TRANSIENT TECHNIQUES IN NMR OF SOLIDS

**An Introduction
to Theory and Practice**

TRANSIENT
TECHNIQUES
IN NMR OF SOLIDS
An Introduction
to Theory and Practice

B. C. GERSTEIN
AMES LABORATORY, MINERALS AND MINING
   RESOURCES RESEARCH INSTITUTE
AND
DEPARTMENT OF CHEMISTRY
IOWA STATE UNIVERSITY
AMES, IOWA

C. R. DYBOWSKI
DEPARTMENT OF CHEMISTRY
UNIVERSITY OF DELAWARE
NEWARK, DELAWARE

1985

ACADEMIC PRESS, INC.
Harcourt Brace Jovanovich, Publishers

Orlando San Diego New York Austin
London Montreal Sydney Tokyo Toronto

52955680

PHYSICS

ACADEMIC PRESS, INC.
Orlando, Florida 32887

United Kingdom Edition published by
ACADEMIC PRESS INC. (LONDON) LTD.
24–28 Oval Road, London NW1 7DX

Library of Congress Cataloging in Publication Data

Gerstein, B. C. (Bernard C.), Date
    Transient techniques in NMR of solids.

    Bibliography: p.
    Includes index.
    1. Solids—Spectra.   2. Nuclear magnetic resonance
spectroscopy.   I. Dybowski, C. R.   II. Title.   III. Title:
Transient techniques in N.M.R. of solids.
QC176.8.O6G47   1985   543'.0877   85-1231
ISBN 0–12–281180–1 (alk. paper)

PRINTED IN THE UNITED STATES OF AMERICA

85 86 87 88   9 8 7 6 5 4 3 2 1

CONTENTS

PREFACE   ix
ACKNOWLEDGMENTS   xi

Chapter 1   **Magnetic Moments, Magnetic Fields, and a Classical Picture of Resonant Absorption**

I.	Introduction	1
II.	The Relation between Magnetic Moment and Angular Momentum	2
III.	The Force on, and the Resulting Motion of, a Dipole in a Magnetic Field	4
IV.	Macroscopic Moments	11
V.	The Bloch Equations: Relaxation	12
VI.	The Rotating Frame	16
VII.	The Pulse NMR Experiment: Pulse and Fourier-Transform NMR	21
VIII.	Relaxation in the Rotating Frame: $T_{1\rho}$	30
IX.	The Magnetic Field of a Solenoid	31
X.	The Production of Radio-Frequency Fields: A Brief Introduction to Alternating-Current Circuit Theory	35
	Problems	39
	References	41

Chapter 2   **Quantum Mechanics of Spin States, the Density Matrix, Interaction Frames, and the Polarization Vector**

I.	Introduction	42
II.	Dirac's Picture of Quantum Mechanics	42
III.	Observables: Physical Interpretation	44
IV.	Angular Momentum Operators and Eigenstates	46
V.	Spin-$\frac{1}{2}$	49
VI.	The Schrödinger Equation for Spin-$\frac{1}{2}$ with a Zeeman Interaction	51
VII.	The Single-Particle Probability Operator	53
VIII.	The Spin-$\frac{1}{2}$ Problem with an Arbitrary Hamiltonian: The Density Operator	54
IX.	The Equilibrium Density Matrix for Spin-$\frac{1}{2}$	57
X.	The Total Density Operator	59
XI.	Time Evolution of the Density Operator: Methods of Solution of the Liouville–von Neumann Equation	61

XII.   Exponential Operators   65
XIII.   Interaction Frames: The $\pi/2$ Pulse Described Quantum Mechanically
        for Spin-$\frac{1}{2}$ 69
XIV.   The Calculation of T_1 79
XV.   The Polarization Vector and Spin-$\frac{1}{2}$: A Quantum Analog of the
        Magnetization Vector   85
XVI.   The Equation of Motion of the Vector **P**: The Quantum Analog of the
        Bloch Equation   87
XVII.   Macroscopic Thermodynamics and the Density Operator   88
        Problems   89
        References   90

Chapter 3 Internal Hamiltonians and Their Spectra

I.   Introduction   91
II.   The Zeeman Interaction   92
III.   The Direct Dipole–Dipole Interaction   94
IV.   The Chemical Shift   103
V.   Indirect Nuclear–Nuclear Interactions   119
VI.   The Quadrupolar Interaction   121
VII. T_1 for Randomly Modulated Hamiltonians   132
VIII.   Concluding Remarks   135
        Problems   135
        References   136

Chapter 4 Exponential Approximations for Evolution Operators: The BCH Formula, the Magnus Expansion, and the Dyson Expression

I.   Introduction   137
II.   The Baker–Campbell–Hausdorff Formula, the Magnus Expansion, and
        the Dyson Expression   138
III.   Interaction Frames Revisited: Examples of the Use of the Magnus
        Expansion and the Dyson Expression   151
IV.   Secular Perturbations   160
        Problems   163
        References   163

Chapter 5 Homonuclear Pulse NMR Experiments

I.   Introduction   164
II.   Multiple-Pulse NMR and the Average Hamiltonian   166
III.   The Symmetry of $\mathcal{H}_{int}$ in the Frame of the Radio Frequency   171

IV.   Treatment of Nonideal RF Perturbations   175
V.   Sample Calculations   182
VI.   The Effect of Motion upon Removal of Dipolar Interactions by
       Multiple-Pulse Techniques   207
VII.   Second Averaging   209
VIII. Combined Rotation and Multiple-Pulse Spectroscopy: Addition of
       Magic-Angle Spinning   213
IX.   Tuning the Spectrometer for Multiple-Pulse Experiments   215
X.   Experimental Considerations   221
       Problems   223
       References   226

Chapter 6 **Heteronuclear Pulse Experiments**

I.   Introduction   227
II.   Pulse Decoupling   229
III.   Continuous-Wave Decoupling   234
IV.   Signal Enhancement by Cross-Polarization   240
V.   Magic-Angle Spinning   255
VI.   Transient Dipole–Dipole Oscillations   267
VII.   Effect of Coupling of Quadrupolar Nuclei to Spin-$\frac{1}{2}$ Nuclei   271
VIII. Concluding Remarks   273
       Problems   274
       References   275

Appendix 1 **The Field of a Current Loop: A Classical Model**   277

Appendix 2 **Units and Physical Constants**   282

Appendix 3 **Vectors, Tensors, and Transformations**   286

BIBLIOGRAPHY   289

INDEX   291

PREFACE

Nuclear magnetic resonance (NMR) is truly a remarkable phenomenon. *Remarkable* can imply different things to different people. From the point of view of a physicist, spin dynamics is an elegant example of the use of time-dependent quantum mechanics, and NMR absorption of energy is a prototype for spectroscopic transitions. From the point of view of the practicing chemist and materials scientist, NMR spectroscopy is an invaluable tool for the identification of chemical species and structures.

Had NMR spectroscopic techniques commercially available in the early 1960s been the only result of investigations of this phenomenon, it would have had a major impact on the course of chemical analysis. The study of liquids and solutions for chemical shifts and couplings of protons had produced a rapid means of identifying chemical species nondestructively. The study of dynamical properties also could be addressed by study of temperature dependence of the spectra or of the saturation of the resonance by high-power irradiation.

Even at that time, however, studies of the spin dynamics had already begun to indicate that there were many interesting facets of the NMR phenomenon left to exploit. For example, the Fourier-transform relationship of the free-induction decay and the absorption spectrum had been shown and the basis of the cross-polarization experiment was being investigated. A number of chemists had begun to study the spin–lattice relaxation times of species by pulse NMR techniques by utilizing methods that were not familiar at that time to the typical chemist but that are now commonly employed in NMR analysis.

The principal characteristic of the NMR technique that makes it so useful for chemical analysis of liquids and solutions is the high resolution that allows one to observe very small interactions such as the chemical shift and the spin–spin coupling. These weak interactions are quite sensitive to the local environment of the spin and therefore may be used as a diagnostic for the environment. The connectivity of chemical structure is often mimicked closely in the NMR connectivity of the spectrum, and quantitative information is relatively easy to obtain.

Nuclear magnetic resonance spectra of solids exhibit such resolution only in special cases. The primary (although not the exclusive) reason for the lack of resolution in the spectrum of a typical solid is the presence of the dipole–dipole interaction, which dominates the NMR spectroscopy of solids that have been of interest to chemists. One solution (no pun intended) to the problem of obtaining chemical-shift information about such solids is to dissolve them and to study them in solution. However, if the solid is insoluble or otherwise intractable or if the analysis involves questions about the properties of the substance *in the solid state,* then there arises a need for techniques to study the weaker interactions in the presence of the dipole–dipole interaction or other overwhelming interactions. This volume describes the means devised by a number of very clever spectroscopists to achieve this goal.

Understanding, like *remarkable,* can imply different things to different people. We have tried to speak to the graduate student who earnestly wishes to learn about NMR spectroscopy of the solid state. A knowledge of quantum mechanics such as one might get in a junior-level physical chemistry course is presumed. Since many graduate students in chemistry characteristically have not been exposed to physics beyond the level of a survey course, very little prior knowledge of the basic theory of electromagnetism is assumed. The reader who is already familiar with NMR of the solid state may thus find that our explanations are long and circuitous. We have strived to provide a background sufficient to allow the student to understand the results at hand. We have also intended this volume as an introduction to concepts of time-dependent quantum mechanics as they apply to NMR spectroscopy of the solid state. As such, we have chosen to discuss certain topics that represent broad applications of this theory to the NMR of solids. Certain other topics, e.g., two-dimensional NMR techniques and multiquantum studies, have been given little exposition here. This choice certainly does not represent any bias on the parts of the authors about the relevance of these techniques to NMR spectroscopy of the solid state.

In the sense that we have provided information for the reader that may help to ease the introduction into this subject, we hope we have contributed to a bridging of the gap of understanding that often frustrates the new student of this field when he encounters the literature, the more advanced books, or a recalcitrant spectrometer. It is our hope that after having read this volume, the student will continue studies of more advanced works but that this work will serve as a reference on some of the machinations that NMR spectroscopists use to explain their experiments on solids.

ACKNOWLEDGMENTS

As with all projects of this magnitude, there are many people who contribute to the final result by offering suggestions, by proofreading, and by doing legwork for the authors. In our case, we have had the good fortune to have been associated with a warm, encouraging group of such persons: Professor Thomas Apple, Dr. Jo-Anne Bonesteel, Dr. Anita J. Brandolini, Mr. Charles Fry, Dr. Mary Kaiser, Mr. Taka Sogabe, and Professor Toshihiko Taki as well as the many students and friends in the NMR community who have helped us in many ways to improve this effort. We also acknowledge the tremendous debt we owe to the late Professor Robert W. Vaughan, whose enthusiasm for the subject of NMR spectroscopy of the solid state and whose unending quest for the deeper meaning in each new experiment not only produced many new avenues of experimentation, but also engendered in those who knew him such a fascination with the subject that two of them might write a volume such as this one.

CHAPTER 1

MAGNETIC MOMENTS,
MAGNETIC FIELDS,
AND A CLASSICAL PICTURE
OF RESONANT ABSORPTION

I. Introduction

The resonant absorption and emission of energy by a system irradiated by light is a fascinating subject. The use of this technique provides an enormous amount of useful information to the chemist whose chief interest is in the determination of the composition, geometry, and reactive characteristics of materials in which he is interested. Although this utilitarian aspect may be the impetus for many practical uses of spectroscopy, the study of the processes underlying these resonant transitions is equally important. It is through the careful study of the details of these processes that more powerful techniques are uncovered, and these allow an even more intimate detailing of molecular behavior, as well as providing means of studying such behavior in realms once thought out of reach. A good example of such developments is the topic of this book—the development of pulse techniques in nuclear magnetic resonance. The novel techniques in NMR that spark so much enthusiasm and whose genesis is no more than a decade old at the time of writing (1981) is the outgrowth of the efforts of a number of investigators who returned to the basic theory describing in detail the interaction of quantum-mechanical systems with resonant radiation. The result of their careful, creative work, together with the happy coupling of theory with experiment, led to results that indicated a still more powerful role for the use of NMR in solving chemical problems in fields as diverse as those of genetics, polymer structure, and heterogeneous catalysis.

In order to understand the fundamentals of the technique used to solve such a broad range of chemical problems, we have recourse to at least two views of the basic mechanism involved, the absorption of resonant radiation by an ensemble of nuclei with magnetic moments. The first is a classical view in which one talks about this ensemble of spins as if it were a single magnet and inquires into how this magnet behaves in an external field (or in fields supplied by other neighboring magnets). The second is a quantum-mechanical view in which the main focus is on a description of how the states of the system are changed in the presence of strong resonant radiation. Both ways of thinking about the nuclear magnetic resonance phenomenon are useful in helping to visualize the effect, and both are developed in this volume.

In this chapter, we focus upon a classical description of magnetic resonance. To this end, we derive the classical equations of motion of a magnetic moment in a field, the classical picture of resonance with a radio-frequency field, and the relations describing relaxation processes by which equilibrium in a magnetic field is established after a resonant perturbation. The result to be immediately obtained is that a magnetic moment [which can be thought of as a circulating charge characterized by a magnetogyric ratio (*vide infra*) γ in a static field **B**] precesses about the field with precession frequency

$$\omega \ (\text{rad sec}^{-1}) = -\gamma \ (\text{rad sec}^{-1} \ G^{-1}) \ \mathbf{B} \ (G) \tag{1.1}$$

This motion is a solution of the equation

$$d\mathbf{M}/dt = \gamma \mathbf{M} \times \mathbf{B} \tag{1.2}$$

The use of (1.2), together with an assumption about the linearity of the differential equations governing relaxation of the x, y, and z components of the spin after an excitation, leads to the Bloch equations and the concept of longitudinal and transverse relaxation, with time constants T_1 and T_2, respectively. It also is seen that solutions of these differential equations describe exactly, in some cases, the observed NMR signal that is the response of a spin system to a resonant radio-frequency excitation.

In addition to discussing the effects of magnetic fields upon classical moments, this chapter deals with the production of radio-frequency fields that are used to act upon the moments and with the spatial distribution of these fields in the circuits used to produce them. The intent of this volume is to provide a source of information on both the theory and the practice of pulse techniques in NMR. Relations that have practical application in the design of circuits for production of resonant radio-frequency pulses are developed and order of magnitude calculations for use in such applications are illustrated.

II. The Relation between Magnetic Moment and Angular Momentum

A concept fundamental to an understanding of the phenomenon of magnetic resonance is that circulating charges give rise both to magnetic moments and to magnetic fields that can be used to probe the environment of

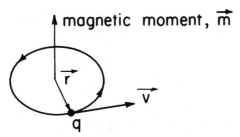

FIG. 1.1. The magnetic moment produced by a current loop. (Arrows indicate vectors in the figures; boldface indicates vectors in the text.)

these moments in matter. In classical mechanics, one learns that equations of motion are obtained from Newton's second law, which deals with the relation between force and a change in linear momentum. Similarly, in discussing the classical equations of motion of magnetic moments in external (or internal) fields, one finds that he is dealing with the relation between torque and a change in *angular* momentum. To start the classical discussion of the magnetic resonance experiment, therefore, it is instructive to investigate the relation between the angular momentum of a circulating charge and its magnetic moment. Consider a charge q moving clockwise[†] on a circle of radius r with velocity V, as illustrated in Fig. 1.1. A moving charge is a current, and a current moving in a circle produces a simply describable magnetic field, the development of which is given in Appendix 1. The spatial description of this field, at distances large compared to r, is similar to that of the electric field produced by an *electric dipole*. In fact, when the extent of the electric dipole is small compared to the distance between the center of the dipole and the point at which the field is measured, the spatial distributions of the fields due to a circulating charge and an electric dipole are identical. A charge whose region of circulation is small compared to the distance at which the field is measured is called a *magnetic dipole* with dipole moment **m** having absolute value

$$m = iA/c \tag{1.3}$$

where A is the area of the current loop and c the speed of light. The period of rotation of such a particle is

$$t = \text{circumference/speed} = 2\pi r/V \tag{1.4}$$

[†] Throughout this text, the term *clockwise* will be taken to mean clockwise when viewed in the direction that a right-hand screw would proceed forward when turned right, i.e., when viewed from below to above the plane of Fig. 1.1. Note that when viewed from above the plane, the circulation appears to be counterclockwise.

For a circulating charge q, this gives an equivalent current of

$$i = q/t = qV/2\pi r \tag{1.5}$$

The magnitude of this magnetic moment is then

$$m = (q/2\mu c)\!\!\!/(\mu V r) \tag{1.6}$$

where μ is the charge's mass. The second term in parentheses is the magnitude of the angular momentum of the circulating particle L. Both the magnetic moment and the angular momentum are vectors. The angular momentum is perpendicular to both the radius vector and the velocity. (See Appendix 1.) For the special case of this model

$$\mathbf{m} = (q/2\mu c)\mathbf{L} \tag{1.7}$$

that is, the magnetic moment is proportional to the angular momentum of the particle.

A circulating particle is not a good representation of a very small nucleus. However, the magnetic interactions of nuclei can be quantified in terms of a magnetic moment that is proportional to the angular momentum of the particle

$$\mathbf{m} = \gamma\mathbf{L} \tag{1.8}$$

where γ is the magnetogyric ratio of the particle.

While this result was derived for the model of an orbiting charged particle, one might similarly derive an expression for a rotating charge distribution in a more general sense; that is, one would find for a charge distribution that the magnetic moment is proportional to the rotational angular momentum of the distribution.

III. The Force on, and the Resulting Motion of, a Dipole in a Magnetic Field[†]

Consider the magnetic dipole's motion in a magnetic field (produced by current loops of one type or another). The fundamental force law for a charge traveling in a magnetic field is

$$\mathbf{F} = q(\mathbf{V}/c) \times \mathbf{B} \tag{1.9}$$

A current loop can be considered to be composed of a number of small segments, each of which has a force exerted on it. If the current is I, then for a small element $d\mathbf{r}_1$

$$q\,d\mathbf{V} = I\,d\mathbf{r}_1$$

[†] See Bleaney and Bleaney [1].

where $\mathbf{r}_1$ is a vector along the loop. The differential force on that part of the loop is (for a constant field)

$$d\mathbf{F} = (I/c)\,d\mathbf{r}_1 \times \mathbf{B} \tag{1.9a}$$

When integrated over the whole loop, the total *force* is

$$\oint d\mathbf{F} = \frac{I}{c} \oint d\mathbf{r}_1 \times \mathbf{B} \tag{1.10}$$

Since $\mathbf{B}$ is uniform across the loop, the integral on the right-hand side is zero. There is therefore no net force on a magnetic moment in a magnetic field. This result may be surprising, but one must remember that, in a uniform magnetic field, the center of mass of a dipole never experiences a translational motion. Infinitesimal forces, however, do exist for small segments of the loop $d\mathbf{r}_1$ and $d\mathbf{r}_2$. It follows from (1.10) that the net forces from various segments must cancel each other exactly, and thus, the combination of two forces on opposite segments causes rotation of the current loop, as indicated in Fig. 1.2.

This action of equal but oppositely directed forces is a *torque* τ and is defined as

$$d\tau = \mathbf{r}_1 \times d\mathbf{F} \tag{1.11}$$

Replacing $d\mathbf{F}$ as in (1.9a) gives

$$d\tau = (I/2c)\mathbf{r} \times d\mathbf{r}_1 \times \mathbf{B} \tag{1.12}$$

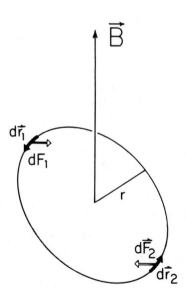

FIG. 1.2. The forces acting on a current loop in a magnetic field $\mathbf{B}$.

Integrating around the whole loop

$$\tau = \left(\frac{I}{2c} \oint \mathbf{r} \times d\mathbf{r}\right) \times \mathbf{B} \tag{1.13}$$

The term in parentheses is the magnetic moment $\mathbf{m}$ for the dipole. [See (A1.21).]
Hence,

$$\tau = \mathbf{m} \times \mathbf{B} \tag{1.14}$$

We know from classical mechanics that *torque* is the time derivative of the *angular momentum*

$$\tau = \partial \mathbf{L}/\partial t \tag{1.15}$$

Therefore, the equations combine to yield the equation of motion

$$\partial \mathbf{L}/\partial t = \mathbf{m} \times \mathbf{B} \tag{1.16}$$

Using the relation (1.8) one arrives at the classical equation of magnetic resonance

$$\partial \mathbf{m}/\partial t = \gamma \mathbf{m} \times \mathbf{B} \tag{1.17}$$

This equation may be solved as follows: with $\mathbf{B}$, defined along the z axis,

$$\mathbf{m} \times \mathbf{B} = \mathbf{i} m_y B - \mathbf{j} m_x B \tag{1.18}$$

Equation (1.17) is really three equations:

$$dm_x/dt = \gamma m_y B \tag{1.19}$$

$$dm_y/dt = -\gamma m_x B \tag{1.20}$$

$$dm_z/dt = 0 \tag{1.21}$$

Equation (1.21) shows that m_z is not coupled to the other components and has the integral result

$$m_z(t) = m_z(0) \tag{1.22}$$

that is, m_z is time independent; m_x and m_y given in (1.19) and (1.20), however, are coupled. The definition

$$m_\pm = m_x \pm i m_y$$

gives

$$dm_\pm/dt = dm_x/dt \pm i\, dm_y/dt = \mp i m_\pm (\gamma B) \tag{1.23}$$

The solutions of (1.23) are

$$m_\pm(t) = m_\pm(0) e^{\mp i\gamma B t} \tag{1.24}$$

which yield

$$m_x(t) = m_x(0) \cos \gamma Bt + m_y(0) \sin \gamma Bt \tag{1.25a}$$

$$m_y(t) = -m_x(0) \sin \gamma Bt + m_y(0) \cos \gamma Bt \tag{1.25b}$$

In a constant field, therefore, the component of the magnetic moment along the field is time independent, whereas those perpendicular to the field oscillate in such a way that the magnitude of the transverse component also remains constant:

$$m_x^2(t) + m_y^2(t) = m_x^2(0) + m_y^2(0)$$

The preceding equations have been solved under the assumption that **B** is static. However, as long as the nonrelativistic limit is considered, **B** may be chosen to be time dependent as well. The solutions may not be so simple, but Eq. (1.17) is certainly valid in the classical limit as long as **B** is spatially uniform over the extent of the dipole. For nuclei, this is always true.

The preceding results suggest a picture in which the magnetic moment **m** is rotating about **B** with rotation frequency $\omega = -\gamma B$ rad sec^{-1}. This angular precession frequency is known as the Larmor frequency. The physical situation is illustrated in Fig. 1.3.

The resonance frequencies of NMR active nuclei are supplied in Table 1.1, at a static field in which protons would resonate at 100 MHz. Also supplied are the values of the spin quantum numbers, the natural abundances, and the relative and absolute sensitivities of detection of these nuclei by NMR at

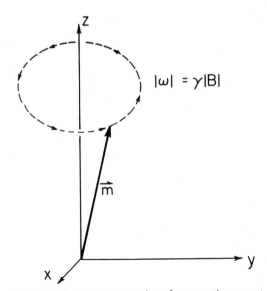

FIG. 1.3. The Larmor precession of a magnetic moment.

TABLE 1.1
Parameters for Selected NMR Active Nuclei

Isotope	Spin	Natural abundance (%)	Quadrupole moment (10^{-28} m^2)	Sensitivity		NMR frequency (MHz) at a field of 2.3488T
				Relative[a]	Absolute[b]	
^{1}H	$\frac{1}{2}$	99.98	—	1.00	1.00	100.000
^{2}H	1	1.5×10^{-2}	2.73×10^{-3}	9.65×10^{-3}	1.45×10^{-6}	15.351
^{3}H	$\frac{1}{2}$	0	—	1.21	0	106.663
^{3}He	$\frac{1}{2}$	1.3×10^{-4}	—	0.44	5.75×10^{-7}	76.178
^{6}Li	1	7.42	-8.0×10^{-4}	8.50×10^{-3}	6.31×10^{-4}	14.716
^{7}Li	$\frac{3}{2}$	92.58	-4.5×10^{-2}	0.29	0.27	38.863
^{9}Be	$\frac{3}{2}$	100	5.2×10^{-2}	1.39×10^{-2}	1.39×10^{-2}	14.053
^{10}B	3	19.58	7.4×10^{-2}	1.99×10^{-2}	3.90×10^{-3}	10.746
^{11}B	$\frac{3}{2}$	80.42	3.55×10^{-2}	0.17	0.13	32.084
^{13}C	$\frac{1}{2}$	1.108	—	1.59×10^{-2}	1.76×10^{-4}	25.144
^{14}N	1	99.63	1.6×10^{-2}	1.01×10^{-3}	1.01×10^{-3}	7.224
^{15}N	$\frac{1}{2}$	0.37	—	1.04×10^{-3}	3.85×10^{-6}	10.133
^{17}O	$\frac{5}{2}$	3.7×10^{-2}	-2.6×10^{-2}	2.9×10^{-2}	1.08×10^{-5}	13.557
^{19}F	$\frac{1}{2}$	100	—	0.83	0.83	94.077
^{21}Ne	$\frac{3}{2}$	0.257	9.0×10^{-2}	2.50×10^{-3}	6.43×10^{-6}	7.894
^{23}Na	$\frac{3}{2}$	100	0.12	9.25×10^{-2}	9.25×10^{-2}	26.451
^{25}Mg	$\frac{5}{2}$	10.13	0.22	2.67×10^{-3}	2.71×10^{-4}	6.1195
^{27}Al	$\frac{5}{2}$	100	0.149	0.21	0.21	26.057
^{29}Si	$\frac{1}{2}$	4.7	—	7.84×10^{-3}	3.69×10^{-4}	19.865
^{31}P	$\frac{1}{2}$	100	—	6.63×10^{-2}	6.63×10^{-2}	40.481
^{33}S	$\frac{3}{2}$	0.76	-5.5×10^{-2}	2.26×10^{-3}	1.72×10^{-5}	7.670
^{35}Cl	$\frac{3}{2}$	75.53	-8.0×10^{-2}	4.70×10^{-3}	3.55×10^{-3}	9.798
^{37}Cl	$\frac{3}{2}$	24.47	-6.32×10^{-2}	2.71×10^{-3}	6.63×10^{-4}	8.156
^{39}K	$\frac{3}{2}$	93.1	5.5×10^{-2}	5.08×10^{-4}	4.73×10^{-4}	4.667
^{41}K	$\frac{3}{2}$	6.88	6.7×10^{-2}	8.40×10^{-5}	5.78×10^{-6}	2.561
^{43}Ca	$\frac{7}{2}$	0.145	-0.05	6.40×10^{-3}	9.28×10^{-6}	6.728
^{45}Sc	$\frac{7}{2}$	100	-0.22	0.30	0.30	24.290
^{47}Ti	$\frac{5}{2}$	7.28	0.29	2.09×10^{-3}	1.52×10^{-4}	5.637
^{49}Ti	$\frac{7}{2}$	5.51	0.24	3.76×10^{-3}	2.07×10^{-4}	5.638
^{50}V	6	0.24	± 0.21	5.55×10^{-2}	1.33×10^{-4}	9.970
^{51}V	$\frac{7}{2}$	99.76	-5.2×10^{-2}	0.38	0.38	26.289
^{53}Cr	$\frac{3}{2}$	9.55	$\pm 3.0 \times 10^{-2}$	9.03×10^{-4}	8.62×10^{-3}	5.652

[a] at constant field for equal numbers of nuclei.
[b] product of relative sensitivity and natural abundance.

(continued)

TABLE 1.1 (Continued)

Isotope	Spin	Natural abundance (%)	Quadrupole moment (10^{-28} m^2)	Sensitivity Relative[a]	Sensitivity Absolute[b]	NMR frequency (MHz) at a field of 2.3488T
^{55}Mn	$\frac{5}{2}$	100	0.55	0.18	0.18	24.664
^{57}Fe	$\frac{1}{2}$	2.19	—	3.37×10^{-5}	7.38×10^{-7}	3.231
^{59}Co	$\frac{7}{2}$	100	0.40	0.28	0.28	23.614
^{61}Ni	$\frac{3}{2}$	1.19	0.16	3.57×10^{-3}	4.25×10^{-5}	8.936
^{63}Cu	$\frac{3}{2}$	69.09	−0.211	9.31×10^{-2}	6.43×10^{-2}	26.505
^{65}Cu	$\frac{3}{2}$	30.91	−0.195	0.11	3.52×10^{-2}	28.394
^{67}Zn	$\frac{5}{2}$	4.11	0.15	2.85×10^{-3}	1.17×10^{-4}	6.254
^{69}Ga	$\frac{3}{2}$	60.4	0.178	6.91×10^{-2}	4.17×10^{-2}	24.003
^{71}Ga	$\frac{3}{2}$	39.6	0.112	0.14	5.62×10^{-2}	30.495
^{73}Ge	$\frac{9}{2}$	7.76	−0.2	1.4×10^{-3}	1.08×10^{-4}	3.488
^{75}As	$\frac{3}{2}$	100	0.3	2.51×10^{-2}	2.51×10^{-2}	17.126
^{77}Se	$\frac{1}{2}$	7.58	—	6.93×10^{-3}	5.25×10^{-4}	19.067
^{79}Br	$\frac{3}{2}$	50.54	0.33	7.86×10^{-2}	3.97×10^{-2}	25.053
^{81}Br	$\frac{3}{2}$	49.46	0.28	9.85×10^{-2}	4.87×10^{-2}	27.006
^{83}Kr	$\frac{9}{2}$	11.55	0.15	1.88×10^{-3}	2.17×10^{-4}	3.847
^{85}Rb	$\frac{5}{2}$	72.15	0.25	1.05×10^{-2}	7.57×10^{-3}	9.655
^{87}Rb	$\frac{3}{2}$	27.85	0.12	0.17	4.87×10^{-2}	32.721
^{87}Sr	$\frac{9}{2}$	7.02	0.36	2.69×10^{-3}	1.88×10^{-4}	4.333
^{89}Y	$\frac{1}{2}$	100	—	1.18×10^{-4}	1.18×10^{-4}	4.899
^{91}Zr	$\frac{5}{2}$	11.23	−0.21	9.48×10^{-3}	1.06×10^{-3}	9.330
^{93}Nb	$\frac{9}{2}$	100	−0.2	0.48	0.48	24.442
^{95}Mo	$\frac{5}{2}$	15.72	±0.12	3.23×10^{-3}	5.07×10^{-4}	6.514
^{97}Mo	$\frac{5}{2}$	9.46	±1.1	3.43×10^{-3}	3.24×10^{-4}	6.652
^{99}Ru	$\frac{3}{2}$	12.72	−0.19	1.95×10^{-4}	2.48×10^{-5}	3.389
^{101}Ru	$\frac{5}{2}$	17.07	7.6×10^{-2}	1.41×10^{-3}	2.40×10^{-4}	4.941
^{103}Rh	$\frac{1}{2}$	100	—	3.11×10^{-5}	3.11×10^{-5}	3.147
^{105}Pd	$\frac{5}{2}$	22.23	−0.8	1.12×10^{-3}	2.49×10^{-4}	4.576
^{107}Ag	$\frac{1}{2}$	51.82	—	6.62×10^{-5}	3.43×10^{-5}	4.046
^{109}Ag	$\frac{1}{2}$	48.18	—	1.01×10^{-4}	4.86×10^{-5}	4.652
^{111}Cd	$\frac{1}{2}$	12.75	—	9.54×10^{-3}	1.21×10^{-3}	21.205
^{13}Cd	$\frac{1}{2}$	12.26	—	1.09×10^{-2}	1.33×10^{-3}	22.182
^{13}In	$\frac{9}{2}$	4.28	1.14	0.34	1.47×10^{-2}	21.866
^{115}In	$\frac{9}{2}$	95.72	0.83	0.34	0.33	21.914
^{15}Sn	$\frac{1}{2}$	0.35	—	3.5×10^{-2}	1.22×10^{-4}	32.699
^{17}Sn	$\frac{1}{2}$	7.61	—	4.52×10^{-2}	3.44×10^{-3}	35.625

(continued)

TABLE 1.1 (Continued)

Isotope	Spin	Natural abundance (%)	Quadrupole moment $(10^{-28}\ m^2)$	Sensitivity Relative[a]	Absolute[b]	NMR frequency (MHz) at a field of 2.3488T
^{119}Sn	$\frac{1}{2}$	8.58	—	5.18×10^{-2}	4.44×10^{-3}	37.272
^{121}Sb	$\frac{5}{2}$	57.25	−0.53	0.16	9.16×10^{-2}	23.930
^{123}Sb	$\frac{7}{2}$	42.75	−0.68	4.57×10^{-2}	1.95×10^{-2}	12.959
^{123}Te	$\frac{1}{2}$	0.87	—	1.80×10^{-2}	1.56×10^{-4}	26.207
^{125}Te	$\frac{1}{2}$	6.99	—	3.15×10^{-2}	2.20×10^{-3}	31.596
^{127}I	$\frac{5}{2}$	100	−0.79	9.34×10^{-2}	9.34×10^{-2}	20.007
^{129}Xe	$\frac{1}{2}$	26.44	—	2.12×10^{-2}	5.60×10^{-3}	27.660
^{131}Xe	$\frac{3}{2}$	21.18	−0.12	2.76×10^{-3}	5.84×10^{-4}	8.199
^{133}Cs	$\frac{7}{2}$	100	-3.0×10^{-3}	4.74×10^{-2}	4.74×10^{-2}	13.117
^{135}Ba	$\frac{3}{2}$	6.59	0.18	4.90×10^{-3}	3.22×10^{-4}	9.934
^{137}Ba	$\frac{3}{2}$	11.32	0.28	6.86×10^{-3}	7.76×10^{-4}	11.113
^{138}La	5	0.089	−0.47	9.19×10^{-2}	8.18×10^{-5}	13.193
^{139}La	$\frac{7}{2}$	99.91	0.21	5.92×10^{-2}	5.91×10^{-2}	14.126
^{141}Pr	$\frac{5}{2}$	100	-5.9×10^{-2}	0.29	0.29	29.291
^{143}Nd	$\frac{7}{2}$	12.17	−0.48	3.38×10^{-3}	4.11×10^{-4}	5.437
^{145}Nd	$\frac{7}{2}$	8.3	−0.25	7.86×10^{-4}	6.52×10^{-5}	3.345
^{147}Sm	$\frac{7}{2}$	14.97	−0.21	1.48×10^{-3}	2.21×10^{-4}	4.128
^{149}Sm	$\frac{7}{2}$	13.83	6.0×10^{-2}	7.47×10^{-4}	1.03×10^{-4}	3.289
^{151}Eu	$\frac{5}{2}$	47.82	1.16	0.18	8.5×10^{-2}	24.801
^{153}Eu	$\frac{5}{2}$	52.18	2.9	1.52×10^{-2}	7.98×10^{-3}	10.951
^{155}Gd	$\frac{3}{2}$	14.73	1.6	2.79×10^{-4}	4.11×10^{-5}	3.819
^{157}Gd	$\frac{3}{2}$	15.68	2.0	5.44×10^{-4}	8.53×10^{-5}	4.774
^{159}Tb	$\frac{3}{2}$	100	1.3	5.83×10^{-2}	5.83×10^{-2}	22.678
^{161}Dy	$\frac{5}{2}$	18.88	1.4	4.17×10^{-4}	7.87×10^{-5}	3.294
^{163}Dy	$\frac{5}{2}$	24.97	1.6	1.12×10^{-3}	2.79×10^{-4}	4.583
^{165}Ho	$\frac{7}{2}$	100	2.82	0.18	0.18	20.513
^{167}Er	$\frac{7}{2}$	22.94	2.83	5.07×10^{-4}	1.16×10^{-4}	2.890
^{169}Tm	$\frac{1}{2}$	100	—	5.66×10^{-4}	5.66×10^{-4}	8.271
^{171}Yb	$\frac{1}{2}$	14.31	—	5.46×10^{-3}	7.81×10^{-4}	17.613
^{173}Yb	$\frac{5}{2}$	16.13	2.8	1.33×10^{-3}	2.14×10^{-4}	4.852
^{174}Lu	1	—	—	—	—	—
^{175}Lu	$\frac{7}{2}$	97.41	5.68	3.12×10^{-2}	3.03×10^{-2}	11.407
^{176}Lu	7	2.59	8.1	3.72×10^{-2}	9.63×10^{-4}	7.928
^{177}Hf	$\frac{7}{2}$	18.5	4.5	6.38×10^{-4}	1.18×10^{-4}	3.120
^{179}Hf	$\frac{9}{2}$	13.75	5.1	2.16×10^{-4}	2.97×10^{-5}	1.869

(continued)

TABLE 1.1 (Continued)

Isotope	Spin	Natural abundance (%)	Quadrupole moment $(10^{-28}\ m^2)$	Sensitivity		NMR frequency (MHz) at a field of 2.3488T
				Relative[a]	Absolute[b]	
^{181}Ta	$\frac{7}{2}$	99.98	3.0	3.60×10^{-2}	3.60×10^{-2}	11.970
^{183}W	$\frac{1}{2}$	14.4	—	7.20×10^{-4}	1.03×10^{-5}	4.161
^{185}Re	$\frac{5}{2}$	37.07	2.8	0.13	4.93×10^{-2}	22.513
^{187}Re	$\frac{5}{2}$	62.93	2.6	0.13	8.62×10^{-2}	22.744
^{187}Os	$\frac{1}{2}$	1.64	—	1.22×10^{-5}	2.00×10^{-7}	2.303
^{189}Os	$\frac{3}{2}$	16.1	0.8	2.34×10^{-3}	3.76×10^{-4}	7.758
^{191}Ir	$\frac{3}{2}$	37.3	1.5	2.53×10^{-5}	9.43×10^{-6}	1.718
^{193}Ir	$\frac{3}{2}$	62.7	1.4	3.27×10^{-5}	2.05×10^{-5}	1.871
^{195}Pt	$\frac{1}{2}$	33.8	—	9.94×10^{-3}	3.36×10^{-3}	21.499
^{197}Au	$\frac{3}{2}$	100	0.58	2.51×10^{-5}	2.51×10^{-5}	1.712
^{199}Hg	$\frac{1}{2}$	16.84	—	5.67×10^{-3}	9.54×10^{-4}	17.827
^{201}Hg	$\frac{3}{2}$	13.22	0.5	1.44×10^{-3}	1.90×10^{-4}	6.599
^{203}Ti	$\frac{1}{2}$	29.5	—	0.18	5.51×10^{-2}	57.149
^{205}Ti	$\frac{1}{2}$	70.5	—	0.19	0.13	57.708
^{207}Pb	$\frac{1}{2}$	22.6	—	9.16×10^{-3}	2.07×10^{-3}	20.921
^{209}Bi	$\frac{9}{2}$	100	-0.4	0.13	0.13	16.069
^{235}U	$\frac{7}{2}$	0.72	4.1	1.21×10^{-4}	8.71×10^{-7}	1.790

constant field. The relative sensitivity is for an equal number of nuclei, and these values vary because the transition probability is proportional to the square of the resonant frequency, among other factors, such that sensitivities decrease with decreasing resonant frequency. The absolute sensitivities are the products of the natural abundances and the relative sensitivities.

IV. Macroscopic Moments

Thus far the main concern has been the motion of *one* magnetic dipole in a field. However, in most experimental situations, one deals with an ensemble of dipoles that emit or absorb radiation, precess, or do whatever dipoles choose to do. In order to describe NMR experiments, it is therefore necessary to obtain the equation of motion for roughly 10^{23} particles with a net moment

$$\mathbf{M} = \sum_k \mathbf{m}_k \qquad (1.26)$$

Here, $\mathbf{M}$ is the total magnetization, and the $\mathbf{m}_k$s are the moments of each of the particles. To the extent that fields produced by individual moments are small compared to the externally produced laboratory field (which is always the case for the systems with which we are concerned here), the equation of motion of the net moment in the field is

$$\frac{d}{dt}\mathbf{M} = \sum_k \frac{d\mathbf{m}_k}{dt} = \sum_k \gamma_k \mathbf{m}_k \times \mathbf{B}(\mathbf{r}_k) \tag{1.27}$$

where $\mathbf{r}_k$ is the position of particle k.

Consider the case in which (1) all the moments are identical ($\gamma_k = \gamma_j = \gamma$) and (2) the field $\mathbf{B}$ is uniform; i.e., it does not depend on space. The equation of motion is then

$$\frac{d\mathbf{M}}{dt} = \gamma \left[\sum_k \mathbf{m}_k \right] \times \mathbf{B} = \gamma \mathbf{M} \times \mathbf{B} \tag{1.28}$$

which is that for a single moment. The solution would be the same if all moments started out at time zero with identical orientations. There is no *a priori* reason for the moments to start out at time zero with the same orientation (unless the experimenter causes them to do so, as will be seen) so a general solution is

$$\mathbf{m}_k = |m_t|[(\cos \phi_k \cos \gamma Bt + \sin \phi_k \sin \gamma Bt)\mathbf{i}$$
$$+ (-\cos \phi_k \sin \gamma Bt + \sin \phi_k \cos \gamma Bt)\mathbf{j}] + m_z^k \mathbf{k} \tag{1.29}$$

where ϕ_k is the initial phase of moment k in the x–y plane. Since each particle is represented by its orientation ϕ_k, the sum $\sum_k$ may be replaced by an integral over all orientations of the density per unit angle; $d\phi_k \rho(\phi_k)$. For random distributions in ϕ_k, $\rho = N/2\pi$, where N is the number of particles. So

$$\mathbf{M}(t) = N m_z \mathbf{k} = M_z \mathbf{k} \tag{1.30}$$

Under these conditions (i.e., random orientation in the x–y components of the small dipoles), the only observable is the *moment along the field*. The other components are observed to be zero.

V. The Bloch Equations: Relaxation[†]

It is known from many experiments that if a magnetizable substance is immersed in a magnetic field for a sufficiently long time, it exhibits magnetization along the magnetic field. Zeeman first proposed that, once immersed in a field, the orientations of the magnetic moments with respect to the field have different energies:

$$E = -\mathbf{m} \cdot \mathbf{B} = -\gamma \mathbf{L} \cdot \mathbf{B} \tag{1.31}$$

[†] For the elementary results, see Farrar and Becker [2]. For a more advanced discussion, see Slichter [3] or Abragam [4].

The energy of a given moment depends upon its orientation with respect to **B**, and the interaction corresponding to Eq. (1.31) is known as the Zeeman interaction. The reason Zeeman chose this form was that certain spectroscopic lines split when the substance is placed in the field (obviously because of interaction with the field). Intuitively, he realized that those molecules which rotate in one direction will have a different energy from those that rotate in the opposite direction. The Zeeman interaction in a static field has nothing to do with perpendicular (i.e., the x and y) components. At equilibrium there is no reason to have any particular direction favored in the transverse plane. Hence, one might expect at equilibrium that

$$\mathbf{M} = M_z^{eq}\mathbf{k}$$

Suppose the system is prepared by soaking in a field aligned along **k**, and then the field (i.e., the magnet) is rotated instantaneously to a new orientation **k'**. After a time, one finds that the magnetization has realigned along the field at **k'**, with no perpendicular component present. The system went from a state $\mathbf{M} = M^{eq}\mathbf{k}$ to a state $\mathbf{M'} = M^{eq}\mathbf{k'}$ in some finite time. The equation of motion derived earlier does not predict this result.

Pulse NMR experiments in particular, and NMR experiments in general, involve just this perturbation and recovery of an ensemble of spins. In many such experiments, there are at least two types of relaxation with which the experimenter must contend in order to accumulate meaningful data with the desired resolution. The first type determines how often one may do an "experiment" (e.g., in pulse NMR, how often one may expose the system to a sequence of one or more pulses of rf power) and still obtain meaningful results. This relaxation rate is that at which the spins repolarize in the external field such that each pulse experiment catches the spin system in the same initial state. The characteristic time for such attainment of equilibrium along the external field is called the spin–lattice or longitudinal relaxation time and is denoted by T_1. With the dc field by convention along the z axis, T_1 characterizes the rate of return to equilibrium along z after a disturbance has taken the system away from equilibrium.

The second type of relaxation determines how rapidly the x and y components of the magnetization attain equilibrium after such a disturbance. The rate of relaxation of these components for many experiments is characterized by a single time constant T_2, called the transverse or spin–spin relaxation time. It will be seen that the decay time of a spin system in response to a single pulse is determined by T_2 and that the observed spectral linewidth in the frequency domain is roughly $(\pi T_2)^{-1}$. The resolution in an NMR experiment, therefore, is determined by the transverse relaxation time T_2.

It is quite common in both solids and liquids to observe T_1s that are of the order of tens of seconds to minutes. This means that in data accumulation for the purpose of signal averaging, the time between experiments must be roughly minutes to tens of minutes. Obviously, in some cases short relaxation times allow faster accumulation. Since the signal-to-noise ratio is proportional to the square root of the number of data accumulations, and the time per data accumulation is proportional to T_1; the experimenter has an interest in knowing the value of T_1 for a given sample in order to obtain an idea of the time and cost involved for a particular signal accumulation. In some cases it is possible to reduce T_1 for a given sample by as much as a factor of 20, with no loss in resolution, by the addition of a material that does not seriously affect the quality of the NMR spectrum, but that allows those processes leading to T_1 relaxation to take place more rapidly. A factor of 20 in reduction of T_1 means a factor of 400 in reduction time of signal averaging for a given signal-to-noise ratio, so such a reduction is indeed nontrivial. The spin–lattice relaxation time T_1 can be useful in understanding molecular motion because T_1 is determined by the rate of transfer of energy to other energy reservoirs in the system. In Chapter 6 we shall see how T_1s associated with a given nuclear species, e.g., ^1H, in a molecule can be used to dominate data accumulation of the signal of another species, such as ^{13}C, in order to reduce signal accumulation times.

While the values of T_1 and T_2 are similar in liquids, the transverse relaxation time T_2 can be as small as microseconds for nuclei in the solid state. This means that linewidths in solids can be as wide as tens of kilohertz. Often this relaxation cannot be described by the simple kinetic scheme that characterizes many liquids. A large portion of this volume deals with techniques designed to counteract those interactions leading to fast transverse relaxation in solids, thus enhancing the resolution of NMR spectra in such systems.

In order to describe transverse and longitudinal relaxation of a spin system in a magnetic field, it is necessary to add the equation describing relaxation to that describing the motion of the moment in a field. To describe transverse and longitudinal relaxation F, Bloch proposed simple first-order kinetics, with rate constants T_2^{-1} and T_1^{-1}, respectively, such that the resulting equation of motion of a moment in a field including relaxation becomes

$$\frac{d}{dt}\mathbf{M} = \mathbf{M} \times \gamma\mathbf{B} - \frac{M_x - M_x^{eq}}{T_2}\mathbf{i} - \frac{M_y - M_y^{eq}}{T_2}\mathbf{j} - \frac{M_z - M_z^{eq}}{T_1}\mathbf{k} \quad (1.32)$$

Under conditions of free decay, equilibrium is that state of completely random phases, i.e., $M_x^{eq} = M_y^{eq} = 0$, and the Bloch equation becomes

$$\frac{d\mathbf{M}}{dt} = \mathbf{M} \times \gamma\mathbf{B} - \frac{M_x}{T_2}\mathbf{i} - \frac{M_y}{T_2}\mathbf{j} - \frac{M_z - M_z^{eq}}{T_1}\mathbf{k} \quad (1.33)$$

This simple form for the decay will be seen to be associated with only the case in which weak interactions between nuclei are present. Such is the case, however, for a wide number of experimental conditions, so it is useful to investigate the form of the solution of Eq. (1.33) in detail.

In terms of individual components of the magnetization, the Bloch equations are

$$dM_z/dt = (1/T_1)(M_z - M_z^{eq}) \tag{1.33a}$$

$$dM_x/dt = \gamma M_y B - (M_x/T_2) \tag{1.33b}$$

$$dM_y/dt = -\gamma M_x B - (M_y/T_2) \tag{1.33c}$$

The first of these is immediately solved to yield

$$M_z(t) = M_z^{eq} + [M_z(0) - M_z^{eq}]\exp(-t/T_1) \tag{1.34}$$

To decouple the equations for M_x and M_y we define $M_\pm$ as we did earlier. The solution of the resulting equations are

$$M_\pm(t) = M_\pm(0)\exp(-t/T_2 \mp i\gamma Bt) \tag{1.35}$$

and $M_x(t)$ and $M_y(t)$ are found by inverting (1.35):

$$M_x(t) = \exp(-t/T_2)[M_x(0)\cos\gamma Bt + M_y(0)\sin\gamma Bt]$$
$$M_y(t) = \exp(-t/T_2)[-M_x(0)\sin\gamma Bt + M_y(0)\cos\gamma Bt] \tag{1.36}$$

If the magnitude of the transverse magnitization at time zero is M_0 and the angle between M_0 and the x axis at time zero is ϕ_0, then

$$M_x(t) = M_0\exp(-t/T_2)\cos(\gamma Bt - \phi_0) \tag{1.37a}$$

$$M_y(t) = -M_0\exp(-t/T_2)\sin(\gamma Bt - \phi_0) \tag{1.37b}$$

The form of these results is shown in Fig. 1.4.

The behavior of M_z is a simple exponential approach to equilibrium, which is not too surprising since the Bloch equation for the z component is based on first-order kinetics with rate constant T_1^{-1}. Representing the decay of the x and y components of the magnetization in the laboratory frame is a bit more difficult since γB_0 is many millions of times larger than $2\pi/T_2$, and on a time scale in which the amplitude of the magnetization would show an appreciable decay, the number of oscillations would be such that no single oscillation could be seen. Fortunately, the experimenter does not detect these oscillations in the laboratory frame, as will be subsequently discussed. The x and y components of **M** are detected in a frame rotating about z with an angular frequency near $\omega_0 = \gamma B_0$ such that differences between the rotating frame of detection and ω_0 are generally between 0 and 100 krad sec^{-1}. On this time scale, the decay of M_x and M_y, with first-order rate constant T_2^{-1}, is easily observed, as shown in the two lower plates of Fig. 1.4. These plates were taken from an actual amplitude versus time trace of the response of the ensemble of proton spins in water to a transient, resonant excitation.

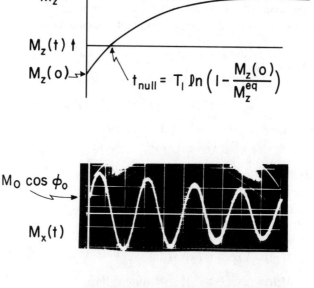

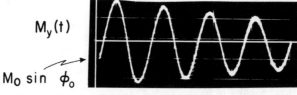

FIG. 1.4. Forms of the solution of the Bloch equations; relaxation in the static field.

VI. The Rotating Frame

Nuclear magnetic resonance signals are detected not in the laboratory frame, but, by the process of demodulation via mixing and filtering, in a frame rotating about z with a frequency ω, such that $(\omega - \omega_0)/\omega_0 \ll 1$. This process is often called "transforming to a rotating frame." The concept of transforming to a frame in which a given interaction either disappears or becomes small compared to its former value is of such importance to an understanding of pulse techniques that we will subsequently devote considerable space to generalizing this concept. After the specific transformation to a frame rotating about z with frequency ω_0 is made (in the absence of other interactions), the magnetization will appear static. Such transformations enormously simplify the description of the motion of the system. Transformation to the frame in which the Zeeman interaction appears to be absent is called transforming to the Zeeman "interaction frame." The operator necessary to perform this transformation must involve the Zeeman interaction directly. More generally, *any* operator may be removed from the description of the motion of the system by transforming to the frame of that particular interaction. The recipe for these transformations will be developed in Chapters 2 and 4.

A. Transformation to the Rotating Frame

The rule for transformation to the frame of the Zeeman interaction is developed by considering a vector **A**, fixed in a coordinate system (cs) rotating with angular velocity ω (rad sec^{-1}) about the z axis of the laboratory frame fixed cs (e.g., **A** might be a line painted on a top). The angular velocity ω is therefore parallel to the z axis of the laboratory frame. Let x, y, and z label the laboratory cs, and x', y', and z' the rotating cs as shown in Fig. 1.5. At time t we choose the lab cs to coincide with the rotating cs. At time $t + \delta t$, the rotating cs has moved clockwise by incremental azimuthal angle $\delta\phi$.

If, as shown in Fig. 1.6, **A** is at fixed angle θ to the axis of rotation, then, with $|\omega| = d\phi/dt$,

$$|\delta\mathbf{A}| = |\omega| \cdot |\mathbf{A}| \sin\theta\, \delta t \tag{1.38}$$

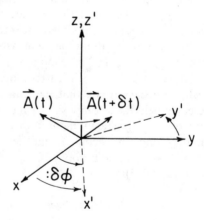

FIG, 1.5. The rotating frame and the vector **A**.

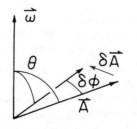

FIG. 1.6. Angular movement of the vector **A**.

or, relative to the fixed laboratory cs,

$$\frac{d\mathbf{A}}{dt} = \lim_{\delta t \to 0} \frac{\delta \mathbf{A}}{\delta t} = \omega \times \mathbf{A} \tag{1.39}$$

Note that (1.39) defines the sense of ω relative to the direction of rotation.

Suppose that $\mathbf{A}$, instead of being a vector fixed in the rotating cs, is a vector from the origin to a peculiar bedbug strolling about in this cs with velocity $D\mathbf{A}/Dt$. Then, relative to the fixed cs

$$d\mathbf{A}/dt = D\mathbf{A}/Dt + \omega \times \mathbf{A} \tag{1.40}$$

The observed motion is just the superposition of the two motions. Identifying $\mathbf{A}$ as the magnetic moment, $\mathbf{M} = \gamma\mathbf{L}$, then

$$d\mathbf{M}/dt = \gamma(d\mathbf{L}/dt) = \gamma\mathbf{M} \times \mathbf{B}_0 \tag{1.41}$$

or the rate of change in the rotating cs is

$$D\mathbf{M}/Dt = \mathbf{M} \times (\gamma\mathbf{B}_0 + \omega) \tag{1.42}$$

Equation (1.42) has two important consequences. The first is that transforming to the rotating cs involves replacing $\mathbf{B}_0$ with an "effective" $\mathbf{B}$, which is $\mathbf{B}_{\text{eff}} = \mathbf{B}_0 + \omega/\gamma$. The second, crucial to this discussion, is that if $\omega = \omega_0 = -\gamma\mathbf{B}_0$, then $D\mathbf{M}/Dt = 0$; i.e., *there is no motion relative to the rotating cs*. An equivalent statement is that, relative to the fixed frame, the moment is precessing about the z axis with angular velocity ω_0. This is the result expressed by Eq. (1.25): a moment with magnetogyric ratio γ will precess about a static field with Larmor precession frequency $\omega_0 = -c\mathbf{B}_0$. Physically, it is easy to see that the transformation to the rotating frame of the Zeeman field $\mathbf{B}_0$ accomplishes two ends. The first is that the equation of motion of the magnetic moment in this frame is simplified; the moment is static in this frame. The second is that the motion of the moment in the laboratory frame is a precession about $\mathbf{B}_0$, a fact perhaps not immediately detectable from Eq. (1.25).

B. Relaxation in the Presence of a Time-Varying Radio-Frequency Field: Saturation and the Continuous-Wave Experiment

The equation of motion of a moment in a stationary magnetic field may be solved directly in the laboratory frame or more simply by transforming to the rotating frame. Moreover, when the complication of a time-varying radio-frequency (rf) is added to the static field, the simplifications resulting from transforming to the rotating frame become even more startlingly clear.

Consider a magnetic moment in a field having two parts: a static field, $\mathbf{B}_0$, and a radio-frequency field $\mathbf{B}_1(t)$. For present purposes, the rf field will be assumed to be simple: it will be perpendicular to $\mathbf{B}_0$ and will have constant magnitude. Its time dependence will arise from a rotation of $\mathbf{B}_1$ about the lab frame z axis, with angular frequency ω. If $\mathbf{B}_1$ has arbitrary phase with respect to the lab cs x and y axes, then the form of $\mathbf{B}_1(t)$ will be

$$\mathbf{B}_1(t) = B_1[\mathbf{i}\cos(\omega t + \phi) + \mathbf{j}\sin(\omega t + \phi)] \tag{1.43}$$

With no loss of generality at this point, we may choose the phase angle ϕ to be zero at some time $t = 0$.

The differential equations describing the behavior of the moment result directly from the fundamental equation (1.17), in which specific account is taken of the fact that the static field $\mathbf{B}_0$ is along the z direction, the time varying $\mathbf{B}_1$ field is given by Eq. (1.43), and relaxation has been added.[†] The resulting equations are

$$dM_x/dt = \gamma B_0 M_y - \gamma B_1 M_z \sin \omega_0 t - (M_x/T_2)$$

$$dM_y/dt = -\gamma B_0 M_x + \gamma B_1 M_z \cos \omega_0 t - (M_y/T_2) \tag{1.44}$$

$$dM_z/dt = \gamma B_1 M_x \sin \omega_0 t - \gamma B_1 M_y \cos \omega_0 t - [(M_z - M_0)/T_1]$$

The equations comprise a coupled set of differential equations in which the rates of change of all three components of $\mathbf{M}$ depend on all three components in a manner complicated by the oscillatory nature of $\mathbf{B}_1$. To make the equations more tractable, it is necessary to transform to a frame rotating about $\mathbf{B}_0$ with frequency ω. We have seen [Eq. (1.42)] that the rule for making this transformation is to replace $\mathbf{B}_0$ with $(\mathbf{B}_0 + \omega/\gamma)$. Since $\mathbf{B}_1$ lies along x at time zero, it will lie along x' at all times in the rotating cs. In the rotating frame there will be an effective field

$$\mathbf{B}_{\text{eff}} = [B_0 + (\omega/\gamma)]\mathbf{k}' + B_1\mathbf{i}' \tag{1.45}$$

In the rotating frame the equation of motion becomes

$$\frac{d\mathbf{M}'}{dt} = \gamma \mathbf{M}' \times \mathbf{B}_{\text{eff}} - \frac{M'_x\mathbf{i}' + M'_y\mathbf{j}'}{T_2} - \frac{(M'_z - M_0)\mathbf{k}'}{T_1} \tag{1.46}$$

where $\mathbf{M}'$ is the magnetization viewed from the rotating frame.

In terms of the components of $\mathbf{M}'$ and the precession frequencies $\omega_i = \gamma B_i$ Eq. (1.46) becomes

$$dM'_x/dt = -(M'_x/T_2) + \delta\omega M'_y$$

$$dM'_y/dt = -\delta\omega M'_x - (M'_y/T_2) - \omega_1 M'_z \tag{1.47}$$

$$dM'_z/dt = \omega_1 M'_y - [(M'_z - M_0)/T_1]$$

[†] An implicit assumption in this treatment is that the presence of the $\mathbf{B}_1$ field does not change the relative rates of relaxation of the x and y components. A subsequent section in this chapter discusses the case when $\mathbf{B}_1$ leads to differing transverse components.

where $\delta\omega = \omega - \omega_0$. A bit of reflection will reveal that Eqs. (1.47) are similar to Eqs. (1.33), which were solved in Section V. Equations (1.47) differ from Eqs. (1.33) in that ω_1 and ω are both present and indicate the presence of an rf field.

Suppose that the system has been allowed to "soak" in this combination of static and time-varying fields. A steady state will ultimately be reached in which none of the components change with time:

$$M'_y \delta\omega - (M'_x/T_2) = 0$$

$$-M'_x \delta\omega - M'_z \omega_1 - (M'_y/T_2) = 0 \tag{1.48}$$

$$M'_y \omega_1 - [M'_z - M_{eq}]/T_1] = 0$$

Equations (1.48) are three equations in three unknowns and are solved to yield the three components of the magnetic moment in the rotating frame under the influence of a rotating field:

$$M'_x = \frac{\gamma B_1 T_2^2 \delta\omega}{1 + (\gamma B_1)^2 T_1 T_2 + (T_2 \delta\omega)^2} M_{eq}$$

$$M'_y = \frac{\gamma B_1 T_2}{1 + (\gamma B_1)^2 T_1 T_2 + (T_2 \delta\omega)^2} M_{eq} \tag{1.49}$$

$$M'_z = \frac{1 + T_2^2 \delta\omega^2}{1 + (\gamma B_1)^2 T_1 T_2 + (T_2 \delta\omega)^2} M_{eq}$$

In the "small-rf" limit, i.e., $(\gamma B_1)^2 T_1 T_2 \ll 1$, which, unlike pulse experiments, which are our primary interest, are the usual conditions in a low-level continuous-wave experiment, Eqs. (1.49) become

$$M'_x \approx (\gamma B_1 M_{eq}) \frac{T_2^2 \delta\omega}{1 + (T_2 \delta\omega)^2} = \gamma B_1 M_{eq} G(\delta\omega)$$

$$\tag{1.50}$$

$$M'_y \approx (\gamma B_1 M_{eq}) \frac{T_2}{1 + (T_2 \delta\omega)^2} = \gamma B_1 M_{eq} F(\delta\omega)$$

These two components are so important that they have special names: M'_y is called the *absorption* and M'_x the *dispersion*. The form of these components for fixed B_1, but with varying B_0, is shown in Fig. 1.7. Here M'_x is the component *in phase* with the driving field, and M'_y is the component $90°$ out of phase with B_1, The functional forms $F(\delta\omega)$ and $G(\delta\omega)$ are the absorption and dispersion for a particular kind of line known as a *Lorentzian* line.

If the rotating components' magnitudes could be measured as the static field B_0 is varied at constant driving frequency ω (or equivalently as ω is varied at constant static field B_0), then a spectrum would be determined. This is the procedure used in *continuous-wave* magnetic resonance experiments.

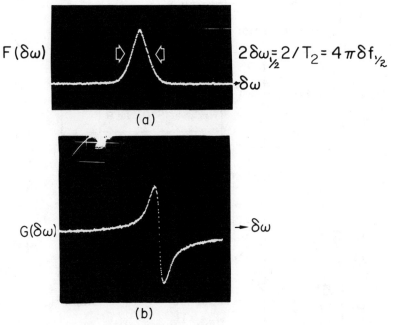

FIG. 1.7. (a) Absorption and (b) dispersion for a Lorentzian line.

In the case of electron spin resonance, it is common to sweep field. In NMR, it is also possible to sweep frequency for high-resolution experiments at fixed field.

VII. The Pulse NMR Experiment: Pulse and Fourier-Transform NMR

In order to initiate discussion of the experiment that is the basis of all subsequent discussions in the text, it is useful to examine carefully a block diagram of an idealized pulse NMR spectrometer. This unit is shown in Fig. 1.8. A stable rf oscillator or frequency synthesizer produces a sinusoidal signal at frequency ω in the megahertz range. This signal is divided into a reference signal and a signal that is fed in pulse mode (by means of a gating switch S controlled by a pulse programmer) to the sample resonant circuit. The pulsed rf excites the sample for a short time. The response of the sample to the pulse, at frequency ω_0, is fed to a high-gain amplifier (a receiver), thence to a device known as a "mixer" or a demodulator. It is at this point that the experimenter steps into the frame rotating at angular frequency ω, by the process known as "demodulation."

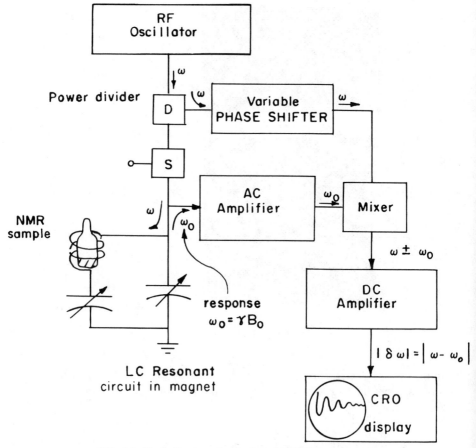

FIG. 1.8. Block diagram of a simple pulse NMR spectrometer.

A mixer is a device that accepts two signals of frequencies ω_1 and ω_2 and yields a signal containing $\omega_1 \pm \omega_2$. In our case, the mixer yields a signal containing frequencies $|\omega \pm \omega_0|$. This signal is fed through an audio amplifier, which has the characteristic of not passing signals of frequency higher than 1 MHz, so the component of the NMR response at $\sim 2\omega$, e.g., 134 MHz, for protons at $B_0 = 1.5 T (15 K G)$ is "filtered out" at the audio amplifier, and only $\delta\omega$ passes. This transient signal, which decays with time constant T_2, can be displayed on a storage oscilloscope or by use of a transient recorder or computer or on an ordinary oscilloscope with a sufficiently long-lasting phosphor. For many pulse experiments in liquids, these transients are simply proportional to the solutions of the Bloch equations in the absence of an rf field but detected in the frame rotating at the frequency of the reference rf.

The solutions are those of (1.37) with γB replaced by $\omega - \omega_0$. The phase of the transient will depend on the phase difference between the reference signal and the transient response at ω_0 arriving at the mixer. If we arbitrarily call the direction of a given B_1 pulse the x direction in the rotating frame, detecting in the x direction *on resonance* (i.e., with ω set exactly at ω_0) will yield a signal of zero.

When we vary the phase of the reference until a zero response is obtained for an on resonance pulse, we know we are "looking" at the component of **M** along $+x$ or $-x$ in the rotating frame. The axes x and y are purely arbitrary, of course; a zero signal on resonance simply indicates that the reference and sample signals are $\pm 90°$ out of phase. We indicate "looking" in the x direction in the rotating frame by an eye gazing at that axis, as shown in Fig. 1.9. Clearly by varying the phase of the reference signal ω with respect to that of the sample signal ω_0, we may "look" anywhere we wish in the $x-y$ plane of the rotating frame.

A "θ pulse" is one that rotates the magnetization about $\mathbf{B}_1$ in the rotating frame by an angle θ. The fundamental relation between the rotation frequency of a moment in a field and the magnetization is given by Eq. (1.1). The rotation angle in radians for a $\mathbf{B}_1$ pulse of duration t seconds is given by

$$\theta \text{ (rad)} = \gamma \text{ (rad sec}^{-1} \text{ G}^{-1}) B_1 \text{ (G)} t \text{ (sec)} \tag{1.51}$$

For example, one calculates from Table 1.1 that γ for ^{1}H is 4.2577 kHz G^{-1}. A $\pi/2$ pulse for a B_1 field of 50 G will take a time of 1.17 μsec.

The relation between the signal seen on an oscilloscope monitoring the output of the audio amplifier and Eqs. (1.37) may be understood with reference to the following example. Suppose we excite the sample with a 90° pulse

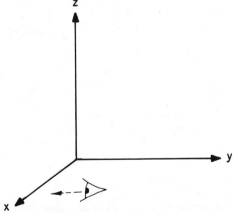

FIG. 1.9. Observation by phase detection along the $+x$ axis in the rotating frame.

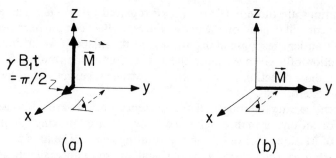

FIG. 1.10. (a) The action of a $\pi/2$ (90°) pulse along the $+x$ axis and (b) a magnetization along $+y$ in the rotating frame immediately after the pulse.

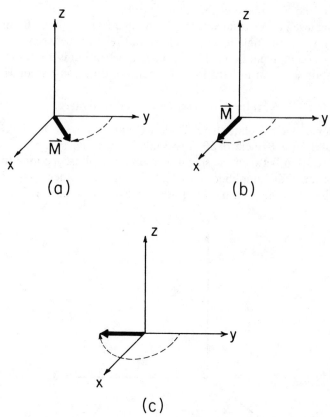

FIG. 1.11. Motion of a magnetic moment with respect to the rotating frame when observed under off-resonance conditions.

along the $+x$ axis, as indicated in Fig. 1.10a, and subsequently observe the magnetization along the $+y$ axis, as indicated in Fig. 1.10b. The offset, $\delta\omega = \omega - \omega_0$, is chosen to be nonzero.

If $\delta\omega \neq 0$, the frame of observation is rotating with respect to $\mathbf{M}$ at an angular frequency $\delta\omega$. The motion of $\mathbf{M}$ relative to the frame of observation is shown in Fig. 1.11. Since the amplitude of the cathode-ray oscilloscope (CRO) display will be proportional to the projection of $\mathbf{M}$ along y as a function of time, the display will appear as shown in Fig. 1.12. In Fig. 1.12, a–c correspond to situations a–c in Fig. 1.11. The response of a magnetic moment to a short pulse, freely decaying in the absence of a driving rf field and inducing a voltage in the inductance coil of the resonant probe circuit, is known as a free-induction decay (FID). An example of the free-induction decay of protons in water, associated with a $\pi/2$ pulse along $-x$, and observation

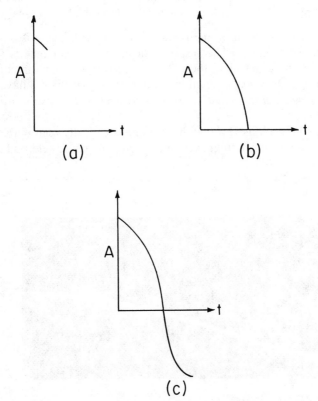

FIG 1.12. The expected time dependence of the signal detected along the $+y$ axis for the moment in Fig. 1.11.

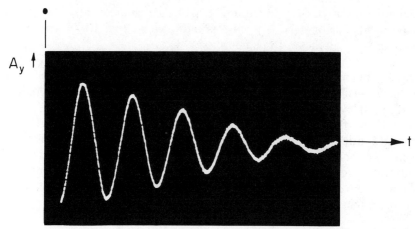

FIG. 1.13. The free-induction decay observed from the protons of water. The exciting pulse is along the $-x$ axis in the rotating frame, with observation along the $+y$ axis.

along $+y$ in the rotating frame is shown in Fig. 1.13. There is a frequency "offset" from resonance. The response detected at resonance is shown in Fig. 1.14. We note that, had the $\pi/2$ pulse been in the $+x$ direction in the rotating frame, the sign of the initially observed signals would have been inverted, *provided that the signal was still observed along the positive y axis in the rotating frame.*

A material such as H_2O, with only one chemically identifiable proton, will have one resonance frequency in a specific magnetic field and will yield

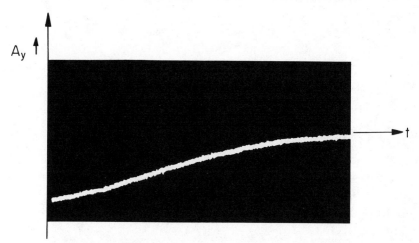

FIG. 1.14. The observed free-induction decay of the protons in water on a longer time scale than shown in Fig. 1.13. The conditions are the same.

a FID of the type shown in Fig. 1.14. A material such as CH_3CO_2H, on the other hand, has both methyl and acid protons. At a resonance frequency of 60 MHz, the different "local fields" about these two kinds of protons lead to Larmor frequencies that differ by about 240Hz. It is this difference that enables the chemist to use NMR as a powerful analytical tool in liquids. Since the methyl and acid proton resonance frequencies differ from each other by about 240 Hz, the FID will be a superposition of both patterns, as shown in Fig. 1.15. The composite is a beat pattern, showing the interference between the two signals. This free-induction decay pattern contains all the information about the protons. However, for complicated systems with many different chemically shifted spins, it is easier to extract information from the frequency spectrum. The functional forms of (1.50) have the property of being related to the FID in the following manner:

$$F(\delta\omega) = \frac{T_2}{1 + T_2^2\,\delta\omega^2} = \int_0^\infty dt\,\exp(-t/T_2)\cos(\delta\omega)t \qquad (1.52)$$

If there exist functions $F(\delta\omega)$ in the frequency domain and $A(t)$ in the time domain, related by the equation [5]

$$F(\delta\omega) = \int_{-\infty}^\infty dt\,A(t)\cos(\delta\omega)t \qquad (1.53)$$

then $F(\delta\omega)$ is said to be the cosine Fourier transform of $A(t)$.

Hence, the absorption spectrum can be obtained from the time decay. With the development of laboratory computers and the fast algorithms [6]

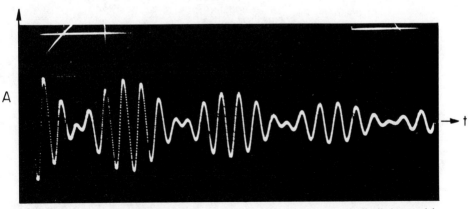

FIG. 1.15. The free-induction decay of a sample containing two chemically distinct nuclei. The proton resonance of acetic acid.

to do this transformation, this procedure has become a viable means to extract the information in NMR spectra of spin systems. The sine transform of $A(t)$ gives the dispersion of a Lorentzian line:

$$\int_0^\infty dt \, \exp(-t/T_2) \sin \delta\omega \, t = G(\delta\omega) \qquad (1.54)$$

Schematically, then, the relation between the free-induction decay observed in a single-pulse NMR experiment and the absorption or dispersion observed in a continuous-wave NMR experiment is shown in Fig. 1.16.

In general the phase of the reference signal ω (Fig. 1.16) will not be exactly zero (leading to the cosine function) or $\pi/2$ (leading to the sine function) with respect to the response signal at ω_0, so an experimentally observed $A(t)$ could appear as shown in Fig. 1.17. The cosine transform of this function would yield neither a pure absorption nor a pure dispersion, but a mixture of the two. In order to obtain the pure absorption and dispersion, a phase correction is applied to the results of Fourier transforming the experimentally observed $A(t)$. This procedure can be done quickly and efficiently by laboratory computers.

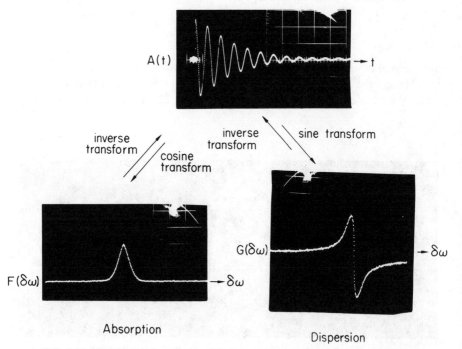

FIG. 1.16. The relation between the free-induction decay and the absorption and dispersion signals.

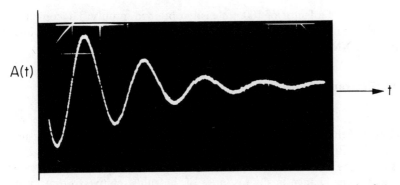

FIG. 1.17. A free-induction decay observed off-axis in the $x-y$ plane of the rotating frame.

The Fourier transform of a beat pattern, resulting from FIDs of nuclei with different ω_0 (due, for example, to chemically shifted nuclei), would result in absorption spectra with peaks for each chemically shifted nucleus. In fact, the Fourier transform of the beat pattern for CH_3CO_2H (in the absence of dipole–dipole interactions, chemical-shift anisotropy, etc.) will be the same as the NMR absorption spectrum of this molecule in solution with two peaks 240 Hz apart and with an area ratio of $3:1$.

In standard Fourier transform (FT) pulse NMR spectroscopy one exposes the sample to an rf field, records the FID, ultimately storing it in a computer, and recovers the NMR absorption spectrum by a Fourier transformation. The experiments may be repeated every five T_1, as a general rule, so with T_1 of the order of seconds (as is found in many liquid samples), a FID may be computer accumulated once every 5–10 sec. The advantage of pulse FT spectroscopy over conventional continuous-wave (cw) spectroscopy is the improvement in the signal-to-noise ratio (S/N) associated with relatively rapid coherent addition of many experiments. The enhancement in S/N is estimated from the ratio $\Delta/v_{1/2}$, where Δ is the total spectral range and $v_{1/2}$ is the width of a typical line in the spectrum. For a spectral range of 1000 Hz, with linewidth of 1 Hz, the pulse FT method gives a factor of 1000 advantage over the cw method, since in the cw method, only during one-thousandth of the scan is information being obtained from a given line. As Farrar and Becker [2] point out, if a cw experiment could utilize 1000 transmitters spread across the spectrum and 1000 detectors, then information could be obtained in one-thousandth the time of a conventional cw experiment, and this advantage of FT NMR over cw experiments would be removed.

In the preceding discussion, it was assumed that the scanning rate for the cw spectrum satisfied the conditions for "adiabatic slow passage" (discussed in standard texts). If a cw scan is made at a faster rate, of course, the advantage of utilizing the pulse FT technique decreases accordingly. In practice, time savings of a factor of 100 are easily obtained in pulse FT compared to cw NMR. This is one of the principal reasons for the success of Fourier transform NMR.

VIII. Relaxation in the Rotating Frame: $T_{1\rho}$

In contrast to the spin system's response to a single 90° pulse of rf, the majority of this text will be concerned with the response of a spin system to sequences of pulses that will take much longer to produce than a $\pi/2$ pulse. In succeeding chapters, we will find that the net effect of such sequences is to produce some change in the response of the spins relative to that found in a one-pulse experiment. A quite powerful, yet simple, experiment that illustrates this effect of these rf fields is that of creating a transverse moment along y with a $\pi/2$ pulse along x in the rotating frame, followed by producing an rf field in the y direction to "spin lock" the net moment along this direction, as shown in Fig. 1.18. If the frequency of the rf field corresponds to the resonance frequency for the spins in question, then the effective field in the

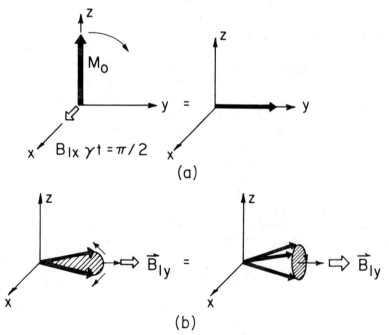

FIG. 1.18. The $T_{1\rho}$ experiment: (a) rotate magnetization to y axis and (b) spin lock along y for time τ.

rotating frame is

$$\mathbf{B}_{\text{eff}} = \mathbf{B}_0 + \omega_0/\gamma + \mathbf{B}_{1y} \cong \mathbf{B}_{1y} \qquad (1.55)$$

Under these circumstances, the natural motion of the spin system will be to precess about the spin locking field $\mathbf{B}_{1y}$ *except* that interactions such as spin–spin relaxation will tend to cause the spins to relax in the x–y plane. To the extent that $\mathbf{B}_{1y}$ is large compared to this dephasing tendency, the "normal" spin–spin relaxation will be affected by the spin locking field, and dephasing will occur less efficiently. The natural tendency to return to equilibrium still causes the magnitude of the magnetization to diminish but at a different rate from that of the magnetization along x. The spin locking field $\mathbf{B}_{1y}$ may be left on for a variable time period τ during which the y component of the spin system decays according to a rate law different from the free precession relaxation. This relaxation rate is called $T_{1\rho}^{-1}$. Assuming that the decay is first order,

$$dM_y/dt = -M_y/T_{1\rho} \qquad (1.56)$$

which at time τ has the solution

$$M_y(\tau) = M_y(0)\exp(-\tau/T_{1\rho}) \qquad (1.57)$$

The initial magnitude of the free-induction decay after removal of the spin locking field is a measure of $M_y(\tau)$, and one could carry out a series of experiments for various values of τ, the results of which could be used with Eq. (1.57) to determine the value of $T_{1\rho}$.

The fact that $T_{1\rho}$ and T_2 do not have to be the same can be seen from the difference in symmetry of the two experiments. In a free-induction decay experiment, there is cylindrical symmetry about the direction of the dc field $\mathbf{B}_0$. It might be expected phenomenologically that a moment would relax toward equilibrium along $\mathbf{B}_0$ with a rate different from the rate at which a perpendicular moment would relax. When the spin-lock field is applied, as in the preceding experiment, space has three unique directions in the rotating frame. If the interactions depend on the components along the various directions, as do Zeeman interactions, the most general way to describe the situation is with three rates, one of which is $T_{1\rho}^{-1}$. As it turns out, for protons in liquids $T_{1\rho} \simeq T_2$, but this is a characteristic of the frequency spectrum of the motion of the nuclei relative to each other in liquids, rather than a requirement. On the other hand, in solids T_2 and $T_{1\rho}$ can differ by orders of magnitude.

IX. The Magnetic Field of a Solenoid

Until now consideration of the production of the time-varying $\mathbf{B}_1$ fields needed to enable the experimenter to partially control ensembles of spins has been delayed. In keeping with the intent of this volume to be a mechanism for exposing the student to both the theory and the practice of NMR, we find it useful to close this chapter with two sections dealing with practical matters. In this section, we show the spatial dependence of the field of a cylindrical inductor. This result will be particularly important when we consider some solid-state experiments that require quite homogeneous radio-frequency $\mathbf{B}_1$ fields over the sample volume. In the section following, a brief introduction to alternating current circuit theory is given. It is hoped that this introduction will serve as a springboard from which the interested and venturesome student may launch into a meaningful understanding of some of the basic circuits used in the production and detection of radio-frequency fields.

To obtain the spatial dependence of the $\mathbf{B}_1$ field of the inductor commonly used to irradiate spin systems, we consider the properties of a cylindrical solenoid of radius a and length D carrying current I (see Fig. 1.19). A solenoid may be considered to be a number of coaxial current loops, the properties of which are considered in Appendix 1. For a solenoid with N total turns, the charge per length per second (i.e., the current density) is NI/D. The equations relating the magnetic intensity vector $\mathbf{B}_1(\mathbf{r}_2)$ ($\mathbf{r}_2$ being the vector from the origin to the point at which the field is measured) to the vector potential $\mathbf{A}(\mathbf{r}_2)$ and the current density vector $\mathbf{J}(\mathbf{r}_1)$, as given in Appendix 1, are

$$\mathbf{B}_1(\mathbf{r}_2) = \text{curl}\,\mathbf{A}(\mathbf{r}_2) = \mathbf{V}_2 \times \mathbf{A}(\mathbf{r}_2) \qquad (1.58)$$

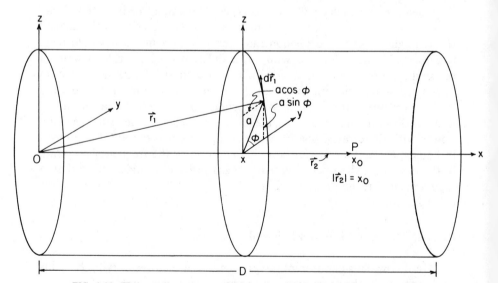

FIG. 1.19. The coordinate system used for describing the field $\mathbf{B}_1$ of a solenoid.

and

$$A(r_2) = \frac{1}{c} \int_v \frac{J(r_1)}{|r_2 - r_1|} d^3r_1 \qquad (1.59)$$

or

$$B_1(r_2) = \frac{\nabla_2}{c} \times \int_v \frac{J(r_1)}{|r_2 - r_1|} d^3r_1 \qquad (1.60)$$

The coordinate system chosen is indicated in Fig. 1.19.

A relatively easily solved problem is that of determining the field along the axis of the solenoid. We inquire about the field at point P along the axis, at a distance x_0 from the origin. By the vector identity

$$\nabla \times (fA) = (\nabla f) \times A + f(\nabla \times A),$$

Eq. (1.60) becomes

$$B_1(r_2) = -\frac{1}{c} \int_v d^3r_1 \, J(r_1) \times \left[\nabla_2 \left(\frac{1}{|r_2 - r_1|} \right) \right] \qquad (1.61)$$

and

$$\nabla_2 \left(\frac{1}{|r_2 - r_1|} \right) = -\frac{r_2 - r_1}{|r_2 - r_1|^3} \qquad (1.62)$$

leading to

$$B_1(r_2) = \frac{1}{c} \int_v \frac{d^3r_1 \, J(r_1) \times (r_2 - r_1)}{|r_2 - r_1|^3} \qquad (1.63)$$

The field at point P is evaluated as follows. From the choice of axes shown in Fig. 1.19 we see that

$$r_2 - r_1 = i(x_0 - x) - ja \cos \phi - ka \sin \phi$$

and

$$|r_2 - r_1| = a \left[1 + \left(\frac{x_0 - x}{a} \right)^2 \right]^{1/2}$$

Also, since $J(r_1)$ is parallel to dr_1 and of magnitude NI/D, then

$$J(r_1) = \frac{NI}{D} (k \cos \phi - j \sin \phi)$$

Thus,

$$J(r_1) \times (r_2 - r_1) = \frac{NIa}{D} \left(i + j \frac{x_0 - x}{a} \cos \phi + k \frac{x_0 - x}{a} \sin \phi \right) \qquad (1.64)$$

changing $\int_v d^3\mathbf{r}_1 \to \int_0^{2\pi} a\,d\phi \int_0^D dx$ in (1.63) and observing that the integration over ϕ leaves only the $\mathbf{i}$ component, gives

$$\mathbf{B}_1(\mathbf{r}_2) = \mathbf{B}_1(x_0) = \frac{2\pi NI}{D}\,\mathbf{i} \int_0^D \frac{d\eta}{[1 + (\eta - \eta_0)^2]^{3/2}} \tag{1.65}$$

where $\eta = x/a$. The usual trigonometric substitution $\eta - \eta_0 = \tan\theta$, results in

$$\mathbf{B}_1(x_0) = (2\pi NI/D)(\sin\theta_2 - \sin\theta_1)\mathbf{i} \tag{1.66}$$

where θ_1 and θ_2 are given by

$$\sin\theta_1 = \frac{-x_0/a}{[(1 + x_0^2)/a^2]^{1/2}}$$

and

$$\sin\theta_2 = \frac{(D - x_0)/a}{\{1 + [(D - x_0)/a]^2\}^{1/2}}.$$

Aside from the functional form, it is important to note that the magnetic field points *along the axis of the coil*. Furthermore, if the current in the coil is time varying, so is the magnetic field. Such a solenoid is the mechanism for producing the time-varying magnetic field needed for nuclear magnetic resonance experiments.

The calculation of fields off the axis of the solenoid is much more difficult. It suffices to say that the field (deep in the coil) is of the form

$$\mathbf{B}_1(x_0) = 2\pi(NI/Dc)f(x_0)\cos\omega t\,\mathbf{i} \tag{1.67}$$

This is not exactly a rotating field, but it is time dependent. A rotating field can be obtained by "adding zero":

$$\begin{aligned} \mathbf{B}_1(x_0) = {}& (\pi NI/Dc)f(x_0)[\cos(\omega t)\mathbf{i} + \sin(\omega t)\mathbf{j}] \\ & + (\pi NI/Dc)f(x_0)[\cos(-\omega t)\mathbf{i} + \sin(-\omega t)\mathbf{j}] \end{aligned} \tag{1.68a}$$

or

$$\mathbf{B}_1(x_0) = \mathbf{B}_1^+(\omega) + \mathbf{B}_1^-(\omega) \tag{1.68b}$$

The linearly polarized field of a coil has the character of two oppositely rotating components.

In a rotating frame from which the interaction is viewed, one of the two components will appear stationary, while the other will rotate at 2ω. The component traveling at 2ω has a negligible effect on the time evolution of magnetic moments. The reason is simple: during one half of its cycle it tries to rotate any magnetic moment in one direction; during the second half, the effect is in the opposite sense. Over times slower than $1/2\omega$, then, the moments appear to ignore the component rotating at 2ω. The problem of magnetic dipoles interacting with a linearly polarized field may then be treated as though they interacted with only one of the rotating components.

Before proceeding, consider the function $f(x_0)$ that characterizes the magnitude of $\mathbf{B}_1$ as the sample for NMR experiments is moved in and out of the coil [Eqs. (1.68)]. The intensity of the field as a function of x is shown in Fig. 1.20.

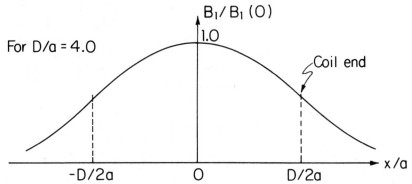

FIG. 1.20. The intensity of the field $\mathbf{B}_1$ on the axis of the solenoid as a function of position.

One sees that, at least inside the center third of the coil, the variation of B_1 with position is less than 10%. Even so, the accumulation of this error over long strings of pulses will be a factor that must be faced in the solid-state NMR experiments discussed in later chapters.

X. The Production of Radio-Frequency Fields: A Brief Introduction to Alternating-Current Circuit Theory

To produce the necessary fields in a coil, alternating current (ac) is required. The necessary ac circuit theory is rather fundamental, and the results are similar to Ohm's law for dc circuits. To obtain a current, a voltage must be imposed across an impedance. However, the voltage varies in time and is produced by a *transmitter* (or a *power amplifier*). To solve completely the current equation in an experiment in which the voltage oscillates at frequency ω and is switched (pulsed) involves solving complicated differential equations. This results because voltages across capacitors, indicators, and resistors are different functions of the current:

$$V_{capacitor}(t) = \frac{1}{C} \int_0^t I(t)dt \qquad (1.69)$$

$$V_{inductor}(t) = L\frac{d}{dt}I(t) \qquad (1.70)$$

$$V_{resistor}(t) = RI(t) \qquad (1.71)$$

with units of volts, farads, amps, ohms, and henries.

An inductor such as is needed in magnetic resonance must be in some way connected to the transmitter and that connection must, of necessity, involve other electrical elements. For example, consider the circuit in Fig. 1.21, where X is a radio-frequency transmitter driving an inductance L, a capacitance C, and a resistance R in parallel. An exact solution of the response of this network to a transient ac voltage is difficult to obtain.

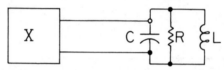

FIG. 1.21. A simple probe circuit, showing a transmitter driving a tuned RLC network.

For the present, however, the steady-state solution to the case of a sinusoidal variation in the voltage applied is all that is required. For an example of a solution involving square-wave excitation, see the paper by Vaughan et al. [7].

For the rest of this section, we will consider the steady-state equations under an ac excitation of the form

$$V(t) = V_0 e^{i\omega t} \qquad (1.72)$$

For a capacitor the relation between voltage and current is from (1.69):

$$V_c(t) = (-i/\omega C)I(t) \qquad (1.73)$$

This equation states that the voltage across a capacitor may be obtained by operating on the current with the operator $-i/\omega C$. The significance of the $-i$ is that the voltage across capacitors, inductors, and resistors may be represented on an Argand diagram, with voltage across resistors along the real axis. The voltage across a capacitor lags the voltage across a resistor in series with the capacitor by 90°. For an inductor, the relation between voltage and current is

$$i\omega L I(t) = V_i(t) \qquad (1.74)$$

which is to say that operating on the current with $i\omega L$ yields the voltage across the inductor. This voltage leads the voltage across a resistor in series with the inductor by 90°.

In steady state, the current–voltage relationships are similar to Ohm's law, but with the resistance R replaced by an impedance operator Z. The units of Z are ohms. Equation (1.71) now becomes

$$V(t) = ZI(t) \tag{1.75}$$

The impedance is in general a complex operator.

The impedance of every component may be defined, and these impedances may be treated like those of resistance in a dc circuit for inductive, capacitive, and resistive impedance, respectively:

$$Z_L = i\omega L \quad \text{ohms} \tag{1.76}$$

$$Z_C = -i/\omega C \quad \text{ohms} \tag{1.77}$$

$$Z_R = R \quad \text{ohms} \tag{1.78}$$

For elements in parallel,

$$1/Z_{\text{TOTAL}} = 1/Z_1 + 1/Z_2 + \cdots \tag{1.79}$$

And for elements in series

$$Z_{\text{TOTAL}} = Z_1 + Z_2 + \cdots \tag{1.80}$$

Thus, the circuit impedance of our model capacitor, resistor, and inductor in parallel (Fig. 1.21) can be expressed in terms of a lumped circuit impedance:

$$\frac{1}{Z_{\text{TOTAL}}} = \frac{1}{Z_C} + \frac{1}{Z_R} + \frac{1}{Z_L} = \frac{-iR(1 - \omega^2 LC) + \omega L}{\omega LR} \tag{1.81}$$

Suppose that we wish to make this circuit behave like a power resistor. In this case, the imaginary component must equal zero. It is obvious that for a given L and C there is only *one* frequency for which this condition is true:

$$\omega^2 = 1/LC$$

and the value of the impedance is just the resistance R. At such frequencies the circuit is said to be *resonant*. The sharpness of the resonance is a measure of how well *energy* is stored in the circuit.

Power is the rate at which energy is transferred from one circuit to another (for example, a transmitter to a resonant circuit containing the NMR coil that produces the $\mathbf{B}_1$ field). For a dc circuit, the power transfer is

$$P = IV = I^2 R = V^2/R \tag{1.82}$$

In the circuit under discussion [i.e., having voltage (1.72)], the power is transferred only to the *resistive* part of the impedance, and the instantaneous power (energy transfer) is

$$P(t) = [\text{Re}\{I(t)\}]^2 \, \text{Re}\{Z\} \tag{1.83}$$

One way to consider power in a continuous power transfer is the *average power* per cycle, which is

$$\bar{P} = \tfrac{1}{2}[\{\text{Re}\,I_0\}]^2 \, \text{Re}\{Z\} \tag{1.84}$$

This is a measure of how fast power is removed from the circuit of the transmitter, smoothed out over any variations that may occur. For the model circuit, the average power is

$$\bar{P} = \frac{1}{2} I_0^2 \, \frac{\omega^2 L^2 R}{\omega^2 L^2 + (1 - \omega^2 LC)^2 R^2} \tag{1.85}$$

A plot of $\bar{P}$ versus ω is shown in Fig. 1.22.

Also indicated in Fig. 1.22 is Q, the quality factor of a resonant circuit, which is a measure of the ability of the circuit to store energy. A useful relation between rf field intensity [oersted (Oe)] in a coil with volume V (in cubic centimeters) contained in a resonant circuit with quality factor Q at frequency f_0 (in megahertz) is

$$B_1 \, (\text{Oe}) \leq 3(PQ/f_0 V)^{1/2} \tag{1.86}$$

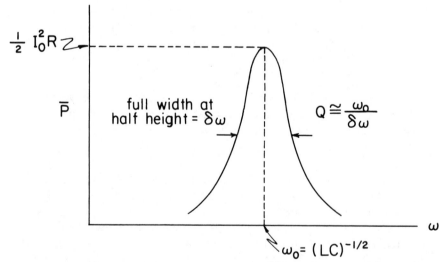

FIG. 1.22. The variation of average power absorption with frequency for a radio-frequency network tuned at ω_0.

for transmitter power P (watts). Practice in the use of this equation is given in Problem 1.8. The circuit theory for some useful probe circuits is given by Murphy and Gerstein [8].

From Fig. 1.22, it is clear that the amount of energy transferred per unit time depends on the value of L and C. A rotation of the magnetic moment requires a transfer of energy, and this can only come from the moments interaction with the field $\mathbf{B}_1$, which is a consequence of the presence of the current. Because of the sharpness of $\bar{P}$ in Fig. 1.22, we must utilize *tuned circuits* (for which $\omega = 1/\sqrt{LC}$) to efficiently transfer energy into the system.

PROBLEMS

1.1 The pulse width for a 90° pulse on a proton system is found to be 1.25 μsec.

(a) How many gauss is B_1 for this pulse?

(b) How long would a 90° pulse for a ^{19}F spin system be in the same spectrometer, utilizing the same transmitter power and the same frequency of operation?

1.2 Why does one prefer to wait $5T_1$ between scans in a pulse NMR experiment? What would be the quantitative effect of waiting $3T_1$?

1.3 Consider the FID shown in Fig. 1.15. What two pieces of information are available from this pattern?

1.4 Prove that the half-width at half-height of a Lorentzian absorption line is given by

$$\delta f = \frac{1}{2\pi T_2} \quad \sec^{-1}$$

1.5 A Gaussian lineshape function is given by

$$g(\delta\omega) = (T_2/\sqrt{\pi}) \exp(-T_2^2 \, \delta\omega^2/2)$$

(a) How is T_2 related to the half-width at half-height for this function? (Express your answer both in terms of $\delta\omega$ and δf.)

(b) Suppose you obtained T_2 for a given sample, assuming that the lineshape was Lorentzian, utilizing the half-width at half-height. What would be the percentage error you would have made in T_2 if the line were really Gaussian?

(c) One sometimes defines $2\delta f_{pp}$ for a given lineshape function, which is the full width of the line evaluated at the points at which the slope of the absorption spectrum changes sign. How is this value related to T_2 for a Gaussian line?

1.6 One of the methods of measuring T_1 is the application of a 180° pulse along the x axis in the rotating frame, followed by a time interval τ, and the subsequent application of a 90° pulse along x, followed by observation of the signal along the y axis of the rotating frame. The initial value of this signal is then plotted in an appropriate manner against τ and T_1 is extracted from this plot.

(a)   What is the differential equation to be solved for this situation?

(b)   Why is the experimental observation made along the y axis?

(c)   Clearly, at $\tau = 0$, the initial value of the magnetization is $-M_0$ along z, where M_0 is the initial polarization in the z direction before the 180° pulse. With this boundary condition, show that the solution to the differential equation of part a is

$$M_z(\tau) = M_0[1 - 2\exp(-\tau/T_1)].$$

1.7   Initial amplitudes of decay, following the sequence $\pi/2_x$, spin lock along y for time τ, observed along y were found as follows for a sample of polyethylene that was estimated to be 65% crystalline from its density.

Duration of locking pulse (τ) (msec)	$A_0(\tau)$	Duration of locking pulse (τ) (msec)	$A_0(\tau)$
0.20	2783	20.0	582
0.40	2730	25.0	461
1.0	2622	30.0	431
1.5	2555	35.0	332
3.0	2511	40.0	330
15.0	690		

What are the values of $T_{1\rho}$ for this sample, and to what does each value physically correspond?

1.8   A rather standard probe circuit used for solid-state NMR is the so-called series tapped tuned circuit shown below.

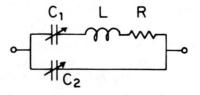

(a)   With $R = 0$, and $L = 0.2\ \mu H$, what will be the value of C_1 that will allow the series resonant circuit consisting of just C_1 and L to resonate at $f = 60 \times 10^6\ \text{sec}^{-1}$ (i.e., $f = 60$ MHz; i.e., $\omega = 377 \times 10^6\ \text{rad sec}^{-1}$)?

(b)   With $R = 0$, how many resonances will the circuit have; i.e., at how many frequencies will the impedance be real?

(c)   This circuit is used with a broadband transmitter, capable of linear operation between 2 and 60 MHz, with power in pulse mode equal to 400 W.

To accommodate a standard 5-mm NMR tube, the inductor is wound to have an inside diameter of 6 mm with a length of 20 mm to ensure a uniform B_1. With $R = 3$ ohms, $L = 0.2 \, \mu H$, $C_1 = 3 \, pF$, and $C_2 = 40 \, pF$, this circuit is found to have a resonance at 56 MHz with Q of 30. What will be the maximum value of the B_1 field attainable?

(d)   What will be the time necessary for a 90° pulse on protons at 56 MHz utilizing this circuit as described?

(e)   What will be the time necessary for a 90° pulse for ^{13}C at 56 MHz using this circuit as described?

REFERENCES

1. B. I. Bleaney and B. Bleaney, "Electricity and Magnetism," par. 20.1, 22.1, and Appendix A. 10. Clarendon Press, Oxford, 1957.
2. T. C. Farrar and E. D. Becker, "Pulse and Fourier Transform NMR," Chapter 1 and par. 2.1. Academic Press, New York, 1971.
3. C. P. Slichter, "Principles of Magnetic Resonance," 2nd ed, Chapter 1, par. 2.1, 2.4, 2.7, and 2.8. Springer-Verlag, Berlin, 1978.
4. A. Abragam, "The Principles of Nuclear Magnetism," Chapter III, Sec. I-III. Clarendon Press, Oxford, 1961.
5. D. C. Champeney, "Fourier Transforms and Their Practical Applications." Academic Press, New York, 1973.
6. L. R. Windmiller, J. B. Ketterson, and J. C. Shaw, "Fast Fourier Transform Using the PDP11." Argonne National Laboratory Report ANL 7907, Argonne, Illinois, (1972).
7. R. W. Vaughan, D. D. Elleman, L. M. Stacey, W-K. Rhim, and J. W. Lee, A simple, low power, multiple pulse NMR spectrometer, *Rev. Sci. Instrumen.* **43**, 1356 (1972).
8. P. Dubois Murphy and B. C. Gerstein, "Analysis and Computerized Design of NMR Probe Circuits." Report IS-4436, National Technical Information Service, U.S. Department of Commerce, Springfield, Virginia.

CHAPTER 2

QUANTUM MECHANICS
OF SPIN STATES, THE DENSITY MATRIX,
INTERACTION FRAMES, AND
THE POLARIZATION VECTOR

I. Introduction

The phenomenon of magnetic resonance (or any resonance spectroscopy utilizing electromagnetic radiation) can be viewed classically, as in Chapter 1. The magnetic resonance phenomenon is, however, quantum mechanical, and a discussion of magnetic resonance requires an understanding of the quantum mechanics of spin. This chapter is devoted to a development of time-dependent quantum mechanics, particularly the time-dependent quantum mechanics of spin systems. Since a physical event is understood with more depth when seen from several points of view, it will be useful for the reader to compare the description from a classical point of view as described in Chapter 1 with the discussion obtained in this chapter.

Time-dependent quantum mechanics is a subject not generally a part of the average chemist's background. This chapter provides a discussion of the time-dependent quantum mechanics that may be used as background for the remainder of the volume. Thus, unlike Chapter 1, the treatment may stress mathematics, with a lack of physical examples. In Chapter 5, the equations of the chapter are applied to concrete examples, after an exposition of interactions (Chapter 3) and the methods of solving the equations (Chapter 4).

II. Dirac's Picture of Quantum Mechanics

Quantum mechanics is often introduced to students via the coordinate representation, in which states of a physical system are functions of the coordinates of the particles that comprise the system and the time. The objective of the exercise is to describe the state of the system by finding the wave-function $\Psi(\mathbf{r}, t)$. In the coordinate representation, this wave function is obtained by solving the Schrödinger equation[†]

$$i \frac{\partial \Psi}{\partial t} (\mathbf{r}, t) = \hat{\mathscr{H}}(\mathbf{r}, t) \Psi(\mathbf{r}, t) \tag{2.1}$$

The Hamiltonian $\hat{\mathscr{H}}(\mathbf{r}, t)$ describes the interactions that specify the system. The equation simplifies when the Hamiltonian is independent of time. The wave function is factorable in this case: $\Psi_n(\mathbf{r}, t) = \phi_n(\mathbf{r}) \exp(-iE_n t)$ and the spatial part of the wave function $\phi_n(\mathbf{r})$ satisfies the equation

$$\hat{\mathscr{H}}(\mathbf{r}) \phi_n(\mathbf{r}) = E_n \phi_n(\mathbf{r}) \tag{2.2}$$

The solutions $\phi_n(\mathbf{r}) \exp(-iE_n t)$ are the *stationary states* of the system, each corresponding to a specific energy E_n. The most general solution of Eq. (2.1) may not be a state of specific energy; but, since the stationary states span the space of solutions to Eq. (2.1), a general solution may be specified as a linear combination of the stationary solutions:

$$\Psi(\mathbf{r}, t) = \sum_n C_n \phi_n(\mathbf{r}) \exp(-iE_n t)$$

where the C_ns are complex numbers, the values of which determine the state.

Often, in discussing a physical situation, it is not necessary to describe the spatial structure of a state, but rather to describe the energy stationary states of the system. Under these circumstances, it is easier to use a notation due to Dirac, in which the stationary states of the system are represented by *kets* and *bras*, which explicitly indicates the dependence on the quantum number, without reference to coordinates. Symbolically, the eigenstate is represented by $|n\rangle$ (or its complex conjugate $\langle n|$). The coordinate representation of this state is just $\phi_n(\mathbf{r})$ [or its complex conjugate $\phi_n^*(\mathbf{r})$]. Thus, one may rewrite the general solution to Schrödinger's equation as

$$|\Psi\rangle = \sum_n C_n |n\rangle \exp(-iE_n t) \tag{2.3}$$

This form suggests the analogy to expansion of a vector in terms of the set of unit vectors that span the space, with $|\Psi\rangle$ being the analog of the vector and the eigenkets $|n\rangle$ being the analogs of the unit vectors that span the space.

The Dirac notation has an additional advantage. The bras and kets represent states, and the product of a bra with a ket represents the total overlap of the states:

$$\langle \Psi_a | \Psi_b \rangle = \int \Psi_a^*(\mathbf{r}) \Psi_b(\mathbf{r}) \, d^3 \mathbf{r}$$

[†] These energies, and those throughout this volume, are expressed in radians per second.

Similarly, *observables* are represented by integrals of the operators $\hat{\theta}$, which in Dirac notation is given, where $|\Psi\rangle$ is the state of the system, by

$$\langle\hat{\theta}\rangle = \langle\Psi|\hat{\theta}|\Psi\rangle \equiv \int \Psi^*(\mathbf{r})\hat{\theta}\Psi(\mathbf{r})\,d^3\mathbf{r}$$

A very useful operator is given by the symbol $|\phi\rangle\langle\chi|$, which may act either as a bra or a ket,

$$\langle a|(|\phi\rangle\langle\chi|) = \langle a|\phi\rangle\langle\chi| = \int a^*\phi d^3\mathbf{r}\,\langle\chi|$$

or

$$(|\phi\rangle\langle\chi|)|b\rangle = |\phi\rangle\langle\chi|b\rangle = |\phi\rangle\int \chi^*bd^3\mathbf{r}$$

One sees that this operator produces a bra or ket multiplied by a number that represents the *overlap* of the bra (or ket) with the ket (or bra). In particular, the operator $|n\rangle\langle n|$, where $|n\rangle$ is one of the eigenstates of the Hamiltonian, is a projection operator with the result

$$(|n\rangle\langle n|)|\Psi\rangle = |n\rangle\langle n|\Psi\rangle = |n\rangle\int \phi_n^*(\mathbf{r})\Psi(\mathbf{r})\,d^3\mathbf{r}$$

The result is the ket $|n\rangle$, multiplied by the overlap integral of the state $|\Psi\rangle$, with the eigenket $|n\rangle$.

III. Observables: Physical Interpretation

All measurements of a physical quantity must yield real numbers. One associates the value of a physical observable with the expectation value of the operator for that observable: $\langle\hat{\theta}\rangle = \langle\Psi|\hat{\theta}|\Psi\rangle$. A real number is equal to its complex conjugate. Hence, the operators associated with physical quantities must have the property that

$$\langle\Psi|\hat{\theta}|\Psi\rangle = (\langle\Psi|\hat{\theta}|\Psi\rangle)^*$$

Such an operator is said to be Hermitian.

One may calculate the value predicted for the observable by noting that any state $|\Psi\rangle$ may be expanded in terms of the eigenstates of the operator

$$|\Psi\rangle = \sum_k C_k|k\rangle$$

The expectation value is then

$$\langle\Psi|\hat{\theta}|\Psi\rangle = \sum_{j,k} C_j^* C_k\langle j|\hat{\theta}|k\rangle$$
$$= \sum_{j,k} C_j^* C_k\theta_k\langle j|k\rangle$$
$$= \sum_k C_k^* C_k\theta_k \qquad (2.4)$$

where it is assumed that the set of eigenkets is orthonormal. If $C_k C_k^*$ is interpreted as the probability p_k of the state $|\Psi\rangle$, having the properties of the eigenket $|k\rangle$, one sees that this expectation value is equivalent to the statistical average $\langle\Psi|\hat\theta|\Psi\rangle = \sum_k \theta_k p_k$.

Very often in such measurements, one does not measure the expectation value of a single small system, but rather the expectation value over some larger system of which the small system is a subsystem. Thus, one observes the total magnetic moment that is a composite of the moments of the subsystems. To treat this situation, one has to define an ensemble of systems. Each system in the ensemble is prepared in exactly the same manner, and the observable is the average over the ensemble of the expectation value of the observable in each system. In particular, when the systems in the ensemble are all loosely coupled to a thermal bath of very large (or infinite) heat capacity for a sufficiently long time that thermal equilibrium has ensued, the distribution of values of the probability $C_k C_k^*$ is very sharp and may be replaced in calculations over the ensemble by its most probable value

$$p_k = \frac{\exp(-\hbar\epsilon_k/kT)}{Z} = \overline{C_k C_k^*}$$

where the bar indicates an average over the systems in the ensemble and Z is the partition function for the subsystem

$$Z = \sum_{\substack{\text{system} \\ \text{energy} \\ \text{states}}} \exp(-\hbar\epsilon_k/kT)$$

The partition function for the ensemble of N identical *but distinguishable* weakly coupled subsystems is a product

$$Z_{\text{ensemble}} = Z_{\text{subsystem}}^N$$

where each subsystem, although identical to all the rest, is distinguishable from the rest. The expectation value averaged over the ensemble is found by taking the expectation value over a state $|\Psi\rangle$ and averaging over the ensemble to give

$$\overline{\langle\Psi|\hat\theta|\Psi\rangle} = \sum_k \overline{C_k C_k^*}\,\theta_k \tag{2.5}$$

What is important in calculating an expectation value is this average over an ensemble of spin systems.

The determination of the energy states of a system such as a spin is one of the first requirements for understanding the phenomenon. We saw in Chapter 1 that an interaction exists between a rotating charge distribution and a static magnetic field. Zeeman first pointed out that the angular momentum states could have energies that depend on the orientation of the angular momentum and the magnetic field Eq. (1.31). This effect corresponds to the existence of a Hamiltonian that depends on the orientation of angular momentum and the magnetic field. For spins, it will be necessary to identify the energy states, which will be the spin angular momentum states. Thus, the quantum mechanics of angular momentum is a topic that is necessary to the understanding of NMR experiments.

IV. Angular Momentum Operators and Eigenstates

Classically the angular momentum is given by

$$\mathbf{L} = \mathbf{r} \times \mathbf{p}$$

Expanding the cross product one finds

$$L_z = xp_y - yp_x, \qquad L_y = zp_x - xp_z, \qquad L_x = yp_z - zp_y$$

Utilizing the operator form for the linear momentum $\mathbf{p} = -i\hbar\nabla$, one finds the forms of the angular momentum operators

$$\hat{L}_z = -i\hbar\left(x\frac{\partial}{\partial y} - y\frac{\partial}{\partial x}\right),\ \hat{L}_y = -i\hbar\left(z\frac{\partial}{\partial x} - x\frac{\partial}{\partial z}\right),\ \hat{L}_x = -i\hbar\left(y\frac{\partial}{\partial z} - z\frac{\partial}{\partial y}\right)$$

$$(2.6)$$

This is generally true for any angular momentum. By examining the commutators of the angular momentum operators with each other, one obtains general relations among the components

$$[\hat{L}_x, \hat{L}_y] = i\hbar\hat{L}_z, \qquad [\hat{L}_x, \hat{L}_z] = -i\hbar\hat{L}_y, \qquad [\hat{L}_y, \hat{L}_z] = i\hbar\hat{L}_x \qquad (2.7)$$

One may simplify the notation by moving to a natural notation in which $\hbar = 1$, in which case Eqs. (2.7) would be written as

$$[\hat{L}_x, \hat{L}_y] = i\hat{L}_z, \qquad [\hat{L}_x, \hat{L}_z] = -i\hat{L}_y, \qquad [\hat{L}_y, \hat{L}_z] = i\hat{L}_x \qquad (2.7')$$

This notation is compatible with the definition of energies in frequency units, and so we shall adopt it henceforth.

The general relations (2.7') are true for any angular momentum, be it orbital, spin, or rotational. Nuclear spin angular momentum is usually given the symbol $\mathbf{I}$. Thus, relations (2.7') for nuclear spin angular momentum may be written concisely as

$$[I_m, I_j] = i\epsilon_{mjk}I_k \qquad (2.8)$$

Here, ϵ_{mjk} is the Levi–Civita symbol, which is 0 if any of m, j, or k are identical, unity if m, j, and k are in cyclic order (i.e., x, y, z or z, x, y, etc.), and -1 if m, j, k are in anticyclic order (e.g., x, z, y, etc.). Therefore, for example,

$$[I_y, I_z] = iI_x \qquad (2.9)$$

and

$$[I_x, I_z] = -iI_y \tag{2.10}$$

One may further define the square of the total angular momentum as

$$I^2 = I_x I_x + I_y I_y + I_z I_z \tag{2.11}$$

for which the respective commutators give

$$
\begin{aligned}
[I^2, I_z] &= [I_x I_x, I_z] + [I_y I_y, I_z] \\
&= I_x I_x I_z - I_z I_x I_x + I_y I_y I_z - I_z I_y I_y \\
&= i\{-[I_x, I_y] + [I_x I_y]\} = 0
\end{aligned}
\tag{2.12}
$$

Similarly, one may show that

$$[I^2, I_x] = 0 \tag{2.13}$$

$$[I^2, I_y] = 0 \tag{2.14}$$

The fact that two operators commute implies that a complete set of simultaneous eigenfunctions exists for the two operators. Thus, one sees from Eqs. (2.8) and (2.12)–(2.14) that one may choose a complete set of simultaneous eigenfunctions of I^2 and any one of the operators I_x, I_y, or I_z; but it is not possible to choose a complete set of simultaneous states of any two of the three components I_x, I_y, and I_z. It is conventional to choose a complete set of simultaneous eigenstates of I^2 and I_z, $|I, m\rangle$, where the eigenvalues are specified as

$$I^2 |I, m\rangle = b|I, m\rangle \tag{2.15}$$

$$I_z |I, m\rangle = m|I, m\rangle \tag{2.16}$$

The action of the operators I_x and I_y on these states is best defined in terms of two operators I^+ and I^-:

$$I^\pm = I_x \pm iI_y \tag{2.17}$$

or, conversely,

$$I_x = \tfrac{1}{2}(I^+ + I^-) \tag{2.18a}$$

$$I_y = (-i/2)(I^+ - I^-) \tag{2.18b}$$

The commutation relations of I^+ and I^- with I_z are particularly useful

$$
\begin{aligned}
[I_z, I^\pm] &= [I_z, I_x] \pm i[I_z, I_y] \\
&= iI_y \pm i(-iI_x) \\
&= \pm(I_x \pm iI_y) \\
&= \pm I^\pm
\end{aligned}
$$

Also,

$$I_z I^{\pm} - I^{\pm} I_z = \pm I^{\pm}$$
$$I_z I^{\pm} = I^{\pm}(I_z \pm 1)$$

(2.19)

Consider a state $|I, m\rangle$, which is one of the states that obey the requirements (2.15) and (2.16). Applying the operator of (2.19) to this state gives the result

$$I_z I^+ |I, m\rangle = (m + 1) I^+ |I, m\rangle$$

(2.20)

that is, $I^+ |I, m\rangle$ is also an eigenstate of I_z, but with a different eigenvalue, namely, $(m + 1)$. Similarly one may show that $I^- |I, m\rangle$ is an eigenstate of I_z, but with the eigenvalue $(m - 1)$. Hence, we see the utility of these operators: Knowing one eigenstate of I_z, one can repetitively generate other eigenstates of I_z.

Here $I^+ |I, m\rangle$ need not be a normalized function, but it is convenient to define all of the eigenstates of I_z and I^2 as normalized functions. Most generally, then, the action of $I^{\pm}$ on $|I, m\rangle$ results in a product

$$I^+ |I, m\rangle = \alpha_m |I, m + 1\rangle$$

(2.21)

Taking the complex conjugate of this function gives

$$\langle I, m| I^- = \alpha_m^* \langle I, m + 1|$$

(2.22)

Taking the product of these two functions, one has

$$\langle I, m| I^- I^+ |I, m\rangle = \alpha_m^* \alpha_m \langle I, m + 1| I, m + 1\rangle$$

But, by expansion, one may show that

$$I^- I^+ = I_x^2 + I_y^2 + i[I_x, I_y]$$
$$= I^2 - I_z^2 - I_z$$

(2.23)

Thence

$$\langle I, m| I^- I^+ |I, m\rangle = |\alpha_m|^2$$
$$\langle I, m| I^2 - I_z^2 - I_z |I, m\rangle = \{b - m^2 - m\} \langle I, m| I, m\rangle = |\alpha_m|^2$$

Thus

$$|\alpha_m| = \{b - m(m + 1)\}^{1/2}$$

(2.24)

That gives

$$I^+ |I, m\rangle = \{b - m(m + 1)\}^{1/2} |I, m + 1\rangle$$

(2.25)

By an analogous argument, one may show that

$$I^- |I, m\rangle = \{b - m(m - 1)\}^{1/2} |I, m - 1\rangle$$

(2.26)

Since I^+ and I^- are Hermitian operators, $|\alpha_m|^2$ must be a positive number; thus

$$b - m^2 - m \geq 0$$

for every allowable value of m.

Call the value of m that gives the equality $|I|$. Then one sees that $b = I(I + 1)$ and $|m| \leq I$. Thus, one must have

$$-I \leq m \leq I \tag{2.27}$$

and one may rewrite the equations as

$$I^+|I,m\rangle = \{I(I + 1) - m(m + 1)\}^{1/2}|I, m + 1\rangle \tag{2.25'}$$

$$I^-|I,m\rangle = \{I(I + 1) - m(m - 1)\}^{1/2}|I, m - 1\rangle \tag{2.26'}$$

Equation (2.27) shows that the eigenvalue spectrum of I_z is bounded, having only $2I + 1$ values. From Eq. (1.31) one sees that each orientation of the spin angular momentum corresponds to a different energy. By the correspondence of classical and quantum properties, one may write the Hamiltonian for this interaction as

$$\hat{\mathscr{H}}_z = -\gamma \mathbf{I} \cdot \mathbf{B} \tag{2.28}$$

This Hamiltonian is the fundamental quantum-mechanical operator of the magnetic resonance phenomenon and will be used in many forms and in many situations in magnetic resonance.

V. Spin-$\frac{1}{2}$

For each particle with spin, the value of m goes from $-I$ to $+I$ in integer steps. The lowest possible number I for which this is true is $I = \frac{1}{2}$. The states that have this angular momentum are $|\frac{1}{2}, \frac{1}{2}\rangle$ and $|\frac{1}{2}, -\frac{1}{2}\rangle$, where the second number indicates the eigenvalue of I_z. For a pure Zeeman interaction then, with the total Hamiltonian being $\mathscr{H}_0$, the Hamiltonian in zero field, plus the Zeeman Hamiltonian $\mathscr{H}_z$,

$$E_{1/2} = -\gamma B_0/2 + E_0, \qquad E_{-1/2} = \gamma B_0/2 + E_0$$

The energy-level diagram for such a system is shown in Fig. 2.1.

It is conventional to choose the magnetic field direction to be the z direction of the laboratory coordinate system. Hence the Hamiltonian for the Zeeman interaction is

$$\mathscr{H}_z = -\gamma B_0 I_z \tag{2.29}$$

The Zeeman Hamiltonian and I_z can have the same eigenstates. Hence, the results of Fig. 2.1.

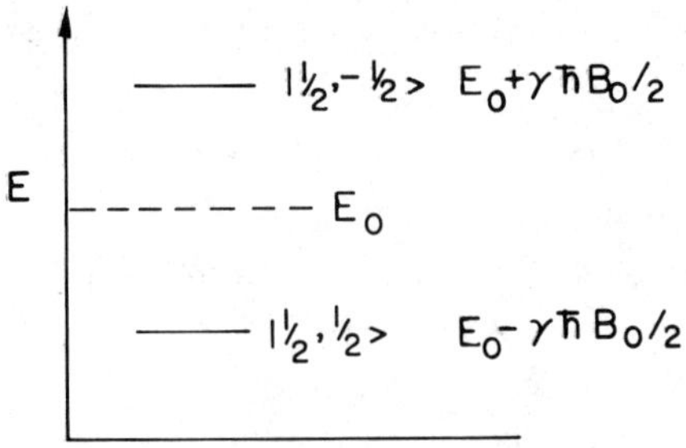

$$|\tfrac{1}{2}, -\tfrac{1}{2}\rangle \quad E_0 + \gamma \hbar B_0/2$$

$$E_0$$

$$|\tfrac{1}{2}, \tfrac{1}{2}\rangle \quad E_0 - \gamma \hbar B_0/2$$

FIG. 2.1. The energy-level diagram for a spin-$\tfrac{1}{2}$ nucleus in a magnetic field.

It is customary to write the operator's braket in matrix form. Thus, the matrix form for I_z, designated by the symbol $\langle I_z \rangle$, is

$$\langle I_z \rangle = \begin{bmatrix} \langle \tfrac{1}{2}, \tfrac{1}{2}|I_z|\tfrac{1}{2}, \tfrac{1}{2}\rangle & \langle \tfrac{1}{2}, \tfrac{1}{2}|I_z|\tfrac{1}{2}, -\tfrac{1}{2}\rangle \\ \langle \tfrac{1}{2}, -\tfrac{1}{2}|I_z|\tfrac{1}{2}, \tfrac{1}{2}\rangle & \langle \tfrac{1}{2}, -\tfrac{1}{2}|I_z|\tfrac{1}{2}, -\tfrac{1}{2}\rangle \end{bmatrix}$$

$$= \frac{1}{2}\begin{bmatrix} 1 & 0 \\ 0 & -1 \end{bmatrix}$$

Similarly,

$$\langle I_x \rangle = \frac{1}{2}\begin{bmatrix} 0 & 1 \\ 1 & 0 \end{bmatrix}$$

and

$$\langle I_y \rangle = \frac{i}{2}\begin{bmatrix} 0 & -1 \\ 1 & 0 \end{bmatrix}$$

One sees, therefore, that the vector angular momentum may be written as

$$\mathbf{I} = \frac{1}{2}\left\{ \begin{bmatrix} 0 & 1 \\ 1 & 0 \end{bmatrix}\mathbf{i} + \begin{bmatrix} 0 & -i \\ i & 0 \end{bmatrix}\mathbf{j} + \begin{bmatrix} 1 & 0 \\ 0 & -1 \end{bmatrix}\mathbf{k} \right\} = \frac{\boldsymbol{\sigma}}{2} \tag{2.30}$$

which defines the *Pauli spin vector* $\boldsymbol{\sigma}$.

The matrices that are coefficients are useful and have a role to play in the quantum mechanics of spin. They are the *Pauli spin matrices*. Since the spin operator is directly proportional to them, their mathematical properties are similar to that of the spin operator.

For example, the commutator is

$$[\sigma_x, \sigma_y] = i\sigma_z \tag{2.31}$$

or cyclic permutations of that equation.

The Pauli spin matrices are normalized in the sense that

$$\sigma_x \sigma_x = \sigma_y \sigma_y = \sigma_z \sigma_z = \begin{bmatrix} 1 & 0 \\ 0 & 1 \end{bmatrix}$$

One sees that the Pauli matrices may be used, particularly in computer calculations, as a representation of spin operators, and the effects of operators, as well as relations among the matrices, may be calculated.

One may generate spin matrices for spins higher than $\frac{1}{2}$ that will have analogous properties to the spin operators (see Problem 2.15). For the moment, we shall concern ourselves with study of spin-$\frac{1}{2}$.

VI. The Schrödinger Equation for Spin-$\frac{1}{2}$ with a Zeeman Interaction

The states $|\frac{1}{2},\frac{1}{2}\rangle$ and $|\frac{1}{2}, -\frac{1}{2}\rangle$ are the only eigenstates for spin-$\frac{1}{2}$. These two states are usually given the shorthand notation $|\alpha\rangle$ for $|\frac{1}{2},\frac{1}{2}\rangle$ and $|\beta\rangle$ for $|\frac{1}{2}, -\frac{1}{2}\rangle$. Any state of the spin-$\frac{1}{2}$ system may be described as a linear combination of these two states:

$$|\Psi_k(t)\rangle = C_{k\alpha}(t)|\alpha\rangle + C_{k\beta}(t)|\beta\rangle \tag{2.32}$$

Thus, to describe the state of the system, one needs only to give the values of $C_{k\alpha}(t)$ and $C_{k\beta}(t)$. The equation for $C_{k\alpha}$ and $C_{k\beta}$ may be found by solving Eq. (2.1), the time-dependent Schrödinger equation. This gives the result

$$\frac{\partial C_{k\alpha}}{\partial t}|\alpha\rangle + \frac{\partial C_{k\beta}}{\partial t}|\beta\rangle = -iC_{k\alpha}\mathcal{H}|\alpha\rangle - iC_{k\beta}\mathcal{H}|\beta\rangle \tag{2.33}$$

Since $|\alpha\rangle$ and $|\beta\rangle$ are orthogonal and normalized, multiplication by $\langle\alpha|$ and $\langle\beta|$, respectively, gives

$$\frac{\partial C_{k\alpha}}{\partial t} = -iC_{k\alpha}\langle\alpha|\mathcal{H}|\alpha\rangle - iC_{k\beta}\langle\alpha|\mathcal{H}|\beta\rangle$$

$$\frac{\partial C_{k\beta}}{\partial t} = -iC_{k\alpha}\langle\beta|\mathcal{H}|\alpha\rangle - iC_{k\beta}\langle\beta|\mathcal{H}|\beta\rangle \tag{2.34}$$

which must be solved for $C_{k\alpha}$ and $C_{k\beta}$.

Let us suppose, as a first example, that the only Hamiltonian is the Zeeman Hamiltonian, which has $|\alpha\rangle$ and $|\beta\rangle$ as eigenstates. Under these conditions, the preceding reduce to

$$\frac{\partial C_{k\alpha}}{\partial t} = -iE_\alpha C_{k\alpha} = +i\frac{\gamma B_0}{2}C_{k\alpha}$$

$$\frac{\partial C_{k\beta}}{\partial t} = -iE_\beta C_{k\beta} = -i\frac{\gamma B_0}{2}C_{k\beta} \tag{2.35}$$

The solutions of these are straightforward: with $C_{k\alpha}(0)$ being the value of the coefficient at time zero (see Problems 2.11 and 2.12),

$$C_{k\alpha}(t) = |C_{k\alpha}(0)|e^{-i\phi_\alpha}e^{i\gamma B_0 t/2}$$

$$C_{k\beta}(t) = |C_{k\beta}(0)|e^{+i\phi_\beta}e^{-i\gamma B_0 t/2} \tag{2.36}$$

where ϕ_α and ϕ_β are phase angles associated with the respective states. The state at time t is therefore

$$|\Psi_k(t)\rangle = |C_{k\alpha}(0)|e^{-i\phi_\alpha}e^{i\gamma B_0 t/2}|\alpha\rangle + |C_{k\beta}(0)|e^{i\phi_\beta}e^{-i\gamma B_0 t/2}|\beta\rangle \qquad (2.37)$$

Requiring $|\Psi_i(t)\rangle$ to be always normalized,

$$\langle\Psi_k(t)|\Psi_k(t)\rangle \equiv 1$$

so

$$|C_{k\alpha}(0)|^2 + |C_{k\beta}(0)|^2 = 1 \qquad (2.38)$$

and

$$|\Psi_k(t)\rangle = |C_{k\alpha}(0)|e^{-i\phi_\alpha}e^{i\gamma B_0 t/2}|\alpha\rangle + (1 - |C_{k\alpha}(0)|^2)^{1/2}e^{i\phi_\beta}e^{-i\gamma B_0 t/2}|\beta\rangle \qquad (2.39)$$

for which we calculate, for example, the single-particle energy *at any time*:

$$\langle E(t)\rangle = -|C_{k\alpha}(0)|^2 \tfrac{1}{2}\gamma B_0 + (1 - |C_{k\alpha}(0)|^2)\tfrac{1}{2}\gamma B_0 \qquad (2.40)$$

Under these conditions, the energy does not change with time!

Similarly

$$\langle M_z(t)\rangle = \frac{\gamma\hbar}{2}|C_{k\alpha}(0)|^2 - \frac{\gamma\hbar}{2}|C_{k\beta}(0)|^2 \qquad (2.41)$$

The z magnetization does not change with time!

Let us examine the transverse components

$$\langle M_x\rangle = \gamma\hbar\langle I_x\rangle = \frac{\gamma\hbar}{2}\{\langle I^+\rangle + \langle I^-\rangle\}$$

$$= \frac{\gamma\hbar}{2}\{\langle\Psi_k(t)|I^+|\Psi_k(t)\rangle + \langle\Psi_k(t)|I^-|\Psi_k(t)\rangle\}$$

$$= \frac{\gamma\hbar}{2}\{|C_{k\alpha}(0)||C_{k\beta}(0)|e^{i\phi_\alpha}e^{i\phi_\beta}e^{-i\gamma B_0 t}$$

$$+ |C_{k\alpha}(0)||C_{k\beta}(0)|e^{-i\phi_\alpha}e^{-i\phi_\beta}e^{i\gamma B_0 t}\}$$

$$= |C_{k\alpha}(0)||C_{k\beta}(0)|\frac{\gamma\hbar}{2}\{e^{-i(\gamma B_0 t - \phi_\alpha - \phi_\beta)} + e^{i(\gamma B_0 t - \phi_\alpha - \phi_\beta)}\}$$

$$= \gamma\hbar|C_{k\alpha}(0)||C_{k\beta}(0)|\cos(\gamma B_0 t - \phi_\alpha - \phi_\beta) \qquad (2.42)$$

Similarly one may show that

$$\langle M_y(t)\rangle = \gamma\hbar|C_{k\alpha}(0)||C_{k\beta}(0)|\sin(\gamma B_0 t - \varphi_\alpha - \varphi_\beta) \qquad (2.43)$$

That is, in agreement with the classical discussion of Chapter 1, under the condition that the system experiences only Zeeman interactions, the magnetic moment does not change, except for an oscillation caused by rotation about the field. However, the system does not decay away to an equilibrium state when it is in a state that is a linear combination of the eigenstates of the Hamiltonian acting upon it.

VII. The Single-Particle Probability Operator

It is apparent from the simple single-spin-$\frac{1}{2}$ example that the most important parameters in determining the state are the values of the coefficients. Since *any* state, transient or not, may be expanded in a complete set, one would like to find some compact way of expressing this relationship. We define a particular operator P arbitrarily such that

$$\langle k|P|j\rangle = C_j^*(t)C_k(t) \tag{2.44}$$

The braket of P is just the product of the coefficients that are utilized to calculate expectation values, where

$$C_k(t) = |C_k(0)|\exp(-i\phi_k)\exp(i\gamma B_0 t/2)$$

This is the single-particle *probability operator*. It is a measure of the state $\Psi(t)$ because it depends on $C_k(t)$ and $C_j^*(t)$. One may rewrite the expectation values of, for example, M_z in terms of the probability operator for the spin-$\frac{1}{2}$ case

$$\langle M_z(t)\rangle = \frac{\gamma\hbar}{2}\langle\alpha|P|\alpha\rangle - \frac{\gamma\hbar}{2}\langle\beta|P|\beta\rangle$$

$$= \langle\alpha|M_z|\alpha\rangle\langle\alpha|P|\alpha\rangle + \langle\beta|M_z|\beta\rangle\langle\beta|P|\beta\rangle \tag{2.45}$$

But one knows

$$\langle\alpha|M_z|\beta\rangle = \langle\beta|M_z|\alpha\rangle = 0 \tag{2.46}$$

So one may add terms to Eq. (2.45):

$$\langle M_z(t)\rangle = \langle\alpha|M_z|\alpha\rangle\langle\alpha|P|\alpha\rangle + \langle\alpha|M_z|\beta\rangle\langle\beta|P|\alpha\rangle$$
$$+ \langle\beta|M_z|\alpha\rangle\langle\alpha|P|\beta\rangle + \langle\beta|M_z|\beta\rangle\langle\beta|P|\beta\rangle$$

$$= \sum_{k=\alpha}^{\beta}(\langle\alpha|M_z|k\rangle\langle k|P|\alpha\rangle + \langle\beta|M_z|k\rangle\langle k|P|\beta\rangle)$$

$$= \sum_{k,j=\alpha}^{\beta}\langle j|M_z|k\rangle\langle k|P|j\rangle \tag{2.47}$$

It can be shown that, in a sum over complete orthonormal states,

$$\sum_k |k\rangle\langle k| = 1 \tag{2.48}$$

Hence

$$\langle M_z(t) \rangle = \sum_j \langle j | M_z P | j \rangle$$

Such a sum over states is independent of the set of states chosen as basis and is called the trace (or spur):

$$\langle M_z(t) \rangle = \text{Tr } M_z P(t). \tag{2.49}$$

Since traces are independent of the states one chooses, we could have chosen any orthonormal linear combination of $|\alpha\rangle$ and $|\beta\rangle$ to represent the spin-$\frac{1}{2}$ system; i.e., observables are independent of unitary transformations of the coordinate system.

While we have chosen to represent the result only for M_z for spin-$\frac{1}{2}$, the operation is quite generally true for any operator θ,

$$\langle \theta(t) \rangle = \text{Tr } \theta P(t) \tag{2.50}$$

One may then transfer the problem from one of finding the $C_i(t)$ to finding $P(t)$, the operator.

VIII. The Spin-$\frac{1}{2}$ Problem with an Arbitrary Hamiltonian: The Density Operator

We found above that a spin-$\frac{1}{2}$ particle did not change its state with time except for some oscillatory character when subject to the Zeeman interaction. We now suppose that the spin-$\frac{1}{2}$ particle is subject to a Hamiltonian, the spin-dependent part of which contains the Zeeman interaction $\mathscr{H}_0$ and other additional terms $\mathscr{H}_1(t)$. The instantaneous state may still be expanded as before:

$$\Psi_k(t) = C_{k\alpha}(t) | \alpha \rangle + C_{k\beta}(t) | \beta \rangle$$

since $|\alpha\rangle$ and $|\beta\rangle$ span the complete set of spin states for spin-$\frac{1}{2}$. Once again, the Schrödinger equation yields Eq. (2.34), but

$$\mathscr{H} = \mathscr{H}_0 + \mathscr{H}_1 \tag{2.51}$$

We assume that we may project out the part of $\mathscr{H}_1$ that commutes with $\mathscr{H}_0$, $\mathscr{H}_1^s$, and that which does not, $\mathscr{H}_1^n$:

$$[\mathscr{H}_0, \mathscr{H}_1^s] = 0, \qquad [\mathscr{H}_0, \mathscr{H}_1^n] \neq 0 \tag{2.52}$$

The fact that two operators commute is a requirement for the two operators to have simultaneous eigenfunctions. Thus,

$$\mathscr{H}_1^s | \alpha \rangle = \epsilon_\alpha | \alpha \rangle, \qquad \mathscr{H}_1^s | \beta \rangle = \epsilon_\beta | \beta \rangle \tag{2.53}$$

Thus we may define $\mathcal{H}_{01} = \mathcal{H}_0 + \mathcal{H}_1^s$ with eigenvalues E'_α and E'_β, where

$$E'_\alpha = \frac{-\gamma B_0}{2} + \epsilon_\alpha, \qquad E'_\beta = \frac{\gamma B_0}{2} + \epsilon_\beta \qquad (2.54)$$

Then Eq. (2.34) reduces to

$$\frac{\partial}{\partial t} C_{k\alpha} = -iC_{k\alpha}E'_\alpha - iC_{k\beta}\langle\alpha|\mathcal{H}_1^n(t)|\beta\rangle$$

$$\qquad (2.55)$$

$$\frac{\partial}{\partial t} C_{k\beta} = -iC_{k\alpha}\langle\beta|\mathcal{H}_1^n(t)|\alpha\rangle - iC_{k\beta}E'_\beta$$

This set of equations is not independent of representation. However, the single-particle probability operator may be used for this purpose. In terms of the states $|\alpha\rangle$ and $|\beta\rangle$, Eq. (2.44) reads

$$\langle\alpha|P|\beta\rangle = C_\beta^* C_\alpha \qquad (2.44')$$

where C_β and C_α are the coefficients of Eq. (2.55). By differentiation of Eq. (2.44'), one gets an equation for the time dependence of the matrix elements of the single-particle probability operator:

$$\frac{\partial}{\partial t}\langle\alpha|P|\beta\rangle = \frac{\partial C_\beta^*}{\partial t}C_\alpha + C_\beta^*\frac{\partial C_\alpha}{\partial t}$$

$$= i[C_\beta^* E'_\beta + C_\alpha^*\langle\alpha|\mathcal{H}_1^n(t)|\beta\rangle]C_\alpha$$
$$\quad - iC_\beta^*[C_\alpha E'_\alpha + C_\beta\langle\alpha|\mathcal{H}_1^n(t)|\beta\rangle]$$
$$\equiv i[\langle\alpha|P|\beta\rangle\langle\beta|\mathcal{H}_{01}|\beta\rangle + \langle\alpha|P|\alpha\rangle\langle\alpha|\mathcal{H}_1^n(t)|\beta\rangle$$
$$\quad - \langle\alpha|P|\beta\rangle\langle\alpha|\mathcal{H}_{01}|\alpha\rangle - \langle\beta|P|\beta\rangle\langle\alpha|\mathcal{H}_1^n(t)|\beta\rangle] \qquad (2.56)$$

Both the off-diagonal matrix elements of $\mathcal{H}_{01}$ and the diagonal matrix elements of $\mathcal{H}_1^n(t)$ are zero. Also, since $\mathcal{H}$ is Hermitian, $\langle\alpha|\mathcal{H}_1^n(t)|\beta\rangle = \langle\beta|\mathcal{H}_1^n(t)|\alpha\rangle$. Using these identities, Eq. (2.56) may be rewritten as

$$\frac{\partial}{\partial t}\langle\alpha|P|\beta\rangle = i[\langle\alpha|P|\beta\rangle\langle\beta|\mathcal{H}_{01}|\beta\rangle + \langle\alpha|P|\alpha\rangle\langle\alpha|\mathcal{H}_{01}|\beta\rangle$$
$$\quad - \langle\alpha|\mathcal{H}_{01}|\beta\rangle\langle\beta|P|\beta\rangle - \langle\alpha|\mathcal{H}_{01}|\alpha\rangle\langle\alpha|P|\beta\rangle]$$
$$\quad + i[\langle\alpha|P|\beta\rangle\langle\beta|\mathcal{H}_1^n(t)|\beta\rangle + \langle\alpha|P|\alpha\rangle\langle\alpha|\mathcal{H}_1^n(t)|\beta\rangle$$
$$\quad - \langle\alpha|\mathcal{H}_1^n(t)|\beta\rangle\langle\beta|P|\beta\rangle - \langle\alpha|\mathcal{H}_1^n(t)|\alpha\rangle\langle\alpha|P|\beta\rangle]$$
$$\equiv -i\sum_{j=\alpha}^{\beta}\{\langle\alpha|\mathcal{H}_{01}|j\rangle\langle j|P|\beta\rangle - \langle\alpha|P|j\rangle\langle j|\mathcal{H}_{01}|\beta\rangle$$
$$\quad + \langle\alpha|\mathcal{H}_1^n(t)|j\rangle\langle j|P|\beta\rangle - \langle\alpha|P|j\rangle\langle j|\mathcal{H}_1^n(t)|\beta\rangle\}$$
$$\equiv i\{\langle\alpha|[P,\mathcal{H}_{01}]|\beta\rangle + \langle\alpha|[P,\mathcal{H}_1^n(t)]|\beta\rangle\} \qquad (2.57)$$

Analogous equations are found for all the other brakets of the single-particle probability operator. If the states are time independent, Eq. (2.57) is the Liouville–von Neumann equation

$$i\frac{\partial P}{\partial t} = [\mathscr{H}, P] \tag{2.58}$$

The importance of this equation is that it gives us a representation-independent means of determining the time evolution of the system. It will be the subject of the rest of the book.

Once again we define an operator to remind us that we are looking at a large number of particles: the density operator ρ,

$$\langle \alpha|\rho|\beta \rangle = \overline{C^*_\beta(t)C_\alpha(t)} \tag{2.59}$$

where the bar indicates an average over all particles in the sample. The density operator behaves in time in a manner perfectly analogous to the probability operator:

$$i\frac{\partial \rho}{\partial t} = [\mathscr{H}, \rho] \tag{2.60}$$

If one further defines a set of coefficients

$$a_{k\alpha} = C_{k\alpha}\exp(iE'_\alpha t), \qquad a_{k\beta} = C_{k\beta}\exp(iE'_\beta t) \tag{2.61}$$

one finds that, for the time dependence of the products, one gets equations of the form

$$\frac{\partial}{\partial t}\overline{a^*_\beta a_\alpha} = i\langle \alpha|[\rho, \mathscr{H}^n_1(t)]|\beta \rangle \exp[-i(E'_\alpha)t] \tag{2.62}$$

Similar results are obtained for the other products.

Defining $\overline{a^*_\beta a_\alpha} = \langle \alpha|\rho^\dagger|\beta \rangle$, one sees

$$\frac{\partial}{\partial t}\langle \alpha|\rho^\dagger|\beta \rangle = i\langle \alpha|\exp(i\mathscr{H}_{01}t)[\rho, \mathscr{H}^n_1(t)]\exp(-i\mathscr{H}_{01}t)|\beta \rangle$$

$$= i\langle \alpha|[\rho^\dagger, \mathscr{H}^{n\dagger}_1(t)]|\beta \rangle \tag{2.63}$$

or

$$i\frac{\partial \rho^\dagger}{\partial t} = [\mathscr{H}^{n\dagger}_1(t), \rho^\dagger] \tag{2.64}$$

where

$$\langle \alpha|\rho^\dagger|\beta \rangle = \langle \alpha|\rho|\beta \rangle \exp[-i(E'_\beta - E'_\alpha)t]$$

$$\mathscr{H}^{n\dagger}_1(t) = \exp(i\mathscr{H}_{01}t)\mathscr{H}^n_1(t)\exp(-i\mathscr{H}_{01}t) \tag{2.65}$$

This is an element of the density operator in the rotating frame

$$\rho^{\dagger}(t) = \exp(i\mathcal{H}_{01}t)\rho(t)\exp(-i\mathcal{H}_{01}t) \tag{2.66}$$

Equation (2.64) is also very important since we detect magnetization in the rotating frame; that is, the observables are

$$\langle M_x(t)\rangle = \operatorname{Tr} \rho^{\dagger}(t)M_x$$

$$\langle M_y(t)\rangle = \operatorname{Tr} \rho^{\dagger}(t)M_y \tag{2.67}$$

$$\langle M_z(t)\rangle = \operatorname{Tr} \rho^{\dagger}(t)M_z$$

and *not* generally the components in the laboratory frame

$$\langle M_x^L(t)\rangle = \operatorname{Tr} \rho(t)M_x$$

$$\langle M_y^L(t)\rangle = \operatorname{Tr} \rho(t)M_y \tag{2.68}$$

$$\langle M_z^L(t)\rangle = \operatorname{Tr} \rho(t)M_z$$

Once one knows how the density operator changes in time, he knows how the state of the system evolves. Thus, the solution of the Liouville–von Neumann equation is important for determining the results of measurements on quantum systems.

IX. The Equilibrium Density Matrix for Spin-$\frac{1}{2}$

The density operator determines the state of a system at any time. Under most conditions it is difficult to find the exact form of this operator, and one resorts to approximations or series expansions. However, the equilibrium density operator for a nuclear spin system in a strong magnetic field can be written down because it is determined by classical thermal energy partitioning. Thus if $\hat{\mathcal{H}}_0$ is the time-independent part of the Hamiltonian for a system

$$\rho_{\text{eq}} = \frac{\exp(-\hbar\hat{\mathcal{H}}_0/kT)}{\sum_k \langle k|\exp(-\hbar\mathcal{H}_0/kT)|k\rangle} \tag{2.69}$$

For a system of spin-$\frac{1}{2}$ particles that are subject only to a Zeeman interaction, the sum in the denominator of Eq. (2.69) can be evaluated. For typical spins-$\frac{1}{2}$ at temperatures above 1 K (see Problem 2.13), $|\hat{\mathcal{H}}_0| \ll kT/\hbar$, and the exponentials may be expanded to lowest order to give

$$\rho_{\text{eq}} \approx \frac{1 + \gamma\hbar B_0 I_z/kT}{\exp(\gamma\hbar B_0/2kT) + \exp(-\gamma\hbar B_0/2kT)}$$

$$\approx \frac{1 + \gamma\hbar B_0 I_z/kT}{2} = \frac{1}{2} + bI_z \tag{2.70}$$

Using this form of ρ_{eq} and Eq. (2.68), one may calculate the equilibrium magnetization

$$
\begin{aligned}
\langle M_z^{eq} \rangle &= \text{Tr } M_z \rho_{eq} \\
&= \sum_{k,j} \langle k|M_z|j\rangle\langle j|\rho_{eq}|k\rangle \\
&= \frac{\gamma\hbar}{2} \sum_{k,j} \langle k|I_z|j\rangle\langle j|k\rangle + \frac{\gamma^2\hbar^2 B_0}{2kT} \sum_{k,j} \langle k|I_z|j\rangle\langle j|I_z|k\rangle \\
&= \frac{\gamma\hbar}{2} \sum_{k} \langle k|I_z|k\rangle + \frac{\gamma^2\hbar^2 B_0}{2kT} \sum_{k} \langle k|I_z^2|k\rangle \\
&= \frac{\gamma^2\hbar^2 B_0}{2kT} \sum_{k} m_k^2 \\
&= \frac{\gamma^2\hbar^2 B_0}{4kT}
\end{aligned}
$$

(2.71)

This result is the Curie–Weiss law

$$\langle M_z^{eq} \rangle = C(B_0/T) \tag{2.72}$$

for spin-$\frac{1}{2}$. More generally, for any spin I at high temperature one may calculate a similar result, with the Curie constant for an ensemble of N spins given by the more general formula

$$C = N[\gamma^2\hbar^2 I(I + 1)/3k] \tag{2.73}$$

The transverse component of magnetization at equilibrium may similarly be calculated for spin-$\frac{1}{2}$:

$$
\begin{aligned}
\langle M_x \rangle &= \text{Tr } \rho_{eq} M_x \\
&= \gamma\hbar \, \text{Tr } \rho_{eq} I_x \\
&= \frac{\gamma\hbar}{2} \{\text{Tr } \rho_{eq} I^+ + \text{Tr } \rho_{eq} I^-\} \\
&= \frac{\gamma\hbar}{4} \sum_{k,j} \{\langle k|j\rangle\langle j|I^+|k\rangle + \langle k|j\rangle\langle j|I^-|k\rangle\} \\
&\quad + \frac{\gamma^2\hbar^2 B_0}{4kT} \sum_{k,j} \{\langle k|I_z|j\rangle\langle j|I^+|k\rangle + \langle k|I_z|j\rangle\langle j|I^-|k\rangle\} \\
&= \frac{\gamma\hbar}{4} \sum_{k} \{\langle k|I^+|k\rangle + \langle k|I^-|k\rangle\} \\
&\quad + \frac{\gamma^2\hbar^2 B_0}{4kT} \sum_{k} \{m_k\langle k|I^+|k\rangle + m_k\langle k|I^-|k\rangle\}
\end{aligned}
$$

(2.74)

The raising and lowering operators do not connect the same state, so each term in each sum is zero. So, consequently

$$\langle M_x^{eq} \rangle = 0 \tag{2.75}$$

Similarly,

$$\langle M_y^{eq} \rangle = 0 \tag{2.76}$$

At equilibrium, the result obtained from classical arguments turns out to be predicted quantum mechanically. That is, there is no transverse magnetization and the component along the field is in agreement with the Curie law.

The result in Eq. (2.71) is the expectation value of M_z for *one* spin-$\frac{1}{2}$ of a weakly coupled ensemble in thermal equilibrium. Of course, the sample consists of N such spins, so the macroscopic magnetization is

$$\langle M_z \rangle_{eq} = (N\gamma^2 \hbar^2 / 4kT) B_0 \tag{2.77}$$

Note that the preceding equation is equivalent to the statement made earlier; $Z_{\text{ensemble}} = Z_{\text{system}}^N$ for a solid.

X. The Total Density Operator

Although we have (with appropriate caveats) talked about the spin system as if it were an isolated piece of the universe, it really exists in constant contact with all the other degrees of freedom of the molecule, clusters of molecules, and, ultimately, the whole universe. It is through these contacts that the spin system takes up and gives off energy, changes state, and wends its way toward equilibrium. The contacts are indicated by Hamiltonians that involve the properties of both the spin system and the remaining parts of the world. For example, the B_1 field really constitutes a manifestation of the interaction of the magnetization with some "distant" (at least compared to nuclear dimensions) circulating charges. Similarly there are couplings to other degrees of freedom—the electrons, the other nuclei, the rotation of molecules, etc. We shall consider the specific form of each one of these in a later chapter. For now, let us assume that the Hamiltonians exist and contribute to the time development of the system. If these external parts of the universe are virtually independent (i.e., "loosely coupled") of the spin system, the density operator may be written as a simple product

$$\rho_T(t) = \rho(t)\rho_R \tag{2.78}$$

where $\rho(t)$ is the density operator for the spins and ρ_R the density operator for the remainder of the system. The physics behind Eq. (2.78) is discussed at the end of Section VII. If one is calculating just the spin properties,

$$\rho(t) = \mathop{\text{Tr}}_{\substack{\text{all other} \\ \text{degrees}}} \rho_T(t) \tag{2.79}$$

So, for example,

$$\langle \mathbf{M}(t) \rangle = \text{Tr } \rho_T(t) \mathbf{M}$$

$$= \underset{\substack{\text{all other spin} \\ \text{degrees}}}{\text{Tr}} \rho_R \text{ Tr } \rho(t) \mathbf{M}$$

Since $\text{Tr } \rho_R$ over all other degrees of freedom gives the average spin density matrix one finds that

$$\langle \mathbf{M}(t) \rangle = \underset{\text{spin}}{\text{Tr}} \rho(t) \mathbf{M} \tag{2.80}$$

Thus, in considering properties of spin, we need only be concerned with the time evolution of ρ, appropriately averaged over the other degrees of freedom. We can unfortunately not always solve the equation in closed form for $\rho(t)$. The physics of Eq. (2.80) will become, it is hoped, more transparent in the example of the calculation of T_1 in Section XIV of this chapter.

All that has been said to this point about the density operator, or the density matrix, has been in reference to spin-$\frac{1}{2}$. That density operator is a 2×2 matrix. One recognizes that there are $(2I + 1)^2 = 4$ components in the density matrix for a spin-$\frac{1}{2}$ system. More generally an ensemble of spin $I \neq \frac{1}{2}$ systems (e.g., an ensemble of quadrupolar nuclei such as the spin-$\frac{5}{2}$ ^{27}Al nucleus or an ensemble of strongly coupled pairs of nuclei making an ensemble of spin-$\frac{1}{2}$ systems, such as the two protons responsible for the Pake doublet in gypsum) will be represented by a matrix, each element of which is an operator, of dimension $(2I + 1)^2$. This matrix represents expectation values of observables for such systems as before, and the time development of this matrix obeys the same differential equations as before. In systems with spin greater than $I = \frac{1}{2}$, however, there is a possibility not mentioned previously, that of multiple-quantum excitations. In the spin-$\frac{1}{2}$ systems, the equilibrium density matrix for an ensemble soaking in a magnetic field had only diagonal elements. The presence of these elements is said to represent "zero-quantum" coherence. When the ensemble is exposed to an rf perturbation, the result for the density matrix was to generate nonzero elements just one row off the diagonal (the only possible nondiagonal matrix elements for spin-$\frac{1}{2}$). The presence of these one-off-the-diagonal elements is referred to as "single-quantum coherence." This single-quantum coherence is the observable in a magnetic resonance experiment. In systems of spin greater than $\frac{1}{2}$, there will be matrix elements greater than one off the diagonal. If these elements of the density matrix are nonzero, one speaks of "multiple-quantum" coherence. For example, in an ensemble of spin-1 systems, such as ^{2}H, which has a 3×3 density matrix, any disturbance that results in a nonzero element two off the diagonal is spoken of as "double-quantum" coherence. This multiple-quantum coherence cannot be directly observed in a magnetic resonance experiment; however, it can be generated and subsequently observed by the conversion of multiple-quantum coherence to single-quantum coherence. Such experiments form the basis of another area of art and science in nuclear magnetic resonance, which is unfortunately beyond the scope of the present work to treat in a manner consistent with the depth of material intended by the authors. The area of multiple-quantum spectroscopy has been recently reviewed [1], and those with an interest are referred to that work for further detail. A lucid discussion of the density operator is supplied by Tolman [2].

XI. Time Evolution of the Density of Operator: Methods of Solution of the Liouville–von Neumann Equation

The Liouville–von Neumann equation is of pivotal importance in the quantum-mechanical description of the dynamics in spectroscopic experiments, since the computation of an observable requires knowledge of ρ at the time of interest. There are a number of ways to solve this problem, most of which are power series expansions in time. The value of each particular method of solution depends on the rapidity of the convergence of the series so that approximate computation of ρ is simplified. In this section we study certain special cases that are related to NMR experiments other than those in which continued strong rf excitation is applied (the latter being the subject of Chapter 4). Here we restrict ourselves to freely evolving systems—that is, to generalized solutions and calculations of T_1 and T_2.

The exchange of energy between the spin system and the remaining parts of the system (called the "lattice" and designated by the density operator ρ_R) tends to drive the spin system toward equilibrium. At equilibrium with the lattice the density operator for the spin system is given by ρ_{eq}.

In this equation $\mathscr{H}$ represents the Hamiltonian that affects the spins. The Hamiltonian contains terms that depend on the state of the "lattice." In the semiclassical approximation that is used here, the Hamiltonian is replaced by the average over the lattice variables as described by the operation

$$\mathscr{H}(t) = \text{Tr}\,\rho_R \mathscr{H}_T \tag{2.81}$$

where $\mathscr{H}_T$ is the total Hamiltonian.

It is assumed that the energy exchange processes never cause the "lattice" to deviate significantly from thermal equilibrium. Thus, one may write

$$i\frac{d}{dt}(\rho - \rho_{eq}) = [\mathscr{H}, (\rho - \rho_{eq})] \tag{2.82}$$

A. Iterative Integration

Let us go through the solution of Eq. (2.82) first in a very formal manner, i.e., by direct integration:

$$\int_0^t \frac{d\rho}{dt'} \, dt' = -i \int_0^t \left[\mathcal{H}(t'), \rho(t') - \rho_{eq} \right] dt' \tag{2.83}$$

$$\rho(t) - \rho(0) = -i \int_0^t \left[\mathcal{H}(t'), \rho(t') - \rho_{eq} \right] dt' \tag{2.84}$$

Equation (2.84) must also be true at $t = t'$. Hence, as an approximation, make the substitution

$$\rho(t') = \rho(0) - i \int_0^{t'} \left[\mathcal{H}(t''), \rho(t'') - \rho_{eq} \right] dt'' \tag{2.85}$$

And, in the second approximation,

$$\rho(t) = \rho(0) - i \int_0^t \left[\mathcal{H}(t'), \rho(0) - i \int_0^{t'} \left[\mathcal{H}(t''), \rho(t'') - \rho_{eq} \right] dt'' - \rho_{eq} \right] dt'$$

$$= \rho(0) - i \int_0^t \left[\mathcal{H}(t'), \rho(0) - \rho_{eq} \right] dt'$$

$$- \int_0^t dt' \int_0^{t'} dt'' \left[\mathcal{H}(t'), \left[\mathcal{H}(t''), \rho(t'') - \rho_{eq} \right] \right] \tag{2.86}$$

In the third approximation, one may substitute for $\rho(t'')$ to give

$$\rho(t) = \rho(0) - i \int_0^t \left[\mathcal{H}(t'), \rho(0) - \rho_{eq} \right] dt'$$

$$+ (-i)^2 \int_0^t dt' \int_0^{t'} dt'' \left[\mathcal{H}(t'), \left[\mathcal{H}(t''), \rho(0) - \rho_{eq} \right] \right]$$

$$+ (-i)^3 \int_0^t dt' \int_0^{t'} dt'' \int_0^{t''} dt'''$$

$$\times \left[\mathcal{H}(t'), \left[\mathcal{H}(t''), \left[\mathcal{H}(t'''), \rho(t''') - \rho_{eq} \right] \right] \right] \tag{2.87}$$

Continued application of this recursion formula results in an infinite summation

$$\rho(t) = \rho(0) - i \int_0^t \left[\mathcal{H}(t'), \rho(0) - \rho_{eq} \right] dt'$$

$$+ (-i)^2 \int_0^t dt' \int_0^{t'} dt'' \left[\mathcal{H}(t'), \left[\mathcal{H}(t''), \rho(0) - \rho_{eq} \right] \right] + \cdots$$

$$+ (-i)^n \int_0^t dt' \int_0^{t'} dt'' \cdots \int_0^{t(n-1)} dt^{(n)}$$

$$\times \left[\mathcal{H}(t'), \left[\mathcal{H}(t''), \left[\ldots, \left[\mathcal{H}(t^{(n)}), \rho(0) - \rho_{eq} \cdots \right] \right] \right] \right] + \cdots \tag{2.88}$$

Thus, to solve the problem explicitly, one needs to know the exact time dependence of $\mathcal{H}(t)$ for *all* times prior to t, the time of interest. This time dependence arises from the averaging over the "other" degrees of freedom. Once known, $\rho(t)$ may be used to calculate a value for the observable by (2.80).

Equation (2.88) appears to be complex at first glance, but in fact it is quite simple to use, as the following example will illustrate. Consider an ensemble of spins in a static field $\mathbf{B}_0$. Let $\rho_{eq} = \frac{1}{2} + bI_z$, (Eq. (2.70)) corresponding to a system in equilibrium with a static field along z. Let $\rho(0) = \frac{1}{2} + bI_y$ correspond to the physical situation immediately following a strong delta function $(\pi/2)_x$ pulse. The physical result one *must* obtain for the time evolution after that pulse is a precession about $\mathbf{B}_0$ with frequency $\omega_0 = \gamma B_0$. Let us examine how Eq. (2.88) gives us this result. The Hamiltonian at all times is $-\omega_0 I_z$.

$$\rho(t) = \frac{1}{2} + bI_y - i \int_0^t dt' \left[-\omega_0 I_z, \left(\frac{1}{2} + bI_y - \frac{1}{2} - bI_z \right) \right]$$

$$+ (-i)^2 \int_0^t dt' \int_0^{t'} dt'' \left[-\omega_0 I_z, \left[-\omega_0 I_z, \left(\frac{1}{2} + bI_y - \frac{1}{2} - bI_z \right) \right] \right] + \cdots$$

$$= \frac{1}{2} + bI_y + \omega_0 tbI_x - \frac{(\omega_0 t)^2}{2!} bI_y - i \frac{(\omega_0 t)^3}{3!} bI_x + \cdots \tag{2.89}$$

which are the first few terms in the expansions leading to the result

$$\rho(t) = \frac{1}{2} + b[I_y \cos \omega_0 t + I_x \sin \omega_0 t] \tag{2.90}$$

The expectation values of the various components of magnetization may be calculated, with the results

$$\langle M_x(t) \rangle = \text{Tr}\, \rho(t) M_x = (\gamma^2 \hbar^2 B_0 / 4kT) \cos \omega_0 t \tag{2.91}$$

$$\langle M_y(t) \rangle = (\gamma^2 \hbar^2 B_0 / 4kT) \sin \omega_0 t \tag{2.92}$$

This is analogous to the classical results, Eqs. (1.25)

B. Solving for the Liouville Operator

One may solve iteratively as above, and we shall return to that form later, but first let us refer to a special case. Suppose for the moment that we have a time-independent Hamiltonian $\mathscr{H}_0$ such as that caused by the Zeeman interaction. The equation we wish to solve is

$$i \frac{\partial \rho}{\partial t} = [\mathscr{H}_0, \rho(t)] = \mathscr{H}_0 \rho(t) - \rho(t) \mathscr{H}_0 \tag{2.93}$$

Let us propose a solution of the form

$$\rho(t) = \exp(-i\mathscr{H}_0 t)\rho(0) \exp(i\mathscr{H}_0 t) \tag{2.93a}$$

the derivative of which is

$$i\frac{d\rho(t)}{dt} = \mathcal{H}_0 \exp(-i\mathcal{H}_0 t)\rho(0)\exp(i\mathcal{H}_0 t) - \exp(-i\mathcal{H}_0 t)\rho(0)\exp(i\mathcal{H}_0 t)\mathcal{H}_0$$

$$= \mathcal{H}_0\rho(t) - \rho(t)\mathcal{H}_0$$

$$= [\mathcal{H}_0, \rho(t)] \tag{2.94}$$

where careful account has been kept of the order of noncommuting operators. One sees that (2.93a) is a solution of (2.93).

Let us try this form of the solution in a general case and define an operator L, called the *Liouville operator*, whose action upon a density operator is such that the result is the density operator at a later time. The defining equation of this operator is

$$\rho(t) = L(t)\rho(0)L^+(t) \tag{2.95}$$

By the Liouville–von Neumann equation,

$$\frac{\partial\rho}{\partial t} = \left(\frac{\partial L}{\partial t}\right)\rho(0)L^+(t) + L(t)\rho(0)\left(\frac{\partial L^+}{\partial t}\right)$$

or

$$\left(\frac{\partial L}{\partial t}\right)\rho(0)L^+(t) + L(t)\rho(0)\left(\frac{\partial L^+}{\partial t}\right)$$

$$= iL\rho(0)L^+ \mathcal{H}(t) - i\mathcal{H}(t)L(t)\rho(0)L^+(t) \tag{2.96}$$

Thus, one sees that the equation reduces to finding a solution for the operator

$$i\frac{\partial L}{\partial t} = \mathcal{H}(t)L \tag{2.97a}$$

$$-i\frac{\partial L^+}{\partial t} = L^+ \mathcal{H}(t) \tag{2.97b}$$

In the case of a static Hamiltonian $\mathcal{H} \equiv \mathcal{H}_0$, the solution is straightforward, since direct integration of Eq. (2.97) yields the exponential form

$$L = \exp(-i\mathcal{H}_0 t) \tag{2.97c}$$

If $\mathcal{H}$ is time dependent, however, direct integration of Eq. (2.97) is not possible. In this case, one must make series approximations to obtain solutions. These series solutions are developed in detail in Chapter 4.

XII. Exponential Operators

The discussion in Section XI has indicated that the Liouville operator can be used to calculate the time dependence of the density operator. Equations (2.97) determine the Liouville operators. The form of the solution to these equations is an exponential of an operator, as already indicated in Eq. (2.93). We therefore take a brief excursion into the mathematics and physics of exponential operators. As a simple example of an exponential operation, consider the complex number

$$\tilde{Z} = re^{i\phi} \tag{2.98a}$$

This complex number can be represented geometrically as shown in Fig. 2.2a.

Multiplication of $\tilde{Z}$ by the exponential form $U = e^{i\theta}$ results in a new complex number Z.

$$Z = re^{i(\phi + \theta)} \tag{2.98b}$$

Geometrically this is represented by the situation in Fig. 2.2b.

The multiplication has clearly changed $\tilde{Z}$ by rotation by the angle θ, leaving the length invariant. In the preceding section we noted a similar result: The action of the *operators*

$$U_0(t) = \exp(-i\mathscr{H}_0 t) \quad \text{and} \quad U_0^+(t) = \exp(i\mathscr{H}_0 t) \tag{2.99}$$

"rotated" the density operator from a time $t = 0$ to a time t; that is, the density operator was transformed from its value at $t = 0$ to its value at $t = t$.

We seek to generalize this operation to complex functions in an N-dimensional function space; that is, we seek to find an operator that transforms an N-dimensional function but that leaves its length invariant. More specifically, we seek an operator that leaves the scalar product of two functions invariant to transformation. Call the operator U and the functions to be transformed $|\tilde{\Psi}_a\rangle$ and $|\tilde{\Psi}_b\rangle$. The rule for transformation is represented

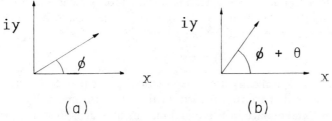

FIG. 2.2. (a) Geometrical representation of a complex number. (b) The effect of rotation on a vector in the complex plane.

by the equation

$$|\Psi\rangle = U|\tilde{\Psi}\rangle \tag{2.100}$$

We require that the scalar product of the N-dimensional functions $|\tilde{\Psi}_a\rangle$ and $|\tilde{\Psi}_b\rangle$ be invariant to transformation, i.e., that

$$\langle\tilde{\Psi}_a|\tilde{\Psi}_b\rangle = \langle\Psi_a|\Psi_b\rangle \tag{2.101}$$

where the bra–ket notation for N-dimensional functions in Hilbert space has been used. Equation (2.101) then requires that

$$\langle\Psi_a|\Psi_b\rangle = \langle U\tilde{\Psi}_a|U\tilde{\Psi}_b\rangle = \langle\tilde{\Psi}_a|U^+U\tilde{\Psi}_b\rangle$$

or that

$$U^+U = UU^+ = 1 \tag{2.102}$$

Equation (2.102) defines a unitary operator.

To anticipate a future discussion, we mention that Eq. (2.100) has a potent physical interpretation. If $|\tilde{\Psi}\rangle$ represents the state of a system at time zero and $|\Psi\rangle$ the state of some subsequent time, then Eq. (2.100) becomes the rule for determining how the state of a system evolves with time:

$$|\Psi(t)\rangle = U|\Psi(t = 0)\rangle \tag{2.103}$$

In density matrix notation, Eq. (2.103) becomes [see Eq. (2.95)]

$$\rho(t) = U\rho(0)U^+ \tag{2.104}$$

where the identity

$$\rho = |\Psi\rangle\langle\Psi| \tag{2.104'}$$

has been used to define the density operator.

Formally, Eqs. (2.103) and (2.104) represent solutions to the time-dependent Schrödinger equation [Eqs. (2.1) and (2.60)]. Since the time evolution of a system is determined by the interactions a system experiences, i.e., by $\mathcal{H}$, it is clear that U and $\mathcal{H}$ must be related.

In order to relate U to a Hermitian operator, we consider the result of transforming in incremental steps from one time to another. The overall transformation is described by the relation

$$\rho(t) = U(t,0)\rho(0)U^+(t,0) \tag{2.104'}$$

Here, we have specifically used the notation $U \equiv U(t,0)$ to indicate that U is an operator that transforms ρ from time zero to time t. For this transformation to take place in two equal incremental steps, one may write

$$U(t,0) = U(t,t_1)U(t_1,0) \tag{2.105}$$

where $2t_1 = t$. Then Eq. (2.104') becomes

$$\rho(t) = U(t, t_1)U(t_1, 0)\rho(0)U^+(t_1, 0)U^+(t, t_1) \tag{2.106}$$

Note that since the transformations take place incrementally in a particular order in time, it may be possible that

$$U(t, t_1)U(t_1, 0) \neq U(t_1, 0)U(t, t_1)$$

or that

$$[U(t, t_1), U(t_1, 0)] \neq 0$$

which simply states that care must be exercised in "time ordering" of these operators.

To pursue the idea behind approximating a finite transformation by incremental steps a bit further, let U consist of a very large number of infinitesimal steps, each by t/n;

$$U(t, 0) = U_n U_{n-1} \cdots U_1 \equiv U(t, (n-1)t/n) \cdots U(t/n, 0) \tag{2.107}$$

Since we anticipate from Eq. (2.94) the idea that there is a one-to-one relation between U and the exponential of a *time-independent* $\mathscr{H}_0$, we let the jth infinitesimal rotation in time *interval* t_j be represented by

$$U_j(t_j) = \exp(-i\mathscr{H}_0 t_j) \cong 1 - i\mathscr{H}_0 t_j \tag{2.108}$$

If U is to be unitary,

$$UU^+ \cong (1 - i\mathscr{H}_0 t_j)(1 + i\mathscr{H}_0^+ t_j) = 1 \tag{2.109}$$

Keeping only first-order terms (which is always allowable if the intervals t_j are made sufficiently small) we find that

$$1 - i\mathscr{H}_0 t_j + i\mathscr{H}_0^+ t_j = 1$$

or that H_0 must be Hermitian:

$$\mathscr{H}_0 = \mathscr{H}_0^+ \tag{2.110}$$

In the limit where the time interval t_j becomes vanishingly small and the number of rotations approaches infinity, we write

$$U(t, 0) \equiv U(t) = \sum_{j=0}^{\infty} \frac{-i\mathscr{H}_0 t_j}{j!} \tag{2.111a}$$

which is just the series representation of the exponential form

$$U(t) = \exp(-i\mathscr{H}_0 t) \tag{2.111b}$$

This is the correct form of U when $\mathscr{H}_0$ is time independent.

The Hermitian operators with which we will be concerned are those that affect the manner in which nuclear spin systems evolve in time. Since many of these may be time *dependent*, the solution Eq. (2.111a) is not applicable. However, an expansion in terms of infinitesimal *time-dependent* operators does turn out to be applicable:

$$U(t) = \prod_{n=0}^{\infty} U_n(t_n) = \prod_{n=0}^{\infty} \exp(-i\mathscr{H}(t_n)t_n) \qquad (2.111c)$$

which may be formally represented by the Dyson expression,

$$U(t) = T \exp\left[-i \int_0^t \mathscr{H}(t')\,dt'\right] \qquad (2.111d)$$

Here T is the Dyson "time-ordering operator," which has the property of making the terms in the series expansion, Eq. (2.111c), increase in time from right to left. This form will be discussed more fully in Chapter 4.

At this point, it is appropriate to point out that both the density operator ρ and the Hamiltonians that affect the time development of this operator are functions of nuclear spin operators $\mathbf{I}$ in consideration of the behavior of nuclear spin systems. One will therefore frequently have occasion to evaluate expressions of the form

$$f(\mathbf{I}) = U\rho U^+ \equiv \exp(i\theta I_j)(aI_x + bI_y + cI_z)\exp(-i\theta I_j) \qquad (2.112a)$$

where I_j is one of I_x, I_y, or I_z. Let us concentrate on just one term in Eq. (2.112a) and attempt to evaluate the form

$$f(\mathbf{I}) = \exp(i\theta I_j)I_k \exp(-i\theta I_j) \qquad (2.112b)$$

where k, j could be any of x, y, or z. Expanding the exponentials in Eq. (2.112b) yields the infinite sum

$$f(\mathbf{I}) = \left(\sum_{n=0}^{\infty} \frac{(i\theta I_j)^n}{n!}\right) I_k \left(\sum_{m=0}^{\infty} \frac{(-i\theta I_j)^m}{m!}\right) \qquad (2.112c)$$

Expanding Eq. (2.112c) to second order yields

$$f(\mathbf{I}) = I_k + i\theta I_j I_k + \frac{i^2\theta^2}{2!} I_j^2 I_k - i\theta I_k I_j + \frac{(-i)^2\theta^2}{2!} I_k I_j^2 + \theta^2 I_j I_k I_j \qquad (2.113)$$

Equation (2.113) may in turn be rewritten in terms of commutators:

$$f(\mathbf{I}) = I_k + i\theta[I_j, I_k] + \frac{i^2\theta^2}{2!} [I_j, [I_j, I_k]] + \cdots$$

$$\equiv (1 + i\theta[I_j, \,] + \frac{i^2\theta^2}{2!} [I_j, [I_j, \,]] + \cdots)I_k \qquad (2.114)$$

The term in parenthesis is an operator that acts as another operator and is sometimes called a "superoperator." By making an appropriate one-to-one correlation between the expansion in Eq. (2.114) and the expansion of an exponential, Eq. (2.114) may be formally written as

$$\exp(i\theta I_j)I_k\exp(-i\theta I_j) = \exp[i\theta(I_j,)]I_k \qquad (2.115)$$

As an example, consider calculating the value of (2.112b) for $I_k = I_x$ and $I_j = I_z$:

$\exp(i\theta I_z)I_x\exp(-i\theta I_z)$

$$= I_x + i\theta[I_z, I_x] + i^2\frac{\theta^2}{2!}[I_z,[I_z,I_x]] + i^3\frac{\theta^2}{3!}[I_z[I_z,I_x]]] + \cdots$$

$$= I_x - \theta I_y + i^2\frac{\theta^2}{2!}[I_z, iI_y] + i^3\frac{\theta^3}{3!}[I_z,[I_z,iI_y]] + \cdots$$

$$= I_x - \theta I_y + i^3\frac{\theta^2}{2!}(-iI_x) + i^4\frac{\theta^3}{3!}[I_z, -iI_x] + \cdots$$

$$= I_x - \theta I_y - \frac{\theta^2}{2!}I_x - i\frac{\theta^3}{3!}(iI_y) + \cdots$$

$$= I_x\left\{1 - \frac{\theta^2}{2!} + \cdots\right\} + I_y\left\{-\theta + \frac{\theta^3}{3!} + \cdots\right\} \qquad (2.116)$$

By extension, it can be shown that the sum multiplying I_x is the $\cos\theta$ expansion and the sum multiplying I_y is $-\sin\theta$. Hence

$$\exp(i\theta I_z)I_x\exp(-i\theta I_z) = I_x\cos\theta - I_y\sin\theta \qquad (2.117)$$

A similar procedure can be carried for each rotation operator acting on the other spin operators. The results are given in Table 2.1. It should be noted that the mathematical results given in part A of Table 2.1 may be visualized by using an appropriate geometrical representation. For example, the form $\exp(i\theta I_x)I_z\exp(-i\theta I_x)$ represents a rotation of I_z about the x axis in the sense that z is rotated toward y by the angle θ. The result will thus be $I_z\cos\theta + I_y\sin\theta$, as given in Table 2.1 and illustrated in Fig. 2.3. As a practical exercise, the reader may similarly wish to evaluate all the entries in Table 2.1A by using this simple physical picture.

XIII. Interaction Frames: The $\pi/2$ Pulse Described Quantum Mechanically for Spin-$\frac{1}{2}$

In Section X the time dependence of the density operator was explicitly given by a unitary transformation. The transformation, in turn, was determined by a particular interaction. In the present section is discussed a view of such transformations that will be routinely used to evaluate the results of

TABLE 2.1[a]
Action of Exponential Operators on Spin Operators

A. $e^{i\mathbf{I} \cdot \mathbf{n}\theta} \mathbf{I} e^{-i\mathbf{I} \cdot \mathbf{n}\theta}$

$e^{i\theta I_x} I_x e^{-\theta I_x} = I_x$

$e^{i\theta I_x} I_y e^{-i\theta I_x} = I_y \cos\theta - I_z \sin\theta$

$e^{i\theta I_x} I_z e^{-i\theta I_x} = I_z \cos\theta + I_y \sin\theta$

$e^{i\theta I_y} I_x e^{-i\theta I_y} = I_x \cos\theta + I_z \sin\theta$

$e^{i\theta I_y} I_y e^{-i\theta I_y} = I_y$

$e^{i\theta I_y} I_z e^{-i\theta I_y} = I_z \cos\theta - I_x \sin\theta$

$e^{i\theta I_z} I_x e^{-i\theta I_z} = I_x \cos\theta - I_y \sin\theta$

$e^{i\theta I_z} I_y e^{-i\theta I_z} = I_y \cos\theta + I_x \sin\theta$

$e^{i\theta I_z} I_z e^{-i\theta I_z} = I_z$

B. $e^{i\mathbf{I} \cdot \mathbf{n}\theta} \mathbf{I}_1 \cdot \mathbf{I}_2 e^{-i\mathbf{I} \cdot \mathbf{n}\theta} = \mathbf{I}_1 \cdot \mathbf{I}_2$

C. $e^{i\theta I_x} I_{z1} I_{z2} e^{-i\theta I_x}$

$\quad = \frac{1}{2}(I_{y1} I_{y2} + I_{z1} I_{z2}) - \frac{1}{2}(I_{y1} I_{y2} - I_{z1} I_{z2}) \cos 2\theta$
$\quad\quad + \frac{1}{2}(I_{y1} I_{z2} + I_{y2} I_{z1}) \sin 2\theta$

D. $e^{i\theta I_y} I_{z1} I_{z2} e^{-i\theta I_z}$

$\quad = \frac{1}{2}(I_{x1} I_{x2} + I_{z1} I_{z2}) - \frac{1}{2}(I_{x1} I_{x2} - I_{z1} I_{z2}) \cos 2\theta$
$\quad\quad - \frac{1}{2}(I_{x1} I_{z2} - I_{z1} I_{x2}) \sin 2\theta$

---

[a] Note that all of the results are simply represented by the general form $e^{i\theta I_j} I_k e^{-i\theta I_j} = I_k \cos\theta - I_l \sin(\theta \epsilon_{jkl})$.

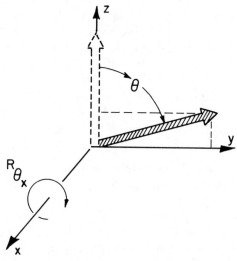

FIG. 2.3. A geometrical representation of the result of the operation $\exp(i\theta I_x) I_z \exp(-i\theta I_x) = I_z \cos\theta + I_y \sin\theta \equiv R_{\theta_x} I_z$.

pulse NMR experiments. To initiate the discussion, recall that in Chapter 1 viewing the motion of a magnetic moment in a magnetic field from the rotating frame was a particularly powerful means of deciding how the moment was moving with respect to the outside world, i.e., the laboratory frame. One found that, if the moment was stationary in the rotating frame, it was precessing at the Larmor frequency in the laboratory frame. We now discuss, from a general point of view, transformations from one frame to another in quantum-mechanical systems. A special example will be the transformation to the rotating frame. The discussions will involve performing unitary transformations on an equation of motion in a representation, in which there exists time dependence in both operators and state vectors, to obtain the Schrödinger representation. Before becoming involved in the algebra, we remind ourselves that the *Schrödinger* representation puts all time dependence in the state vector

$$i \frac{d|\Psi(t)\rangle}{dt} = \mathcal{H}|\Psi(t)\rangle \tag{2.118}$$

The *Heisenberg* representation, on the other hand, puts all time dependence into the operator $\hat{A}$:

$$i \frac{d\langle \hat{A} \rangle}{dt} = \langle [\hat{A}, \mathcal{H}] \rangle + i \left\langle \frac{\partial \hat{A}}{\partial t} \right\rangle \tag{2.119}$$

In the *interaction representation*, or *interaction frame*, there is time dependence in both operators and state vectors. The Hamiltonian, for example, is expressed as a sum of contributions, some or all of which may be time dependent. The game is to move to a frame which there *are* no time-dependent interactions, i.e., back to the Schrödinger representation. To do so, one or more unitary transformations are performed that successively remove or take account of selected portions of the Hamiltonian from the description of motion of the system. An example just considered from a classical point of view is that of a magnetic moment in a magnetic field, part of which is time independent (the dc Zeeman field) and part of which is time dependent (the radiofrequency field causing magnetic resonance absorption). By transforming to the rotating frame of the Zeeman interaction, the Zeeman field is removed from the description of the system in that frame. The rf field is stationary in the transformed frame. Which unitary transformation is applied, of course, depends upon that portion of the Hamiltonian one desires to remove from the description of the system in the "rotated" (transformed) frame. To illustrate the arithmetic, consider a Hamiltonian consisting of two terms $\mathcal{H}_1$ and $\mathcal{H}_2$, either of which may be an explicit function of time. The equation of motion of the system is given by

$$i \frac{d|\Psi\rangle}{dt} = (\mathcal{H}_1 + \mathcal{H}_2)|\Psi\rangle \tag{2.120}$$

If we make the unitary transformation

$$|\Psi(t)\rangle = U_1|\tilde{\Psi}(t)\rangle \tag{2.121}$$

and we insist that U_1 is determined by $\mathscr{H}_1$ via the differential equation

$$\frac{i\,dU_1}{dt} = \mathscr{H}_1 U_1 \tag{2.122}$$

with the boundary condition

$$U_1(t = 0) = U_1^+(t = 0) = 1 \tag{2.123}$$

Then Eqs. (2.120)–(2.122) give

$$\frac{d|\Psi\rangle}{dt} = \frac{\mathscr{H}_1 + \mathscr{H}_2}{i}|\Psi\rangle \equiv \frac{(\mathscr{H}_1 + \mathscr{H}_2)U_1|\tilde{\Psi}\rangle}{i}$$

$$= \frac{dU_1}{dt}|\tilde{\Psi}\rangle + \frac{\mathscr{H}_2 U_1|\tilde{\Psi}\rangle}{i} \tag{2.124}$$

Now

$$\frac{d|\Psi\rangle}{dt} = \frac{d(U_1|\tilde{\Psi}\rangle)}{dt} = \frac{dU_1}{dt}|\tilde{\Psi}\rangle + U_1\frac{d|\tilde{\Psi}\rangle}{dt} \tag{2.125}$$

Comparison of Eqs. (2.124) and (2.125) yields

$$U_1\frac{d|\tilde{\Psi}\rangle}{dt} = \frac{\mathscr{H}_2 U_1|\tilde{\Psi}\rangle}{i} \tag{2.126}$$

Multiplication of Eq. (2.126) on the left by iU_1^+ yields the time-dependent Schrödinger equation for $\tilde{\Psi}$:

$$\frac{i\,d|\tilde{\Psi}\rangle}{dt} = U_1^+\mathscr{H}_2 U_1|\tilde{\Psi}\rangle \equiv \tilde{\mathscr{H}}_2|\tilde{\Psi}\rangle \tag{2.127}$$

We therefore see that the overall result of the transformation (2.121) with the constraints of Eqs. (2.122) and (2.123) is to "transform" Ψ to a frame where its transformed description, called $|\tilde{\Psi}\rangle$, no longer explicitly depends on $\mathscr{H}_1$, the interaction used to perform the transformation. The time dependence of $\tilde{\Psi}$ is only governed by the value of $\tilde{\mathscr{H}}_2 \equiv U_1^+\mathscr{H}_2 U_1$, which in turn, of course, is implicitly governed by $\mathscr{H}_1$ through Eq. (2.122). This is a transformation to the "interaction frame" of $\mathscr{H}_1$.

The process of transforming into successive frames by unitary transformations may be continued. For example, if the transformation

$$|\tilde{\Psi}\rangle = U_2|\tilde{\tilde{\Psi}}\rangle \tag{2.128}$$

is made, with the constraints

$$U_2(t = 0) = 1, \qquad \frac{i\,dU_2}{dt} = \tilde{\mathscr{H}}_2 U_2 \qquad (2.129)$$

then the equation of motion of $|\tilde{\tilde{\Psi}}\rangle$, which is said to be in the interaction frames of $\mathscr{H}_1$ and $\tilde{\mathscr{H}}_2$, is

$$\frac{i|d\tilde{\tilde{\Psi}}\rangle}{dt} = 0 \qquad (2.130)$$

i.e., $|\tilde{\tilde{\Psi}}\rangle$ is stationary.

The reader will wish to verify that Eqs. (2.120)–(2.130) have analogs in density matrix formulation as follows: The density matrix analog of Eq. (2.120) is

$$\frac{i\,d\rho}{dt} = [(\mathscr{H}_1 + \mathscr{H}_2), \rho] \qquad (2.131)$$

Let the transformation

$$\rho = U_1 \tilde{\rho} U_1^+ \qquad (2.132)$$

be made, in which U_1 is determined through $\mathscr{H}_1$ via Eqs. (2.122) and (2.123). Then the equation of motion of $\tilde{\rho}$ is

$$\frac{i\,d\tilde{\rho}}{dt} = [\tilde{\mathscr{H}}_2, \tilde{\rho}] \qquad (2.133)$$

where $\tilde{\mathscr{H}}_2$ is given by Eq. (2.127). Furthermore, if a second transformation

$$\tilde{\rho} = U_2 \tilde{\tilde{\rho}} U_2^+ \qquad (2.134)$$

is made, then, with U_2 determined by Eq. (2.129), the behavior of $\tilde{\tilde{\rho}}$ is given by

$$\frac{i\,d\tilde{\tilde{\rho}}}{dt} = [0, \tilde{\tilde{\rho}}] = 0 \qquad (2.135)$$

and ρ and $\tilde{\tilde{\rho}}$ are related by the transformations

$$\rho = U_2 U_1 \tilde{\tilde{\rho}} U_1^+ U_2^+ \qquad (2.136a)$$

In appropriate circumstances, and in a spirit quite like that used in discussing a physical implication of Eq. (2.100), we shall subsequently see that Eq. (2.136) may be interpreted in terms of a rule for time development; i.e., we shall subsequently have occasion to make the identifications $\rho \equiv \rho(t)$ and $\tilde{\tilde{\rho}} = \rho(0)$. This identification, and the conditions under which it may be usefully made, will be discussed in detail in Chapter 5. It suffices to mention at this point that if the identity

$$\rho \equiv \rho(t_c) = \rho(0) \equiv U_1 U_2 \tilde{\tilde{\rho}} U_2^+ U_1^+ \qquad (2.136b)$$

is obeyed for any particular time t_c, at such a time, observation of the behavior of the spin system will indicate that both U_1 and U_2 will be unity at time t_c. Here U_1 is determined by $\mathscr{H}_1$, and U_2 by $\tilde{\mathscr{H}}_2$, which in turn is determined by $\mathscr{H}_1$. With $\mathscr{H}_1$ being identified as the radio-frequency Hamiltonian, to which the experimenter has manipulative access, we begin to see the possibility of massaging a system with appropriate radio-frequency pulses such that specific interactions $\tilde{\mathscr{H}}_2$ appear at observation times appropriately chosen not to have affected the time evolution of the system. As a first example of the details of the calculation of transformation to an interaction frame, we consider the frame of the Zeeman interaction, the rotating frame. Consider a system subject to three interactions: (1) the time-independent Zeeman interaction $-\omega_0 I_z$, (2) a static interaction $-\delta\omega I_z$, and (3) a time-dependent interaction $v(t)$. The second interaction might represent a chemical-shift Hamiltonian (e.g., of protons in methanol relative to each other). The Hamiltonian, with the dc field along z, is

$$\mathscr{H} = \mathscr{H}_0 + \mathscr{H}_1 \equiv -\gamma I_z(B_0 + \delta B) + v(t) \equiv -I_z(\omega_0 + \delta\omega) + v(t) \quad (2.137)$$

Here,

$$\mathscr{H}_1 = -\delta\omega I_z + v(t) \quad (2.138)$$

The transformation to the frame of the *static* Zeeman interaction is effected as follows

$$|\Psi\rangle = U_0|\tilde{\Psi}\rangle = \exp\{-i\mathscr{H}_0 t\}|\tilde{\Psi}\rangle = \exp\{i\omega_0 I_z t\}|\tilde{\Psi}\rangle \quad (2.139)$$

(or equivalently, $\rho = U_0\tilde{\rho}U_0^{-1}$). The time-dependent Schrödinger equations for $\tilde{\Psi}$ and $\tilde{\rho}$ are then

$$\frac{i\,d|\tilde{\Psi}\rangle}{dt} = \tilde{\mathscr{H}}_1|\tilde{\Psi}\rangle = U_0^+[-\delta\omega I_z + v(t)]U_0|\tilde{\Psi}\rangle \quad (2.140a)$$

$$\frac{i\,d\tilde{\rho}}{dt} = [\tilde{\mathscr{H}}_1, \tilde{\rho}] \quad (2.140b)$$

In this frame, the time dependences of $|\tilde{\Psi}\rangle$ or $\tilde{\rho}$ depend explicitly on $\delta\omega I_z$ and $v(t)$. Since $\delta\omega I_z$ is very much like a Zeeman interaction with frequency $\delta\omega$ (which produces oscillation of the transverse magnetization in the laboratory frame), one might expect the solution in the new frame to have oscillatory character, as indeed it does. Since the experimenter detects signals in the interaction frame of the rf excitation, all chemical shifts are Hamiltonians of this type and our interest will be in solving equations like (2.140).

It is instructive to further illustrate with a calculation of the quantum-mechanical description of a $\pi/2$ pulse, where $v(t)$ in (2.138) specifically describes a circularly polarized rf field. In Chapter 1 we discussed the fact that an ensemble of spins polarized along the z axis could be considered as a classical moment pointing in the z direction. This moment could be rotated to the y axis in the rotating frame by a rf pulse along the x axis in the rotating frame with the magnitude of rf field satisfying the condition

$$\gamma B_1 t = \omega_1 t = \pi/2 \tag{2.141}$$

We have developed a form for the density operator that is proportional to I_z before the pulse and to I_y after the transformation corresponding to the pulse. While this is a useful formulation, it tells us little in detail about the population of states corresponding to the situation immediately after a $\pi/2_x$ pulse. In order to study the population changes, consider a spin-$\frac{1}{2}$ system initially totally polarized along z. The wave function, with $|\alpha\rangle$ and $|\beta\rangle$ representing the two state vectors spanning spin-$\frac{1}{2}$ space, generally is

$$|\Psi\rangle = C_1|\alpha\rangle + C_2|\beta\rangle \tag{2.142}$$

and will have $C_1 = 1$, $C_2 = 0$, at $t = 0$. We inquire about the detailed population of states as a function of time subject to the radio-frequency interaction and the Zeeman interaction from $t = 0$ to the time $t = \pi/2\gamma B_1 = \pi/2\omega_1$.

In the laboratory frame, the magnetic field corresponding to a static field B_0 along z and rf field of magnitude B_1 precessing about z with frequency ω_z is

$$\mathbf{B} = \mathbf{i}B_1 \cos \omega_z t - \mathbf{j}B_1 \sin \omega_z t + \mathbf{k}B_0 \tag{2.143}$$

This field corresponds to a time-dependent Hamiltonian

$$\mathscr{H} = -\omega_0 I_z - \omega_1(I_x \cos \omega_z t - I_y \sin \omega_z t) \tag{2.144}$$

where we have expressed γB_0 as ω_0 and γB_1 as ω_1. Therefore in the laboratory frame

$$i\frac{\partial}{\partial t}|\Psi\rangle = \mathscr{H}|\Psi\rangle$$

or

$$i\frac{\partial}{\partial t}|\Psi\rangle = [-\omega_0 I_z - \omega_1(I_x \cos \omega_z t - I_y \sin \omega_z t)]|\Psi\rangle \tag{2.145}$$

noting that

$$I_x \cos \omega_z t - I_y \sin \omega_z t = \exp(i\omega_z t I_z)I_x \exp(-i\omega_z t I_z) \tag{2.146}$$

Defining

$$|\tilde{\Psi}\rangle = \exp(-i\omega_z t I_z)|\Psi\rangle = U|\Psi\rangle \tag{2.147a}$$

and

$$\tilde{\mathscr{H}} = \exp(-i\omega_z t I_z)\mathscr{H} \exp(i\omega_z t I_z) = U\mathscr{H}U^+ \tag{2.147b}$$

leads to the differential equation

$$i\frac{\partial}{\partial t}|\tilde{\Psi}\rangle = [-(\omega_0 - \omega_z)I_z - \omega_1 I_x]\tilde{\Psi}\rangle \qquad (2.148)$$

On resonance (i.e., when $\omega_0 = \omega_z$), one finds that

$$i\frac{\partial}{\partial t}|\tilde{\Psi}\rangle = -\omega_1 I_x|\tilde{\Psi}\rangle \qquad (2.149)$$

the equation for the wave function in the rotating frame. This result is the analog of Eq. (2.126) for the problem at hand. We therefore find that

$$i\frac{\partial}{\partial t}e^{-i\omega_0 I_z t}[C_1|\alpha\rangle + C_2|\beta\rangle] = -\omega_1 I_x e^{-i\omega_0 I_z t}[C_1|\alpha\rangle + C_2|\beta\rangle] \quad (2.150)$$

determines the differential equations for C_1 and C_2. Using the relations

$$\begin{aligned}
e^{-i\omega_0 I_z t}|\alpha\rangle &= e^{-i\omega_0 t/2}|\alpha\rangle \\
e^{-i\omega_0 I_z t}|\beta\rangle &= e^{-i\omega_0 t/2}|\beta\rangle \\
\langle\alpha|I_x|\alpha\rangle &= 0 \\
\langle\alpha|I_x|\beta\rangle &= \tfrac{1}{2} = \langle\beta|I_x|\alpha\rangle
\end{aligned} \qquad (2.151)$$

we find that multiplication of (2.150) on the left by $\langle\alpha|$ or by $\langle\beta|$ yields

$$-i\frac{d}{dt}e^{-i\omega_0 t/2}C_1 = \frac{\omega_1}{2}e^{i\omega_0 t/2}C_2 \qquad (2.152)$$

$$-i\frac{d}{dt}e^{i\omega_0 t/2}C_2 = \frac{\omega_1}{2}e^{-i\omega_0 t/2}C_1 \qquad (2.153)$$

If we define the *rotating frame* values of C_1 and C_2 to be

$$\tilde{C}_2 = e^{-i\omega_0 t/2}C_1 \qquad \tilde{C}_2 = e^{i\omega_0 t/2}C_2 \qquad (2.154)$$

i.e., if we let $|\tilde{\Psi}\rangle = \tilde{C}_1|\alpha\rangle + \tilde{C}_2|\beta\rangle$, we arrive at the differential equations

$$-i\frac{d\tilde{C}_1}{dt} = \frac{\omega_1}{2}\tilde{C}_2 \qquad (2.155)$$

$$-i\frac{d\tilde{C}_2}{dt} = \frac{\omega_1}{2}\tilde{C}_1 \qquad (2.156)$$

Equations (2.155) and (2.156) are a particularly significant set of relations for purposes of pedagogy, because they are a coupled set of differential equations. Since in problems of the type with which we deal in this text it is more common than not for the differential equations to be coupled, we digress for a moment to talk about uncoupling differential equations.

A. *Uncoupling Differential Equations*

Consider a set of coupled differential equations

$$\dot{y}_1 = C_{11}y_1 + C_{12}y_2 + \cdots + C_{1n}y_n$$
$$\dot{y}_2 = C_{21}y_1 + \cdots + C_{2n}y_n$$
$$\vdots$$
$$\dot{y}_n = C_{n1}y_1 + \cdots + C_{nn}y_n$$

which may be written in matrix form as

$$\dot{\mathbf{Y}} = \mathbf{CY}$$

with $\dot{\mathbf{Y}}$ and $\mathbf{Y}$ being column vectors and $\mathbf{C}$ the $n \times n$ matrix of the transformation of $\dot{\mathbf{Y}}$ into $\mathbf{Y}$. Then let $\mathbf{Y} = \mathbf{PU}$, where $\mathbf{P}$ is some $n \times n$ constant matrix. Therefore, $\dot{\mathbf{Y}} = \mathbf{P\dot{U}}$ and

$$\dot{\mathbf{U}} = \mathbf{P}^{-1}\mathbf{CPU} = \mathbf{DU} \tag{2.157}$$

where $\mathbf{P}^{-1}\mathbf{CP}$ is a diagonal matrix. The Eqs. (2.157) are an uncoupled set that may be solved directly for U, and the value of $\mathbf{Y}$ may be obtained from the relation $\mathbf{Y} = \mathbf{PU}$. To find $\mathbf{P}$, it is necessary to diagonalize $\mathbf{C}$, which is performed by first finding the eigenvalues of $\mathbf{C}$, using the determinantal form

$$|C_{ij} - \lambda\delta_{ij}| = 0$$

yielding the eigenvalues λ_i. The λ_i may then be used to determine the columns of the matrix $\mathbf{P}$ since

$$\mathbf{CP}_i = \lambda_i\mathbf{P}_i \tag{2.158}$$

which is just another statement of the relation $\mathbf{P}^{-1}\mathbf{CP} = \mathbf{D}$, where $\mathbf{D}$ is the diagonal matrix having λ_i as diagonal elements. In (2.158), $\mathbf{P}_i$ is the ith column of the matrix $\mathbf{P}$. The requirement that the columns of $\mathbf{P}$ form orthonormal vectors

$$\sum_k P_{ik}P_{jk} = \delta_{ij}$$

completes the determination of the elements of $\mathbf{P}$. The inverse of $\mathbf{P}$ is found from the relation

$$\mathbf{P}^{-1} = \mathbf{P}^{\text{adj}}/|\mathbf{P}|$$

where $\mathbf{P}^{\text{adj}}$, the adjoint of $\mathbf{P}$, is the transpose of the matrix in which every element of $\mathbf{P}$ has been replaced by its cofactor

$$\mathbf{P}^{\text{adj}} = (C_{h_j})^{\text{T}} = [(-1)^{h+j}|M_{h_j}|]$$

We illustrate this process with Eqs. (2.155) and (2.156). In this case,

$$\mathbf{C} = \begin{bmatrix} 0 & -\dfrac{\omega_1}{2i} \\ -\dfrac{\omega_1}{2i} & 0 \end{bmatrix}$$

with eigenvalues $\pm i\omega_1/2$. $\mathbf{P}$ is then found to be

$$\mathbf{P} = \frac{1}{\sqrt{2}} \begin{bmatrix} 1 & 1 \\ 1 & -1 \end{bmatrix}$$

which in this case is also $\mathbf{P}^{-1}$. Therefore

$$\mathbf{D} = \mathbf{P}^{-1}\mathbf{C}\mathbf{P} = \frac{\omega_1}{2i} \begin{bmatrix} 1 & 0 \\ 0 & -1 \end{bmatrix}$$

The uncoupled equations are

$$\dot{\mathbf{U}} = \mathbf{D}\mathbf{U}$$

or

$$\dot{U}_1 = \frac{\omega_1}{2i} U_1, \qquad \dot{U}_2 = \frac{-\omega_1}{2i} U_2$$

with general solutions

$$U_1 = a\exp(-i\omega_1 t/2), \qquad U_2 = b\exp(i\omega_1 t/2)$$

Inverting back to $\mathbf{Y}$ via $\mathbf{PU} = \mathbf{Y}$ yields

$$y_1 = a\exp(-i\omega_1 t/2) + b\exp(i\omega_1 t/2) \equiv \tilde{C}_1$$

$$y_2 = a\exp(-i\omega_1 t/2) - b\exp(i\omega_1 t/2) \equiv \tilde{C}_2$$

The boundary conditions at $t = 0$, $y_1 = 1$, $y_2 = 0$ yield

$$\tilde{C}_1 \equiv y_1 = \frac{1}{\sqrt{2}} \cos\frac{\omega_1 t}{2}, \qquad \tilde{C}_2 \equiv y_2 = \frac{i}{\sqrt{2}} \sin\frac{\omega_1 t}{2} \qquad (2.159)$$

and

$$\tilde{\Psi}(t) = \cos\frac{\omega_1}{2} t|\alpha\rangle + i\sin\frac{\omega_1}{2} t|\beta\rangle \qquad (2.160)$$

A $\pi/2$ pulse, $\omega_1 t = \pi/2$, gives

$$\tilde{\Psi}(t) = \frac{1}{\sqrt{2}}(\alpha + i\beta) \qquad (2.161)$$

Thus, after a $\pi/2$ pulse on resonance, the populations of the $|\alpha\rangle$ and $|\beta\rangle$ states are equal and a phase coherence is created such that $\langle I_y \rangle = \frac{1}{2}$ (cf. Problem 2.9).

XIV. The Calculation of T_1

A second important calculation that utilizes a transformation to the frame of the Zeeman Hamiltonian, but where the remaining Hamiltonian is not controlled by the experimenter, is that in which the second interaction is a randomly modulated function of time

$$\mathcal{H}_1(t) = h_z(t)I_z + h_x(t)I_x + h_y(t)I_y \qquad (2.162)$$

where $h_i(t)$ is a component of a random time-varying field in the i direction. Note that h has units of frequency. The Zeeman Liouville operator associated with the Zeeman Hamiltonian $\mathcal{H}_0$ is given by

$$U_0(t) = \exp(-i\mathcal{H}_0 t) \qquad (2.163)$$

For our own purposes, we consider P to be the single-particle probability operator.

In the interaction frame of the Zeeman interaction, the equation of motion of P is

$$i\frac{\partial \tilde{P}}{\partial t} = [\tilde{\mathcal{H}}_1(t), \tilde{P}] \qquad (2.164)$$

with

$$\tilde{\mathcal{H}}_1(t) = \exp(i\mathcal{H}_0 t)\mathcal{H}_1(t)\exp(-i\mathcal{H}_0 t) \qquad (2.165)$$

If h_x, h_y, h_z are random and small compared to the Larmor frequency, then (2.164) may be solved to second order in time using the series approximation

$$\tilde{P}(t) = \tilde{P}(0) - i\int_0^t [\tilde{\mathcal{H}}_1(t'), \tilde{P}(0) - \tilde{P}_{eq}]\,dt'$$
$$+ (-i)^2 \int_0^t dt' \int_0^{t'} dt''[\tilde{\mathcal{H}}_1(t'), [\tilde{\mathcal{H}}_1(t''), \tilde{P}(t'') - \tilde{P}_{eq}]]$$

Taking the value of t'' to be $t' - \tau$ for fixed t', then $dt'' = -d\tau$ and

$$\tilde{P}(t) = \tilde{P}(0) - i\int_0^t [\tilde{\mathcal{H}}_1(t'), \tilde{P}(0) - \tilde{P}_{eq}]\,dt'$$
$$+ (-i)^2 \int_0^t dt' \int_0^{t'} d\tau[\tilde{\mathcal{H}}_1(t'), [\tilde{\mathcal{H}}_1(t' - \tau), \tilde{P}(t' - \tau) - \tilde{P}_{eq}]] \qquad (2.166)$$

Taking the derivative with respect to t gives

$$\frac{\partial \tilde{P}}{\partial t} = -i[\tilde{\mathcal{H}}_1(t), \tilde{P}(0) - \tilde{P}_{eq}]$$
$$- \int_0^t d\tau[\tilde{\mathcal{H}}_1(t), [\tilde{\mathcal{H}}_1(t - \tau), \tilde{P}(t - \tau) - \tilde{P}_{eq}]] \qquad (2.167)$$

Substituting

$$-\tilde{\mathcal{H}}_1(t) \equiv \exp(i\mathcal{H}_0 t)\{h_x(t)I_x + h_y(t)I_y + h_z(t)I_z\}\exp(-i\mathcal{H}_0 t)$$
$$= \exp(i\mathcal{H}_0 t)\sum_j h_j(t)I_j\exp(-i\mathcal{H}_0 t)$$

yields

$$\frac{\partial \tilde{P}}{\partial t} = i\sum_j h_j(t)[\exp(i\mathcal{H}_0 t)I_j\exp(-i\mathcal{H}_0 t), \tilde{P}(0) - \tilde{P}_{eq}]$$

$$+ \int_0^t d\tau \sum_{j,m} h_j(t)h_m(t-\tau)[\exp(i\mathcal{H}_0 t)I_j\exp(-i\mathcal{H}_0 t),$$

$$[\exp[i\mathcal{H}_0(t-\tau)]I_m\exp[-i\mathcal{H}_0(t-\tau)], \tilde{P}(t-\tau) - \tilde{P}_{eq}]] \quad (2.168)$$

We now take the ensemble average of both sides of Eq. (2.168), which converts $\tilde{P}$ into $\tilde{\rho}$, the density operator. The result is

$$\frac{\partial \tilde{\rho}}{\partial t} = i\sum_j \overline{h_j}(t)[\tilde{I}_j, \tilde{\rho}(0) - \tilde{P}_{eq}]$$

$$- \int_0^t d\tau \sum_{j,m} \overline{h_j(t)h_m(t-\tau)}[\tilde{I}_j, [\tilde{I}_m, \rho(t-\tau) - P_{eq}]] \quad (2.169)$$

where

$$\tilde{I}_j = \exp(i\mathcal{H}_0 t)I_j\exp(-i\mathcal{H}_0 t)$$

and   (2.170)

$$\tilde{I}_m = \exp[i\mathcal{H}_0(t-\tau)]I_m\exp[-i\mathcal{H}_0(t-\tau)]$$

Since the $h_j(t)$ are random functions, the ensemble average of any $h_j(t)$ vanishes. The first term in (2.169) is zero, and we find

$$\dot{\tilde{\rho}} = \frac{\partial \tilde{\rho}}{\partial t} = -\int_0^t d\tau \sum_{j,m} \overline{h_j(t)h_m(t-\tau)}$$

$$\times [e^{i\mathcal{H}_0 t}I_j e^{-\mathcal{H}_0 t}, [e^{i\mathcal{H}_0(t-\tau)}I_m e^{-i\mathcal{H}_0(t-\tau)}, \tilde{\rho}(t-\tau) - \tilde{\rho}_{eq}]] \quad (2.171)$$

If the h_j are assumed to be uncorrelated

$$\overline{h_j(t)h_m(t-\tau)} = 0 \qquad \text{unless} \quad j = m$$

Therefore

$$\dot{\tilde{\rho}} = -\int_0^t d\tau \sum_j \overline{h_j(t)h_j(t-\tau)}$$

$$\times [e^{i\mathcal{H}_0 t}I_j e^{-i\mathcal{H}_0 t}, [e^{i\mathcal{H}_0(t-\tau)}I_j e^{-i\mathcal{H}_0(t-\tau)}, \tilde{\rho}(t-\tau) - \tilde{\rho}_{eq}]]$$

One may calculate the observed behavior of the longitudinal magnetization from Eq. (2.71):

$$\frac{\partial \langle I_z \rangle}{\partial t} = \frac{\partial}{\partial t} \operatorname{Tr} \tilde{\rho}(t) I_z = \operatorname{Tr} \dot{\tilde{\rho}} I_z$$

Repeated application of the identity $\operatorname{Tr} AB = \operatorname{Tr} BA$ yields

$$\operatorname{Tr} \dot{\tilde{\rho}} I_z = \int_0^t d\tau \sum_j \overline{h_j(t)h_j(t-\tau)} \operatorname{Tr}[[I_z, I_j], e^{i\mathcal{H}_0\tau} I_j e^{-i\mathcal{H}_0\tau}]$$

$$\times (\tilde{\rho}(t-\tau) - \tilde{\rho}_{eq}) \tag{2.172}$$

The approximation of truncating at second order is really only good if the time τ over which $\overline{h_j(t)h_j(t-\tau)}$ is nonzero is short compared to t.

We may replace $\rho(t-\tau)$ by $\rho(t)$ and expand the sum over j, carrying out commutators explicitly, to give

$$\frac{d}{dt} \langle I_z(t) \rangle = i \int_0^t d\tau \{ \overline{h_x(t)h_x(t-\tau)} \operatorname{Tr}[I_y, e^{i\mathcal{H}_0\tau} I_x e^{-\mathcal{H}_0\tau}]$$

$$\times (\tilde{\rho}(t) - \tilde{\rho}_{eq})$$

$$+ \overline{h_y(t)h_y(t-\tau)} \operatorname{Tr}[-I_x, e^{i\mathcal{H}_0\tau} I_y e^{-i\mathcal{H}_0\tau}]$$

$$\times (\tilde{\rho}(t) - \tilde{\rho}_{eq}) \}$$

The exponential operators only induce rotation

$$e^{-i\omega_0\tau I_z} I_x e^{i\omega_0\tau I_z} = I_x \cos \omega_0\tau + I_y \sin \omega_0\tau$$

$$e^{-i\omega_0\tau I_z} I_y e^{i\omega_0\tau I_z} = I_y \cos \omega_0\tau - I_x \sin \omega_0\tau$$

Therefore,

$$\frac{d}{dt} \langle I_z(t) \rangle = - \int_0^t d\tau \, \overline{h_x(t)h_x(t-\tau)} \cos \omega_0\tau \operatorname{Tr} I_z[\tilde{\rho}(t) - \tilde{\rho}_{eq}]$$

$$- \int_0^t d\tau \, \overline{h_y(t)h_y(t-\tau)} \cos \omega_0\tau \operatorname{Tr} I_z[\tilde{\rho}(t) - \tilde{\rho}_{eq}] \tag{2.173}$$

One sees that this equation for the observable has the form

$$\frac{d}{dt} \langle I_z \rangle = -\frac{1}{T_1} [\langle I_z(t) \rangle - \langle I_z \rangle eq] \tag{2.174}$$

which was postulated for the Bloch equations. However, T_1 depends on the ending time t! One of the initial assumptions was that the entire time dependence could be approximated by a second-order truncation of the perturbation expansion. This procedure is possible if the integrals [and hence the correlation functions $\overline{h_j(t)h_j(t-\tau)}$] are nonzero only for a short time. Then, the integral to t has a nonzero value only at very short times. Specifically at times $\ll T_1$, the approximation is made

$$\int_0^t \overline{h_j(t)h_j(t-\tau)} \cos\omega_0\tau \, d\tau \cong \int_0^\infty \overline{h_j(t)h_j(t-\tau)} \cos\omega_0\tau \, d\tau \quad (2.175)$$

Further, it is assumed that, in the other degrees of freedom, the system remained more or less in equilibrium. Consistent with the assumption, we may make the approximation that the value of the average in (2.175) is independent of origin in time and is symmetric in time—the stochastic approximation. We denote these averages as $G_j(\tau)$.

$$\overline{h_j(t)h_j(t-\tau)} = \overline{h_j(t)h_j(t+\tau)} = G_j(\tau) \quad (2.176)$$

One then obtains a form for T_1 in terms of the dynamics of the system

$$\frac{1}{T_1} = \int_0^\infty [G_x(\tau) + G_y(\tau)] \cos\omega_0\tau \, d\tau \quad (2.177)$$

which reduces, by virtue of the symmetric properties of $G_j(\tau)$, to

$$\frac{1}{T_1} = \frac{1}{2}\int_{-\infty}^{+\infty} [G_x(\tau) + G_y(\tau)] \exp(i\omega_0\tau) \, d\tau \quad (2.178)$$

This is a recipe for calculating T_1 in terms of microscopic dynamics. The models for the interaction leading to loss of energy from the Zeeman-energy reservoir and the microscopic dynamics are all one needs to describe T_1 behavior. One model is the dipolar fluctuation model, in which the dipolar interactions between various nuclei are time dependent due to molecular motion, e.g., because of random rotation of the molecule or movement of molecules relative to each other (diffusion). The results are derived in the same manner as in the discussion leading to Eq. (2.178), but they are slightly different in form. Relaxation due to interactions of the nuclei by other mechanisms is also possible. In Chapter 3, where the forms of various interactions are discussed, we present the effects of various mechanisms of spin–lattice relaxation.

For the present, let us turn back to the more general (if slightly more nebulous) form of (2.178) to see whether one may glean more information about the characteristics of relaxation. The usual manner in which the approximate behavior of $G_x(\tau)$ and $G_y(\tau)$ is handled is to assume they decay away in short times characterized by correlation times τ_x and τ_y. Hence

$$G_j(\tau) = G_j(0)\exp(-|\tau/\tau_j|) \quad (2.179)$$

if the correlation decay is modeled as an exponential. Whether this is reasonable depends on the exact dynamics of the many-body system. In what follows we assume that it is correct. Substituting this form into (2.178) gives one a result for T_1 after integration:

$$\frac{1}{T_1} = G_x(0) \frac{\tau_x}{1 + \omega_0^2 \tau_x^2} + G_y(0) \frac{\tau_y}{1 + \omega_0^2 \tau_y^2} \tag{2.180}$$

For purposes of illustration, assume $\tau_x = \tau_y = \tau$. Then one may construct a plot of $1/T_1$ in terms of $\omega_0 \tau$ as shown in Fig. 2.4. The relaxation mechanism is seen to be most effective at or around $\tau \approx 1/\omega_0$ and is only about 10% as effective if the correlation time is an order of magnitude longer or shorter than this value.

One more commonly sees this information displayed in a log–log plot of T_1 versus τ, as in Fig. 2.5. From both of these plots, it is seen that, when correlation times in the other degrees of freedom (often referred to collectively as the *lattice*) are on the order of the fundamental frequencies of the spin system, energy exchange (and return to equilibrium) is rapid. It is as if the spin system had come into resonance with a field with which it could readily exchange energy.

The simplest model of lattice motion is that of the correlation time for the modulation of the Hamiltonian being thermally activated:

$$\tau(T) = \tau_j \exp(E_a/RT) \tag{2.181}$$

Hence, for the simple model given in (2.181), with $\tau_x = \tau_y = \tau$,

$$\frac{1}{T_1} = 2\{G_x(0) + G_y(0)\} \frac{\tau_j \exp(E_a/RT)}{1 + \omega_0^2 \tau_j^2 \exp(2E_a/RT)} \tag{2.182}$$

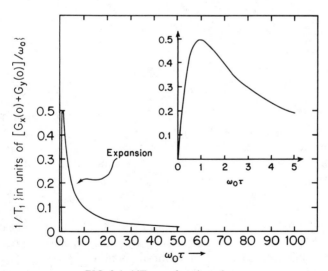

FIG. 2.4. $1/T_1$ as a function of $\omega_0 \tau$.

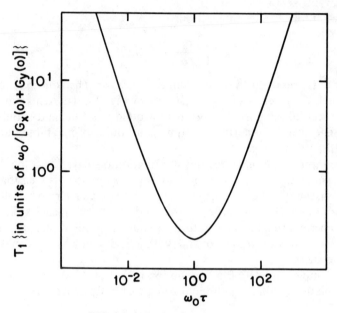

FIG. 2.5. $\ln T_1$ as a function of $\omega_0 \tau$.

In the low-temperature regime $\omega_0^2 \tau_j^2 \exp(2E_a/RT) \gg 1$

$$\frac{1}{T_1} = \frac{2\{G_x(0) + G_y(0)\}}{\omega_0^2 \tau_j^2} \exp(-E_a/RT) \tag{2.183}$$

or

$$\ln T_1 = -\ln \frac{2(G_x(0) + G_y(0))}{\tau_j^2} + 2\ln \omega_0 + \frac{E_a}{RT} \tag{2.184}$$

In other words, a plot of $\ln T_1$ versus $1/T$ at fixed frequency should, in this limit, give a straight line with a slope of E_a/R, where E_a is the activation energy for the motion producing modulation.

In the other limit, $1 \gg \omega_0^2 \tau_j^2 \exp(2E_a/RT)$, a similar procedure to the one above gives

$$\ln T_1 = -\ln 2(G_x(0) + G_y(0))\tau_j - (E_a/R)(1/T) \tag{2.185}$$

Thus, in this high-temperature limit, then, the plot is also linear with a slope $-E_a/R$. In this limit, the intercept does not depend on ω_0, whereas in the low-temperature (slow-correlation-time) limit it does.

XV. The Polarization Vector and Spin-$\frac{1}{2}$: A Quantum Analog of the Magnetization Vector

We discuss further how the density matrix may be used to describe a spin-$\frac{1}{2}$ system. In order to do so, the density matrix is expressed in terms of the Pauli spin matrices:

$$\rho = \frac{1}{2}\left[\begin{bmatrix} 1 & 0 \\ 0 & 1 \end{bmatrix} + P_x\sigma_x + P_y\sigma_y + P_z\sigma_z\right] \tag{2.186}$$

where

$$P_x = 2\operatorname{Re}(C_\alpha^* C_\beta), \qquad P_y = 2\operatorname{Im}(C_\alpha^* C_\beta), \qquad P_z = |C_\alpha|^2 - |C_\beta|^2 \tag{2.187}$$

which are real numbers—the elements of the polarization vector.

Equations (2.186)–(2.187) give the density matrix as

$$\rho = \frac{1}{2}\left\{\begin{bmatrix} |C_\alpha|^2 + |C_\beta|^2 & 0 \\ 0 & |C_\alpha|^2 + |C_\beta|^2 \end{bmatrix} + \begin{bmatrix} P_z & P_x - iP_y \\ P_x + iP_y & -P_z \end{bmatrix}\right\}$$

We may formally represent the density matrix as

$$\rho = \tfrac{1}{2}(1 + \mathbf{P} \cdot \boldsymbol{\sigma}) \tag{2.188}$$

where $\mathbf{1}$ is the unit matrix, $\mathbf{P}$ the polarization vector, and $\boldsymbol{\sigma}$ the Pauli-spin-matrix vector (a slightly peculiar construction because the vector's components are themselves 2×2 matrices).

Before discussing the properties of ρ further, we again recall the properties of matrices which are similar to diagonal matrices.

If a matrix $\mathbf{A}$ is similar to a diagonal matrix $\mathbf{D}$, there exists a transformation

$$\mathbf{P}^{-1}\mathbf{A}\mathbf{P} = \mathbf{D} \tag{2.189}$$

The characteristic roots of $\mathbf{A}$, λ_i, are found from the determinant $|A_{ij} - \lambda\delta_{ij}| = 0$, and these are the diagonal elements of $\mathbf{D}$. Also, by (2.189)

$$\mathbf{A}\mathbf{P} = \mathbf{P}\mathbf{D} = \operatorname{diag}\{\lambda_1, \lambda_2, \ldots, \lambda_n\} \tag{2.190}$$

If $\mathbf{P}_i$ denotes the ith column of $\mathbf{P}$, we have

$$\mathbf{A}\mathbf{P}_i = \lambda_i\mathbf{P}_i \tag{2.191}$$

Hence $\mathbf{A}$ has as one of its eigenstates the vector $\mathbf{P}_i$ and the eigenvalue is λ_i. With respect to the density matrix, the secular equation is

$$\begin{vmatrix} |C_\alpha^2| - \lambda & C_\alpha C_\beta^* \\ C_\alpha^* C_\beta & |C_\beta^2| - \lambda \end{vmatrix} = 0 \tag{2.192}$$

the solution of which is $\lambda = 1$ or zero. The value of the eigenfunction corresponding to $\lambda = 1$ is found from

$$\rho P_1 = \begin{bmatrix} |C_\alpha|^2 & |C_\alpha|e^{i\gamma_\alpha}|C_\beta|e^{-i\gamma_\beta} \\ |C_\alpha|e^{-i\gamma_\alpha}|C_\beta|e^{i\gamma_\beta} & |C_\beta|^2 \end{bmatrix} \begin{bmatrix} |C_\alpha|e^{i\gamma_\alpha} \\ |C_\beta|e^{i\gamma_\beta} \end{bmatrix}$$

$$= \lambda P_1 = \begin{bmatrix} |C_\alpha|e^{i\gamma_\alpha} \\ |C_\beta|e^{i\gamma_\beta} \end{bmatrix}$$

and

$$P_1 = \begin{bmatrix} |C_\alpha|e^{i\gamma_\alpha} \\ |C_\beta|e^{i\gamma_\beta} \end{bmatrix} \equiv \chi \tag{2.193}$$

That is, the eigenfunction is just the instantaneous state. The spinor corresponding to $\lambda = 0$ must be orthogonal to χ and have eigenvalue zero. A useful exercise would be to show this statement is true.

Calling the eigenvector corresponding to $\lambda = 0$ by the name η, we see that the two-by-two space of states is spanned by χ and η. Any function ϕ in this space may be expressed as a linear combination

$$\phi = a_1\chi + a_2\eta \tag{2.194}$$

and

$$\rho\phi = a_1\chi \tag{2.195}$$

Therefore ρ is a projection operator which projects ϕ onto χ. A projection operator is always idempotent; i.e., it has the property $A^2 = A$. This property is easily seen for ρ since

$$\rho = \chi\chi^*$$

and $\chi^*\chi = 1$ so

$$\rho = \chi(\chi^*\chi)\chi^* = \rho^2 \tag{2.196}$$

We inquire as to what the expectation value of $\boldsymbol{\sigma}$ is when the system is described by spinor χ

$$\begin{aligned} \langle\sigma_x\rangle &= \operatorname{Tr}\rho\sigma_x \\ &= \operatorname{Tr}\tfrac{1}{2}[1 + \mathbf{P}\cdot\boldsymbol{\sigma}]\sigma_x \\ &= \tfrac{1}{2}\operatorname{Tr}\sigma_x + \tfrac{1}{2}\operatorname{Tr}P_x\sigma_x^2 \\ &\quad + \tfrac{1}{2}\operatorname{Tr}P_y\sigma_y\sigma_x + \tfrac{1}{2}\operatorname{Tr}P_z\sigma_z\sigma_y \\ &= P_x \end{aligned} \tag{2.197}$$

because $\operatorname{Tr}\sigma_i\sigma_j = 2\delta_{ij}$. Similarly

$$\langle\sigma_y\rangle = P_y, \qquad \langle\sigma_z\rangle = P_z$$

so the polarization vector

$$\mathbf{P} = \langle \boldsymbol{\sigma} \rangle = \mathrm{Tr}(\rho\boldsymbol{\sigma}) = \mathrm{Tr}(\boldsymbol{\sigma}\rho). \tag{2.198}$$

This result gives immediate physical significance to **P**. **P** may be said to be a vector pointing in the direction of the spin of the particle described by χ. In other words, it is the quantum-mechanical analog of the direction of the magnetic moment.

XVI. The Equation of Motion of the Vector P: The Quantum Analog of the Bloch Equation

We may use the Liouville–von Neumann equation to develop the quantum-mechanical analog of the Bloch equation. This result will bring us full circle—the quantum-mechanical equations lead us to the classical equation of motion.

We let the Hamiltonian be represented by the matrix

$$\mathscr{H} = \frac{1}{2}(Q_0 \mathbf{1} + \mathbf{Q} \cdot \boldsymbol{\sigma}) \tag{2.199}$$

We relate this interaction to the classical interaction of a magnetic moment **M** (with quantum analog **P**), and a field **B** (with quantum analog **Q**), $E = -\mathbf{M} \cdot \mathbf{B}$. We see that the second term on the right-hand side of (2.199) is the analog of the classical interaction. The first term on the right-hand side of (2.199) is a constant, added to make the calculation convenient. Adding a constant to the energy is always permissible since only energy changes are measurable. Here $Q_0 = \frac{1}{3}\mathrm{Tr}\,Q$ is a real number. We assert that (2.199) represents $\mathscr{H}$, even if $\mathscr{H}$ is an explicit function of time.

With the realization that $d\boldsymbol{\sigma}/dt = 0$ (since the components of $\boldsymbol{\sigma}$ are just complex numbers), we find

$$\frac{d\mathbf{P}}{dt} = \frac{d\langle \boldsymbol{\sigma} \rangle}{dt} = \frac{\langle \boldsymbol{\sigma}\mathscr{H} - \mathscr{H}\boldsymbol{\sigma} \rangle}{i} \tag{2.200}$$

with $\mathscr{H}$ defined as above. This yields

$$\frac{d\mathbf{P}}{dt} = \frac{\langle \boldsymbol{\sigma}\mathbf{Q} \cdot \boldsymbol{\sigma} - \mathbf{Q} \cdot \boldsymbol{\sigma}\boldsymbol{\sigma} \rangle}{2i} \tag{2.201}$$

One may use the well-known vector identity

$$\mathbf{A} \times (\mathbf{B} \times \mathbf{C}) = \mathbf{B}(\mathbf{A} \cdot \mathbf{C}) - \mathbf{C}(\mathbf{A} \cdot \mathbf{B})$$

to show this reduces to

$$\frac{d\mathbf{P}}{dt} = \frac{\mathbf{Q} \times \boldsymbol{\sigma} \times \boldsymbol{\sigma}}{2i} \tag{2.202}$$

Since $\boldsymbol{\sigma}$ is not a simple three-component vector with each component a scalar, but a vector with each component a Pauli matrix, $\boldsymbol{\sigma} \times \boldsymbol{\sigma}$ is not zero. The standard commutation rules for the Pauli matrices yield $\boldsymbol{\sigma} \times \boldsymbol{\sigma} = 2i\boldsymbol{\sigma}$, so (2.202) becomes

$$d\mathbf{P}/dt = \mathbf{Q} \times \mathbf{P} \tag{2.203}$$

the quantum analog of the Bloch equations.

We see that the equations of motion of systems comprised of spins which obey the quantum-mechanical rules for angular momentum are in analogy to the classical equations derived from the models in Chapter 1. However, with the quantum-mechanical equations, one may relate the quantities in the equations of motion (like T_1) to averages over microscopic quantities like the correlation functions for molecular motions.

XVII. Macroscopic Thermodynamics and the Density Operator

An interesting connection between the equilibrium density operator and macroscopic thermodynamic functions exists. Another way of writing Eq. (2.69) is

$$\rho_{eq} = \{1/Z\} \exp[-\hbar\mathscr{H}_0/kT] \tag{2.69'}$$

where Z is the canonical ensemble partition function. The relation between equilibrium thermodynamic functions and the canonical ensemble partition function is

$$A = -kT \ln Z \tag{2.204}$$

A is the Helmholtz free energy, defined in terms of the internal energy U and entropy $\mathscr{S}$ as

$$A = U - T\mathscr{S} \tag{2.205}$$

The internal energy is the expectation value of the Hamiltonian;

$$U = \langle \hbar\mathscr{H}_0 \rangle \tag{2.206}$$

We may therefore show that the entropy is given by the form

$$\mathscr{S} = -k \operatorname{Tr} \rho \ln \rho \tag{2.207}$$

This is a very powerful result; in manipulations of ensembles of spins, especially in multiple-quantum experiments not discussed here, spins are frequently manipulated so that an observable magnetization is not present for a period of time but is made to appear after this period by further manipulation. Interesting questions may be asked about such a process; for example, what is it that disappears? What is conserved such that it may be "read back" by the application of radio-frequency pulses? Equation (2.207) holds one of the keys to understanding the physical reality in such experiments.

PROBLEMS

2.1   Show that

$$\text{Tr}\,\sigma_k\sigma_j = \delta_{kj}$$

2.2   Show that
   (a) $\langle\sigma_x\rangle = P_x$.
   (b) $\langle\sigma_y\rangle = P_y$.
   (c) $\langle\sigma_z\rangle = P_z$.

2.3   Prove that the spinors corresponding to the eigenvalues $\lambda = 1$ and $\lambda = 0$ are orthonormal.

2.4   Suppose the spinor χ has the form

$$\chi = \begin{bmatrix} e^a \\ e^{-a} \end{bmatrix}$$

What is the polarization vector for such a spinor?

2.5   Show that the commutation relation for the Pauli spin matrices is

$$\sigma \times \sigma = 2i\sigma$$

2.6   For a Hamiltonian of the form

$$\mathscr{H} = \tfrac{1}{2}(Q_0 \cdot \mathbf{1} + \mathbf{Q} \cdot \sigma)$$

what are the eigenfunctions and associated eigenvalues?

2.7   Verify that Eq. (2.66) is the solution to Eq. (2.64).

2.8   (a)   What are the units of γ in Eq. (2.28)?
   (b)   How would this equation have to be changed if the units of E were ergs?

2.9   What are the rotating frame values of $\langle I_y\rangle$, $\langle I_x\rangle$, and $\langle I_z\rangle$ after a $\pi/2_x$ pulse on an ensemble of spins-$\tfrac{1}{2}$? What do you physically infer from these answers?

2.10   Show that Eq. (2.8) follows from Eq. (2.7).

2.11   Show that Eqs. (2.36) are solutions of (2.35) by substitution.

2.12   Ordinary laboratory magnets produce dc fields which are less than 70 kG (7 T in SI units). Using the magnetogyric ratios in Table 1.1 and the usual values for $\hbar$ and k, calculate $\gamma\hbar B_0/kT$ at 1 K at a field of 7 T. Under these conditions, what is the error made in the expansion used in Eq. (2.70)?

2.13   Prove that if the rule for transformation of the wave function Ψ is $\tilde{\Psi} = U^+\Psi$, then the rule for transforming the density operator $\rho = \Psi\Psi^*$ is

$\rho = U\dot{\rho}U^+$. (Keep in mind the fact that both ρ and U are matricies, and consequently the rules that govern the Hermitian conjugate of the product of two matricies.)

2.14   (a)   Generate the matrices for I_x, I_y and I_z for a spin-1 particle.

   (b)   Define Pauli spin matrices for spin-1 consistent with Eq. (2.31).

   (c)   Write out the matrices for I^+ and I^- for spin-1.

   (d)   What is (by matrix multiplication) $[I^+, I^-]$? Does this suggest a property of matrix multiplication?

2.15   (a)   Show that Eq. (2.204) follows from the other equations in Section XVII.

   (b)   Calculate the entropy of a mole of systems with spin quantum number I, weakly interacting as to satisfy the last statement in section IX. Assume any temperature large compared to 0.01 K, which means the high temperature limit.

REFERENCES

1. D. P. Weitekamp, Time domain multiple quantum NMR, *Adv. Magn. Reson.* **11**, 111 (1983).
2. R. C. Tolman, "The Principles of Statistical Mechanics," Chapter 9, Oxford Univ. Press, Oxford, 1950.

CHAPTER 3

INTERNAL HAMILTONIANS
AND THEIR SPECTRA

I. Introduction

In Chapter 2 it was shown that the measured values of observables in NMR spectroscopy, the transverse components of angular momentum (corresponding to single-quantum coherences), depend on the time development of the wave function through equations such as (2.67). The wave function, in turn, depends on interactions experienced by the system under study, which is reflected in the time-dependent Schrödinger equation. From this equation the Hamiltonian of the system is a determining parameter in specifying the time evolution of a system. In this chapter, we derive the forms of Hamiltonians that affect the time evolution of a nuclear spin system in conditions under which a chemist might observe it. These Hamiltonians are quantum-mechanical operators and are often derived by comparison to classical treatments such as that of Chapter 1. Two of the interactions are supplied by the spectroscopist: the static Zeeman interaction and the Zeeman interaction with applied oscillating magnetic fields. In rigid nonmetallic solids, four interactions are supplied by nature: (1) the dipole–dipole interaction among spins, (2) the shielding interaction with electrons, (3) the scalar J or indirect coupling of two or more nuclei, and (4) the quadrupolar interaction with electric-field gradients. In solid materials, one or more of these may dominate the others, producing spectra that contain little information about all the potential interactions. The subjects of Chapters 5 and 6 are the means of manipulating these systems with radio-frequency irradiation to allow observation of these "smaller" interactions.

The focus of much of the work in the field of NMR spectroscopy in the solid state is the suppression of the effects of the dipole–dipole interaction. The dipole–dipole interaction results in *homogeneously* broadened NMR resonances. To say that the resonance is homogeneously broadened means that *each nucleus* in the solid experiences couplings to many other nuclei. Such nuclei exhibit an NMR spectrum in which the linewidth in the absence of any decoupling is much larger than would be expected on the basis of the lifetime of the Zeeman states. For each nucleus in such a solid, e.g., ^{19}F in

CaF_2, T_2^* is much shorter than T_1. This is true even for nuclei in a single crystal. On the other hand, a nucleus whose dominant interaction is the chemical shift or the quadrupolar interaction may have a very broad line in a powder because the nuclei in crystallographically inequivalent sites have slightly different resonance positions. However, the resonance of *each nucleus* subject to these interactions is sharp. The broadness results from the super-position of many sharp lines. Such spectra are said to be *inhomogeneously broadened*.

Rotations in the appropriate manner, either in spin space or in "real" space, can produce suppression of the effects of specific internal Hamiltonians while retaining the effects of others. A question that arises is how fast does one need to perform the requisite suppression? The answer is simple: one must perform the averaging by whatever process faster than the interaction can cause significant changes in the system. This time is roughly of the order of the inverse of the inherent linewidth due to that interaction. Inhomoge-neously broadened lines, being a superposition of inherently sharp spectra, need only to be affected by rotational processes that are slow compared to the total *inhomogeneous* linewidth, but that are fast compared to the inherent linewidth. An example that we discuss later is the "slow sample rotation" that breaks a line broadened by chemical-shift anisotropy into a set of sharp lines even though the inhomogeneous linewidth is much greater. Homogeneously broadened lines, because the total width of the spectrum represents the line-width of each spin's spectrum, can only be narrowed by appropriate rota-tional manipulation if that manipulation may be done at a rate greater than the inverse of the entire line. The reader will want to keep in mind that, of all the interactions discussed, the dipole–dipole interaction is the only one for which many-body effects lead to homogeneous broadening in solids and so is special in that regard.

II. The Zeeman Interaction

The Hamiltonians associated with the Zeeman and radio-frequency fields have been derived in Chapter 2 [cf. Eqs. (2.28) and (2.143)]. These Hamil-tonians represent the energy of a moment in a field $-\mathbf{M} \cdot \mathbf{B}$. In terms of the spin angular momenta $\mathbf{I}_k$ of the nuclei in the sample, the Hamiltonians are

$$\mathcal{H}_Z = -\gamma \mathbf{I} \cdot \mathbf{B}_0 = -\omega_0 \sum_k I_{zk} \tag{3.1}$$

and, for a particular choice of phase ϵ

$$\mathcal{H}_{rf} = -\omega_1 \sum_k I_{xk} \cos[(\omega_0 + \delta)t + \epsilon]$$
$$- I_{yk} \sin[(\omega_0 + \delta)t + \epsilon] \tag{3.2}$$

where δ is the resonance offset in radians per second. We recall from Chapter 2 that, in the frame of the Zeeman Hamiltonian, i.e., in the frame in which the NMR response is observed, a pulse along the k axis has a Hamiltonian of the form

$$\mathscr{H}_{1k} = -\omega_1 I_k \qquad (3.2a)$$

such that, for an x pulse,

$$\mathscr{H}_{1x} = -\omega_1 I_x \qquad (3.2b)$$

The forms of (3.2) remind us that the symbol ω indicates both an *intensity* (the magnitudes of the Zeeman and radio-frequency fields are measured by ω_0/γ and ω_1/γ, respectively) and a *frequency of oscillation* (the Larmor frequency ω_0 and the frequency of oscillation of the radio-frequency field in the laboratory frame $\omega_0 + \delta$). These Hamiltonians are written as an intensity times a product of space and spin operators. For example, the Zeeman Hamiltonian of Eq. (3.1) may be written as

$$\mathscr{H}_Z = -\omega_0(I_x \sin\theta \cos\phi + I_y \sin\theta \sin\phi + I_z \cos\theta) \qquad (3.1a)$$

where (θ, ϕ) represent the polar angles of the magnetic field relative to the coordinate system. The internal Hamiltonians may be similarly represented as products of intensity factors with functions of space and spin variables.

As examples of intensity factors for Zeeman and rf Hamiltonians, in a 14 kG (1.4 T) field, the magnitude of the static Zeeman interaction for a proton with $\gamma = 2.675 \times 10^4$ rad sec^{-1} is

$$\omega_0^{^1H} = |\gamma^{^1H} B_0| = 3.72 \times 10^8 \text{ rad sec}^{-1} = 59.2 \text{ MHz.}$$

For ^{13}C, $\gamma^{^{13}C} = 6.727 \times 10^3$ rad sec^{-1} G^{-1}, and the interaction is $\omega_0^{^{13}C} = 9.418 \times 10^7$ rad sec$^{-1} = 14.99$ MHz; and for ^{2}H, $\gamma^{^2H} = 4.107 \times 10^3$ rad sec^{-1} G^{-1}, yielding a magnitude of interaction $\omega_0^{^2H} = 5.749 \times 10^7$ rad sec$^{-1} = 9.150$ MHz.

The magnitude of the interaction with the time-dependent field in an NMR experiment is determined by the magnitude of the field. We have previously calculated that, for the currents achievable with a small coil, the magnitude of the radio-frequency field produced in the coil can be as high as 60 G. This seems to be a value one might expect to generate in a NMR experiment, so we shall use it as an estimate of the field in such a coil. The magnitude of the interaction of protons with such a field is $\omega_1^{^1H} = |\gamma_H B_1| = 1.605 \times 10^6$ rad sec^{-1}. For ^{13}C, interaction with a field of a similar size gives a magnitude $\omega_1^{^{13}C} = 4.036 \times 10^5$ rad sec^{-1}.

The Zeeman interaction is often used as a model for all the other possible interactions to which spins may be subject. For example, the dipolar interaction is often described in terms of a *local field*, as if each spin interacted with a field produced by all other spins. Although not strictly correct, it is often useful to picture other interactions as Zeeman interactions with effective fields.

III. The Direct Dipole–Dipole Interaction

The presence of other spins may be observed in a number of interactions. One of the potentially major interactions is that of a spin at point $\mathbf{r}$ with the dipolar field produced by other spins $\mathbf{m}_j$ at other points $\mathbf{r} - \mathbf{r}_j$ in space. This field is found in Appendix 1 to be

$$\mathbf{B(r)} = \sum_j \left[\frac{\mathbf{m}_j}{|\mathbf{r} - \mathbf{r}_j|^3} + \frac{3\mathbf{m}_j \cdot (\mathbf{r} - \mathbf{r}_j)}{|\mathbf{r} - \mathbf{r}_j|^5} (\mathbf{r} - \mathbf{r}_j) \right] \tag{3.3}$$

The Hamiltonian for the overall dipole–dipole interaction is segregated into interactions with like nuclei (e.g., proton–proton interactions), with unlike nuclei (e.g., proton–carbon), and with electronic moments. The total interaction between a given moment $\mathbf{m}_j = \hbar \gamma_j \mathbf{I}_j$ and N moments $\mathbf{m}_k = \gamma_k \mathbf{S}_k \hbar$ is

$$\mathcal{H}_D = \sum_{jk} \left(\frac{\gamma_j \gamma_k \hbar \mathbf{I}_j \cdot \mathbf{S}_k}{|\mathbf{r}_j - \mathbf{r}_k|^3} - \frac{3\gamma_j \gamma_k \hbar [\mathbf{I}_j \cdot (\mathbf{r}_j - \mathbf{r}_k)][\mathbf{S}_k \cdot (\mathbf{r}_j - \mathbf{r}_k)]}{|\mathbf{r}_j - \mathbf{r}_k|^5} \right) \tag{3.4}$$

where $\mathbf{m}_k$ may represent like nuclei, other types of nuclei, or electron moments. For electrons, the moment is generally expressed in terms of the electron g factor and the Bohr magneton $\gamma \hbar = g\beta$.

For protons, which are relatively abundant (99.8%), the like nuclei (homonuclear) term is largest, unless there are unpaired electrons present. Let us calculate the magnitude of the homonuclear proton–proton interaction, assuming that a reasonable distance of separation for protons is about 2 Å. For example, this distance in water is 1.58 Å:

$$\left| \mathcal{H}_D^{H-H} \right| = \frac{\gamma^2 \hbar}{r^3}$$

$$= \frac{(2.675 \times 10^4 \quad \text{rad sec}^{-1} \text{ G}^{-1})^2 (1.05 \times 10^{-27} \quad \text{erg sec rad}^{-1})}{(2.0 \times 10^{-8} \quad \text{cm})^3}$$

$$= 9.39 \times 10^4 \quad \text{rad sec}^{-1} \equiv 14.95 \quad \text{kHz}$$

Let us calculate the magnitude of the dipolar interaction between a ^{13}C nucleus and its adjacent proton. Bond lengths are on the order of 1.0 to 1.5 Å in hydrocarbons. Let us use 1.0 Å:

$$\left| \mathcal{H}_D^{C-H} \right| = [(2.675 \times 10^4 \quad \text{rad sec}^{-1} \text{ G}^{-1})(6.727 \times 10^3 \quad \text{rad sec}^{-1} \text{ G}^{-1})$$

$$\times (1.05 \times 10^{-27})](1.0 \times 10^{-8} \quad \text{cm})^{-3}$$

$$= 1.88 \times 10^5 \quad \text{rad sec}^{-1} \equiv 30 \quad \text{kHz}$$

The magnitude of the dipolar interaction between protons and carbons is roughly comparable to that among protons, primarily because a carbon nucleus often lies closer to its nearest proton neighbor than a proton does. The difference in gyromagnetic ratio should make the magnitude less for comparably situated protons and carbons.

In terms of the vector $\mathbf{r}$ equal to $\mathbf{r}_2 - \mathbf{r}_1$, with magnitude r and components along the coordinate axes x, y, z, the dipole–dipole Hamiltonian for like nuclei may be rewritten as

$$\mathcal{H}_D = \frac{\gamma^2\hbar}{r^3}[I_{x_1}, I_{y_1}, I_{z_1}]\begin{bmatrix} 1 - 3x^2/r^2 & -3xy/r^2 & -3xz/r^2 \\ -3xy/r^2 & 1 - 3y^2/r^2 & -3yz/r^2 \\ -3xz/r^2 & -3yz/r^2 & 1 - 3z^2/r^2 \end{bmatrix}\begin{bmatrix} I_{x_2} \\ I_{y_2} \\ I_{z_2} \end{bmatrix}$$

$$= \mathbf{I}_1 \cdot \mathbf{D} \cdot \mathbf{I}_2 \tag{3.5}$$

Here $\mathbf{D}$ is known as the dipolar coupling tensor. In (3.5) this tensor is written in Cartesian coordinates. One may apply a transformation to spherical coordinates (r, θ, ϕ), where z is chosen as the polarization axis. The result is an expansion of the "alphabet soup" form:

$$\mathcal{H}_D = \frac{\gamma^2\hbar}{r^3}\{A + B + C + D + E + F\}$$

$$\equiv \omega_D\{A + B + C + D + E + F\} \tag{3.6}$$

where

$$\begin{aligned}
A &= (1 - 3\cos^2\theta)I_{z_1}I_{z_2} \\
B &= -\tfrac{1}{4}(1 - 3\cos^2\theta)[I_1^+ I_2^- + I_1^- I_2^+] \\
C &= -\tfrac{3}{2}\sin\theta\cos\theta e^{-i\phi}[I_{z_1}I_2^+ + I_1^+ I_{z_2}] \\
D &= -\tfrac{3}{2}\sin\theta\cos\theta e^{i\phi}[I_{z_1}I_2^- + I_1^- I_{z_2}] \\
E &= -\tfrac{3}{4}\sin^2\theta e^{-i2\phi}[I_1^+ I_2^+] \\
F &= -\tfrac{3}{4}\sin^2\theta e^{i2\phi}[I_1^- I_2^-] \\
\omega_D &= \gamma^2\hbar/r^3 \quad \text{(the dipolar frequency)}
\end{aligned} \tag{3.7}$$

This form is quite useful because each of the spin operators I^+, I^-, and I_z has a well-defined action on the basis states of the Zeeman Hamiltonian, as derived in Chapter 2. For example, consider the effect of a stationary perturbation by the dipolar Hamiltonian on a system consisting of two spin-$\tfrac{1}{2}$ particles. The system has four possible independent two-spin states; one form of these is

$$(|\alpha_1\rangle|\alpha_2\rangle;\ |\alpha_1\rangle|\beta_2\rangle;\ |\beta_1\rangle|\alpha_2\rangle;\ |\beta_1\rangle|\beta_2\rangle)$$

These states have Zeeman energies of $-\gamma B_0$, 0, 0, and $+\gamma B_0$, respectively.

The linear combinations of these product functions that are eigenfunctions of I^2 and I_z for the triplet t and singlet s spin-one systems are

$$|t_1\rangle = |\alpha_1\rangle|\alpha_2\rangle$$
$$|t_{-1}\rangle = |\beta_1\rangle|\beta_2\rangle$$

$$|t_0\rangle = \frac{1}{\sqrt{2}}(|\alpha_1\rangle|\beta_2\rangle + |\alpha_2\rangle|\beta_1\rangle)$$

$$|s_0\rangle = \frac{1}{\sqrt{2}}(|\alpha_1\rangle|\beta_2\rangle - |\alpha_2\rangle|\beta_1\rangle)$$

The effect of the dipolar interaction (~ 30 kHz) is small compared to that of the Zeeman interaction (~ 60 MHz in a 1.4-T field), so first-order perturbation theory is used to calculate the effect of $\mathcal{H}_D$ on the eigenstates of $\mathcal{H}_z$. It is therefore necessary to calculate the matrix elements $\langle i|\mathcal{H}_D|j\rangle$. The A term contributes an energy shift both to $|t_1\rangle$ and $|t_{-1}\rangle$ of $\omega_D(1 - 3\cos^2\theta)/4$. The A and B terms shift the $|t_0\rangle$ level by the amount $-\omega_D(1 - 3\cos^2\theta)/2$. The results are shown in the energy-level diagram of Fig. 3.1. Note that the center of gravity of the levels is conserved under $\mathcal{H}_D$, a useful check on the result.

Transitions caused by an oscillating applied field $\mathbf{B}_1(t)$ occur only between triplet levels because of the selection rule $\Delta I = 0$. In the "Zeeman-only" spectrum, this fact means that the two transitions are coincident. In the "Zeeman + dipolar" spectrum, the transitions are not coincident except

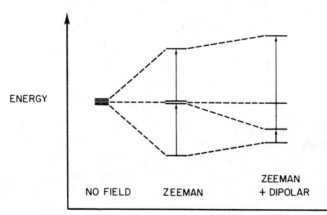

FIG. 3.1. Effect of Zeeman and Zeeman plus dipolar interactions on two spin-$\frac{1}{2}$ particles coupled by a homonuclear dipolar interaction.

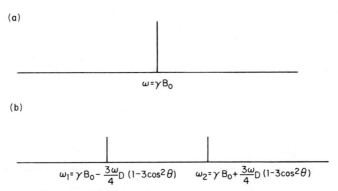

FIG. 3.2. (a) The absorption spectrum for a two spin-$\frac{1}{2}$ system with only a Zeeman interaction; (b) the absorption spectrum for a two spin-$\frac{1}{2}$ system subjected to a Zeeman and a dipole–dipole interaction.

under special circumstances. The stick spectra for the two cases are shown in Fig. 3.2.

The exact splitting of the line depends upon the spatial variables θ and r. Dipolar interactions can, therefore, be used to infer internuclear distances for pairs of interacting spins, provided that other interactions such as shielding anisotropy or scalar-coupling anisotropy (*vide infra*) are appropriately taken into account. If a single crystal is available, a study of the dipolar splitting as a function of orientation can supply not only the internuclear distance of a pair of interacting spins, but also the orientation of the internuclear vector with respect to the crystal axes. If only a powder sample is available, the dipolar powder pattern (*vide infra*) can be used to infer the internuclear distance. Note that there is a particular value of the angle between the static field and the internuclear vector for which the dipolar interaction is zero (see Problem 3.4). The value of this angle is easy to remember because it is the angle between the side of a cube and the cube diagonal.

A randomly oriented, or "powder," sample is one that contains microcrystallites oriented with respect to the field at all possible angles with equal probabilities. The NMR spectrum of such a sample consists of a band of lines as shown in Fig. 3.3. This is a rather characteristic spectrum, termed a "powder pattern of Pake doublets." Such a spectrum of a two-spin system is closely approximated by the powder spectra of such solids as $CaSO_4 \cdot 2H_2O$ and Cl_3CCOOH [Fig. 3.3(b)], which are examples of isolated two-spin systems. Note that these powder spectra are examples of *inhomogeneously* broadened lines.

In addition to the single dipolar interaction, there may be other protons (or other nuclei) close-by. In this case, all mutual dipolar interactions among

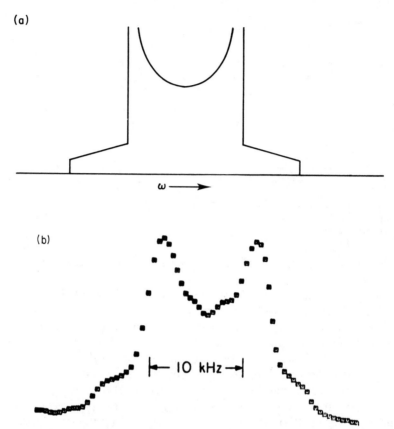

FIG. 3.3. (a) The calculated spectrum for a powder of randomly oriented crystallites having two spin-$\frac{1}{2}$ mutual dipole–dipole interactions; (b) the experimental spectrum for the protons in a randomly oriented powder of trichloroacetic acid. This spectrum is a close approximation to the ideal spectrum of (a).

the spins must be considered to solve the problem, and the spectrum becomes much more complicated. Consider the problem of three spins, each of which, in addition to the Zeeman interaction, has a pairwise interaction with the others. The energy-level scheme of such a three-spin system in the presence of pairwise dipolar interactions and the possible transitions are indicated in Fig. 3.4.

One sees that, when dipolar interactions are added, the three-spin system has a more complicated spectrum than a two-spin system because of the additional dipolar interactions. If one could unravel the transitions completely, he would have a complete description of the geometry of the system. For a

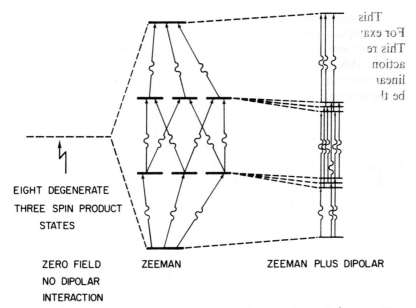

EIGHT DEGENERATE
THREE SPIN PRODUCT
STATES

ZERO FIELD   ZEEMAN   ZEEMAN PLUS DIPOLAR
NO DIPOLAR
INTERACTION

FIG. 3.4. Depiction of the effect of dipolar interactions in a three spin-$\frac{1}{2}$ system. Note the complication of the transition scheme when compared to the system of spins-$\frac{1}{2}$ interacting in pairs (Fig. 3.1).

two-spin system, such a description can be carried out. An unraveling of the spectrum may be feasible even for a six-spin system. However, in most systems there are roughly $10^{23} \times 10^{23}$ interactions that must be considered if there is a mole of nuclei present. These produce a spectrum for each nucleus with a marvelous complexity of structure. Lines may overlap, be degenerate, and produce a band of absorptions that is in general difficult to associate with any particular transition; i.e., the spectrum for each nucleus in the sample is homogeneously broadened.

It is possible, however, to obtain some information about the average value of the dipolar interactions in the sample from such a band of frequencies. To accomplish this, it is necessary to use an expansion technique originally suggested by van Vleck [1]. This is the expansion in terms of moments of the line. The nth moment of a line is defined by

$$M_n = \frac{\partial^n \langle M \rangle}{\partial t^n}\bigg|_{t=0} \bigg/ M_0 \tag{3.8}$$

where M_0 is the area under the absorption spectrum (which is the zeroth moment).

This expansion is an infinite-series approximation to the exact line shape. For example, consider a free-induction decay after a single-pulse experiment. This response is caused by a Hamiltonian such as that of the dipolar interaction. This decay can be completely described in terms of an appropriate linear combination of a complete set of functions. Such a complete set might be the integral powers of the time

$$\langle M_x(t) \rangle = \left\{ \sum_{n=0}^{\infty} M_n \frac{t^n}{n!} \right\} M_0 \qquad (3.9)$$

All that is needed is the set of coefficients M_n to describe the line shape. To see the relationship to observables, consider the derivatives of M_x:

$$\frac{\partial \langle M_x \rangle}{\partial t} = \left\{ \sum_{n=1}^{\infty} M_n \frac{t^{n-1}}{(n-1)!} \right\} M_0 \qquad (3.10)$$

$$\frac{\partial^2 \langle M_x \rangle}{\partial t^2} = \left\{ \sum_{n=2}^{\infty} M_n \frac{t^{n-2}}{(n-2)!} \right\} M_0 \qquad (3.11)$$

or generally

$$\frac{\partial^k \langle M_x \rangle}{\partial t^k} = \left\{ \sum_{n=k}^{\infty} M_n \frac{t^{n-k}}{(n-k)!} \right\} M_0 \qquad (3.12)$$

These derivatives may be evaluated at any time, but the time zero is particularly important. At $t = 0$, the derivatives have the simple form

$$\left. \frac{\partial^k \langle M_x \rangle}{\partial t^k} \right|_{t=0} = M_k M_0 \qquad (3.13)$$

However, the expectation value of M_x as a function of time is given by

$$\langle M_x(t) \rangle = \text{Tr}\, \rho(t) M_x$$

An iterative integration yields a solution similar to (2.91). The observed value of the magnetization along the x axis of the rotating frame is

$$\langle M_x(t) \rangle = \text{Tr} \left\{ \left(\rho(0) - i \int_0^t dt' [\mathscr{H}(t'), \rho(0) - \rho_{\text{eq}}] \right) M_x \right.$$

$$+ (-i)^2 \int_0^t dt' \int_0^{t'} dt'' [\mathscr{H}(t'), [\mathscr{H}(t''), \rho(0) - \rho_{\text{eq}}]] M_x$$

$$\left. + \cdots M_x \right\} \qquad (3.14)$$

The value of the derivative of $\langle M_x(t) \rangle$ with respect to time is

$$\frac{\partial \langle M_x \rangle}{\partial t} = -i \operatorname{Tr}[\mathcal{H}(t), \rho(0) - \rho_{eq}]M_x$$

$$+ (-i)^2 \operatorname{Tr} \int_0^t dt'[\mathcal{H}(t), [\mathcal{H}(t'), \rho(0) - \rho_{eq}]]M_x$$

$$+ \cdots \qquad (3.15)$$

The value at $t = 0$ is

$$\left. \frac{\partial \langle M_x \rangle}{\partial t} \right|_{t=0} = -i \operatorname{Tr}[\mathcal{H}, \rho(0)]M_x$$

$$= -i \operatorname{Tr}[\mathcal{H}, M_x]\rho(0) \qquad (3.16)$$

The first moment M_1 is thus found to be $M_1 = -i \operatorname{Tr}[\mathcal{H}, M_x]\rho(0)/M_0$, the expectation value of the commutator of $\mathcal{H}$ with M_x.

We may continue the differentiation process. The second derivative, evaluated at $t = 0$, is

$$\left. \frac{\partial^2 \langle M_x \rangle}{\partial t^2} \right|_{t=0} = (-i)^2 \operatorname{Tr}[\mathcal{H}, [\mathcal{H}, \rho(0)]]M_x$$

$$= (-i)^2 \operatorname{Tr}[\mathcal{H}, [\mathcal{H}, M_x]]\rho(0) \qquad (3.17)$$

Hence

$$M_2 = (-i)^2 \operatorname{Tr}[\mathcal{H}, [\mathcal{H}, M_x]]\rho(0)/M_0 \qquad (3.18)$$

This process of differentiation may be continued to calculate any higher-order moments.

The relationship between the time-dependent magnetization $\langle M_x(t) \rangle$ and the frequency spectrum $M_x(\omega)$ is well known. It is just a Fourier transform:

$$\langle M_x(t) \rangle = \int_{-\infty}^{+\infty} \cos(\omega t) M_x(\omega)\, d\omega + i \int_{-\infty}^{+\infty} \sin(\omega t) M_x(\omega)\, d\omega \qquad (3.19)$$

For a symmetric line, the second integral vanishes and this leads to some important consequences. The expansion for the cosine is given by

$$\cos \omega t = \sum_{n=0}^{\infty} \frac{(\omega t)^{2n}}{(2n)!} (-1)^n \qquad (3.20)$$

Collecting terms in powers of t yields, for a symmetric line,

$$\langle M_x(t) \rangle = \int_{-\infty}^{+\infty} M_x(\omega)\, d\omega - \frac{1}{2!} \int_{-\infty}^{+\infty} \omega^2 M_x(\omega)\, d\omega\, t^2$$

$$+ \frac{1}{4!} \int_{-\infty}^{+\infty} \omega^4 M_x(\omega)\, d\omega\, t^4 + \cdots \qquad (3.21)$$

From the first expansion [Eq. (3.9)], the following identification is made for the coefficients M_n, $n = 0, 2, \ldots$,

$$M_0 = \int_{-\infty}^{+\infty} M_x(\omega)\, d\omega \qquad (3.22a)$$

$$M_2 = -\frac{1}{2!} \int_{-\infty}^{+\infty} \omega^2 M_x(\omega)\, d\omega \qquad (3.22b)$$

$$M_4 = \frac{1}{4!} \int_{-\infty}^{+\infty} \omega^4 M_x(\omega)\, d\omega \qquad (3.22c)$$

and similarly for higher powers. We see that Eqs. (3.22b) and (3.22c) provide a means of calculating the spread of frequencies in a given spectrum.

For a Lorentzian line, the integrals diverge. However, truncation of the integral to the range $-\alpha < \omega < \alpha$ can give a value for a pseudosecond moment

$$M_2\big|_{\text{Lorentzian}} = \frac{2\alpha}{\pi T_2}$$

and a pseudofourth moment

$$M_4\big|_{\text{Lorentzian}} = \frac{2\alpha^3}{3\pi T_2}$$

and so

$$\frac{M_4}{M_2^2}\bigg|_{\text{Lorentzian}} = \frac{\pi}{6} \alpha T_2$$

Hence the values and ratio depend on the end points chosen to limit the region of interest.

The method of moments is a reasonable method of describing magnetic resonance lines where no exact shape is known and is approximate in the sense that only a finite number (usually two) of the moments can be measured from the frequency spectrum with reasonable accuracy. All higher-order moments depend more and more heavily on the intensity away from the center of the line. Since the intensity is weaker, the signal-to-noise ratio is less "in the wings," resulting in less accurate values for higher moments. Timewise, this means that one must carefully evaluate the form of $M_x(t)$ very near $t = 0$ to obtain accurate values for higher moments. Because of instrumental problems, this has been a difficult process to accomplish by observation of the free-induction decay. A recent technique developed by Lowe called "zero-time-resolution NMR" has dramatically improved one's capability to measure these higher-order moments and should result in more accurate values of line shapes in solids [2]!

One utility of the second moment is that, in the absence of interactions other than dipole–dipole interactions, it is a direct measure of internuclear distances in the sample. The second moment may be used to check the validity of a proposed crystal structure. Without derivation, we quote van Vleck's result for the second moment of a line homogeneously broadened by homonuclear dipolar interactions among spins-I at internuclear distances r_{jk} in a powdered sample [1]:

$$M_2 = \frac{3}{5} \gamma^4 \hbar^2 I(I + 1) \sum_{jk} r_{jk}^{-6}/N \tag{3.23a}$$

For protons with $I = \frac{1}{2}$ and $\gamma = 4.257 \times 10^3 \text{ sec}^{-1} \text{ G}^{-1}$, this result reads

$$M_2 = 1.644 \times 10^8 \frac{1}{N} \sum_{jk} r_{jk}^{-6}/\text{Hz}^2 \tag{3.23b}$$

with r_{jk} in angstroms. For a simple cubic lattice of protons separated by a distance d, the result is

$$M_2 = 1.397 \times 10^9 \, d^{-6}/\text{Hz}^2 \tag{3.23c}$$

It is also useful to recall that for a Gaussian line $M_2 = T_2^{-2}$.

IV. The Chemical Shift[†]

Consider next the indirect coupling of nuclei to the static magnetic field by interaction with the electrons. This interaction leads to shifts in resonance frequencies that are a reflection of the chemical environment of a nucleus in an atom or molecule and are therefore important in the analytical application of NMR, as well as in testing theoretical descriptions of molecules.

Electrons, in addition to possible interactions with nuclei, have a number of interactions with the applied magnetic field. They have the direct spin interaction because they are spin-$\frac{1}{2}$. However, since they are delocalized in space, electrons may have an energy due to the coupling of the orbital angular momentum to the static magnetic field. As a result, the electronic wave function has slightly more orbital angular momentum about the static field in

[†] See Pople *et. al.* [3]

one direction than in the other. This slight rotation produces a magnetic field that alters the effective field at the site of the nucleus.

The nucleus interacts directly with the *total field* at its site by the Zeeman interaction. Since in NMR we perceive different magnetic fields by the different absorption frequencies of nuclei located in different electronic environments, this electronic field is observed as a shift in the resonance frequency of a nucleus surrounded by electrons relative to the value for the free nucleus.

To solve the problem of the effect of the angular momentum of an electron on the spectrum of a nuclear spin, one must first determine the orbital states of the electron in the presence of a static field B_0 *and* the nuclear spin. The Hamiltonian for this situation is composed of a kinetic energy term and a potential energy term.

To construct the kinetic energy operator, it is necessary to recall that the motion of the electron is a result of its linear momentum and the circulation caused by the magnetic fields. The kinetic energy of the electron may be classically expressed as

$$T = \tfrac{1}{2}m\mathbf{v} \cdot \mathbf{v} = (1/2m)(m\mathbf{v}) \cdot (m\mathbf{v}) \tag{3.24}$$

In the presence of magnetic fields, $m\mathbf{v}$ and the linear momentum $\mathbf{p}$ are *not* equivalent. An extra term must be added to account for the circulation caused by the fields:

$$\mathbf{p} = m\mathbf{v} + (e/c)\mathbf{A} \tag{3.25}$$

where $\mathbf{A}$ is the vector potential describing the field.

As discussed in Appendix 1, the relation between the field $\mathbf{B}$ and the vector potential $\mathbf{A}$ is

$$\mathbf{B} = \nabla \times \mathbf{A}$$

Of course, since $\nabla \times (\nabla f)$, where f is some scalar function, is always zero, the value of $\mathbf{A}$ may always be changed by the addition of a term ∇f without changing the physical value of $\mathbf{B}$. The transformation in which a term ∇f is added to the vector potential is called a *gauge transformation*. The choice of the function f is called a choice of gauge.

The kinetic energy of an electron with linear momentum $\mathbf{p}$ in a magnetic field with vector potential $\mathbf{A}$ is

$$T = (1/2m)[\mathbf{p} - (e/c)\mathbf{A}] \cdot [\mathbf{p} - (e/c)\mathbf{A}] \tag{3.26}$$

and the overall classical Hamiltonian is

$$\mathscr{H} = (1/2m\hbar)[p^2 - (e/c)(\mathbf{A} \cdot \mathbf{p} + \mathbf{p} \cdot \mathbf{A}) + (e/c)^2\mathbf{A} \cdot \mathbf{A}] + V \tag{3.27}$$

There are two sources of fields at the site of the electron: the static magnetic field B_0 and the nuclear field. The vector potential then has two components

A_0 and A_1 corresponding to each of these, respectively. Expanding the Hamiltonian

$$\mathcal{H} = (1/2m\hbar)[p^2 - (e/c)(A_0 \cdot p + p \cdot A_0) + (e^2/c^2)A_0 \cdot A_0] + V$$
$$-(e/2m\hbar c)(A_1 \cdot p + p \cdot A_1) + (e^2/2m\hbar c^2)(A_0 \cdot A_1 + A_1 \cdot A_0)$$
$$+ (e^2/2m\hbar c^2)A_1 \cdot A_1 \tag{3.28}$$

With both A_0 and A_1 being zero, one has the problems of an electron moving in the potential describing the structure of the molecule. This is the problem for which, e.g., molecular orbital treatments can give an answer. Assuming the solutions to that problem to be available, the question is then how the terms in $\mathcal{H}$ that contain A_0 and A_1 alter that solution.

By first-order time-independent perturbation theory, the solution to the problem of a one-electron molecule perturbed by the presence of the static field is

$$\mathcal{H}_1 = -(e/2m\hbar c)(A_0 \cdot p + p \cdot A_0) \tag{3.29}$$

The first-order correction to the wave function in this case is

$$|\Psi^{(1)}\rangle = \sum_{m \neq 0} \frac{|\Psi_m\rangle\langle\Psi_m|\mathcal{H}_1|\Psi_0\rangle}{E_m - E_0}$$
$$= -\frac{e}{2m\hbar c} \sum_{n \neq 0} |\Psi_n\rangle \frac{\langle\Psi_n|(A_0 \cdot p + p \cdot A_0)|\Psi_0\rangle}{E_n - E_0} \tag{3.30}$$

where the set $\{|\Psi_n\rangle\}$ is the set of solutions to the unperturbed problem and the energy E_n corresponds to the unperturbed state $|\Psi_n\rangle$ of the electron moving in the potential V.

Addition of terms like

$$\mathcal{H}_2 = -(e/2mc\hbar)(A_1 \cdot p + p \cdot A_1)$$
$$+ (e^2/2m\hbar c^2)(A_0 \cdot A_1 + A_1 \cdot A_0) + (e^2/2m\hbar c^2)A_1 \cdot A_1 \tag{3.31}$$

correct for the effect of the magnetic field of the nucleus on the electronic motion. With respect to the nuclear states, the average of this Hamiltonian over the perturbed electronic state will be an operator:

$$\mathcal{H}_N = \langle\Psi_{el}|\mathcal{H}_2|\Psi_{el}\rangle$$
$$= \{\langle\Psi_0| + \langle\Psi_1|\}\mathcal{H}_2\{|\Psi_0\rangle + |\Psi_1\rangle\}$$
$$= \langle\Psi_0|\mathcal{H}_2|\Psi_0\rangle + \langle\Psi_1|\mathcal{H}_2|\Psi_0\rangle + \langle\Psi_0|\mathcal{H}_2|\Psi_1\rangle + \langle\Psi_1|\mathcal{H}_2|\Psi_1\rangle$$
$$= \langle\Psi_0|\mathcal{H}_2|\Psi_0\rangle + \frac{-e}{2m\hbar c} \sum_{n \neq 0} \left\{ \frac{\langle\Psi_0|A_0 \cdot p + p \cdot A_0|\Psi_n\rangle}{E_n - E_0} \langle\Psi_n|\mathcal{H}_2|\Psi_0\rangle \right.$$
$$+ \frac{\langle\Psi_0|\mathcal{H}_2|\Psi_n\rangle}{E_n - E_0} \langle\Psi_n|A_0 \cdot p + p \cdot A_0|\Psi_0\rangle \bigg\} \tag{3.32}$$

where we have only kept terms to first order in nuclear-spin variables. Before proceeding further, it is necessary to choose the gauges of $\mathbf{A}_0$ and $\mathbf{A}_1$. One possible choice of gauge is

$$\mathbf{A}_0 = \frac{1}{2} \mathbf{B}_0 \times (\mathbf{r}_{el} - \mathbf{r}_{nuc}) \equiv \frac{1}{2} \mathbf{B}_0 \times \mathbf{r} \tag{3.33}$$

and

$$\mathbf{A}_1 = \gamma \mathbf{I} \times \frac{(\mathbf{r}_{el} - \mathbf{r}_{nuc})}{|\mathbf{r}_{el} - \mathbf{r}_{nuc}|^3} = \gamma \mathbf{I} \times \frac{\mathbf{r}}{|\mathbf{r}|^3} \tag{3.34}$$

where $\mathbf{r}$ has been used to represent $\mathbf{r}_{el} - \mathbf{r}_{nuc}$. Keeping only those terms linear in B_0 and I yields

$$
\begin{aligned}
\mathscr{H}_N &= \frac{e^2}{2mc^2\hbar} \langle \Psi_0 | \mathbf{A}_0 \cdot \mathbf{A}_1 + \mathbf{A}_1 \cdot \mathbf{A}_0 | \Psi_0 \rangle \\
&+ \left(\frac{e}{2m\hbar c} \right) \sum_{n \neq 0} \left\{ \frac{\langle \Psi_0 | \mathbf{A}_0 \cdot \mathbf{p} + \mathbf{p} \cdot \mathbf{A}_0 | \Psi_n \rangle \langle \Psi_n | \mathbf{A}_1 \cdot \mathbf{p} + \mathbf{p} \cdot \mathbf{A}_1 | \Psi_0 \rangle}{E_n - E_0} \right. \\
&+ \left. \frac{\langle \Psi_0 | \mathbf{A}_1 \cdot \mathbf{p} + \mathbf{p} \cdot \mathbf{A}_1 | \Psi_n \rangle \langle \Psi_n | \mathbf{A}_0 \cdot \mathbf{p} + \mathbf{p} \cdot \mathbf{A}_0 | \Psi_0 \rangle}{E_n - E_0} \right\} \\
&= \frac{\gamma e^2}{4mc^2\hbar} \left\langle \Psi_0 \left| \frac{(\mathbf{B}_0 \times \mathbf{r}) \cdot (\mathbf{I} \times \mathbf{r})}{|r|^3} \right| \Psi_0 \right\rangle \\
&+ \frac{\gamma e}{4m\hbar c} \sum_{n \neq 0} \left\{ \langle \Psi_0 | (\mathbf{B}_0 \times \mathbf{r} \cdot \mathbf{p} + \mathbf{p} \cdot \mathbf{B}_0 \times \mathbf{r} | \Psi_n \rangle \right. \\
&\cdot \left\langle \Psi_n \left| \frac{\mathbf{I} \times \mathbf{r} \cdot \mathbf{p} + \mathbf{p} \cdot \mathbf{I} \times \mathbf{r}}{|r|^3} \right| \Psi_0 \right\rangle \frac{1}{E_n - E_0} \\
&+ \left. \left\langle \Psi_0 \left| \frac{\mathbf{I} \times \mathbf{r} \cdot \mathbf{p} + \mathbf{p} \cdot \mathbf{I} \times \mathbf{r}}{|r|^3} \right| \Psi_n \right\rangle \left\langle \Psi_n \left| \frac{\mathbf{B}_0 \times \mathbf{r} \cdot \mathbf{p} + \mathbf{p} \cdot \mathbf{B}_0 \times \mathbf{r}}{E_n - E_0} \right| \Psi_0 \right\rangle \right\}
\end{aligned}
\tag{3.35}
$$

At this point, we may utilize the two vector identities:

$$(\mathbf{A} \times \mathbf{B}) \cdot \mathbf{C} = \mathbf{A} \cdot (\mathbf{B} \times \mathbf{C}) \tag{3.36}$$

$$\mathbf{A} \times (\mathbf{B} \cdot \mathbf{C}) = \mathbf{B}(\mathbf{A} \cdot \mathbf{C}) - \mathbf{C}(\mathbf{A} \cdot \mathbf{B}) \tag{3.37}$$

Then

$$(\mathbf{B}_0 \times \mathbf{r}) \cdot (\mathbf{I} \times \mathbf{r}) = (\mathbf{r} \cdot \mathbf{r})\mathbf{I} \cdot \mathbf{B}_0 - (\mathbf{I} \cdot \mathbf{r})(\mathbf{r} \cdot \mathbf{B}_0) \tag{3.38}$$

and

$$\mathbf{B}_0 \times \mathbf{r} \cdot \mathbf{p} = \mathbf{B}_0 \cdot \mathbf{r} \times \mathbf{p} = \hbar \mathbf{B}_0 \cdot \mathbf{L} \tag{3.39}$$

Therefore,

$$\mathcal{H}_N = \frac{\gamma e^2}{4mc^2} \left\langle \Psi_0 \left| \frac{\mathbf{r} \cdot \mathbf{r}}{|r|^3} \right| \Psi_0 \right\rangle \mathbf{I} \cdot \mathbf{B}_0 - \frac{\gamma e^2}{4mc^2} \left\langle \Psi_0 \left| \frac{(\mathbf{I} \cdot \mathbf{r})(\mathbf{r} \cdot \mathbf{B}_0)}{|r|^3} \right| \Psi_0 \right\rangle$$

$$+ \frac{\gamma e^2 \hbar}{4mc^2} \sum_{n \neq 0} \left\{ \langle \Psi_0 | \mathbf{B}_0 \cdot \mathbf{L} | \Psi_n \rangle \right.$$

$$\times \left\langle \Psi_n \left| \frac{\mathbf{I} \cdot \mathbf{L}}{|r|^3} \right| \Psi_0 \right\rangle \frac{1}{E_0 - E_n}$$

$$\left. + \left\langle \Psi_0 \left| \frac{\mathbf{I} \cdot \mathbf{L}}{|r|^3} \right| \Psi_n \right\rangle \langle \Psi_n | \mathbf{B}_0 \cdot \mathbf{L} | \Psi_0 \rangle \frac{1}{E_n - E_0} \right\} \tag{3.40}$$

The matrix form of Eq. (3.40) is

$$\mathcal{H} = \gamma \mathbf{I} \cdot \boldsymbol{\sigma} \cdot \mathbf{B}_0 \tag{3.40a}$$

Here, $\boldsymbol{\sigma}$ is the "shielding tensor" [note the similarity of the form of the shielding interaction and the dipolar coupling interaction, Eq. (3.5)]. We see that the shielding tensor, as with the dipolar coupling tensor, is a 3×3 matrix. While not symmetric in principle, the shielding tensor is symmetric for many nuclei of chemical interest, and in our subsequent discussions, we shall take it to be symmetric; i.e., $\sigma_{ij} = \sigma_{ji}$. Such is the case for ^{1}H, ^{13}C, ^{15}N, ^{27}Al, ^{29}Si but not the very heavy elements such as ^{209}Tl.

If $\mathbf{B}_0$ is chosen to be along the z axis, the terms that survive in $\boldsymbol{\sigma}$ are

$$\sigma_{zz} = \frac{-e^2}{2mc^2} \left\langle \Psi_0 \left| \frac{x^2 + y^2}{|r|^3} \right| \Psi_0 \right\rangle + \frac{e^2 \hbar}{2mc^2} \sum_n \left\{ \frac{\langle \Psi_0 | L_z | \Psi_n \rangle}{E_n - E_0} \left\langle \Psi_n \left| \frac{L_z}{|r|^3} \right| \Psi_0 \right\rangle \right.$$

$$\left. + \left\langle \Psi_0 \left| \frac{L_z}{|r|^3} \right| \Psi_n \right\rangle \left\langle \Psi_n \left| \frac{L_z}{E_n - E_0} \right| \Psi_0 \right\rangle \right\} \tag{3.41}$$

$$\sigma_{xz} = \frac{-e^2}{4mc^2} \left\langle \Psi_0 \left| \frac{xz}{|r|^3} \right| \Psi_0 \right\rangle + \frac{e^2 \hbar}{8m^2c^2} \sum_n \left\{ \frac{\langle \Psi_0 | L_z | \Psi_n \rangle}{E_n - E_0} \left\langle \Psi_n \left| \frac{L_x}{|r|^3} \right| \Psi_0 \right\rangle \right.$$

$$\left. + \left\langle \Psi_0 \left| \frac{L_x}{|r|^3} \right| \Psi_n \right\rangle \frac{\langle \Psi_n | L_z | \Psi_0 \rangle}{E_n - E_0} \right\} \tag{3.42}$$

and

$$\sigma_{yz} = \frac{-e^2}{2mc^2} \left\langle \Psi_0 \left| \frac{yz}{|r|^3} \right| \Psi_0 \right\rangle + \frac{e^2 \hbar}{2mc^2} \sum_n \left\{ \frac{\langle \Psi_0 | L_z | \Psi_n \rangle}{E_n - E_0} \left\langle \Psi_n \left| \frac{L_y}{|r|^3} \right| \Psi_0 \right\rangle \right.$$

$$\left. + \left\langle \Psi_0 \left| \frac{L_y}{|r|^3} \right| \Psi_n \right\rangle \frac{\langle \Psi_n | L_z | \Psi_0 \rangle}{E_n - E_0} \right\} \tag{3.43}$$

We see that the shielding components depend on two characteristics of the electronic structure: (a) the spatial distribution of the electronic charge and (b) the orbital angular momentum of the electron. In most treatments of the theory the two contributions are given names. The term that depends only on spatial operators is the *diamagnetic* part of the chemical shift, and the one that depends on the orbital angular momentum and spatial operators is the *paramagnetic* part. This terminology is due originally to Lamb and is used extensively in the literature.

For purposes of calculation of the magnitude of the first term, we see that

$$|\sigma_d| = (e^2/4mc^2)\langle 1/r \rangle \tag{3.44}$$

where $\langle 1/r \rangle$ is the expectation value of $1/r$ for the electron.

As an example, take the hydrogen atom with an electron in a 1s orbital: $\langle 1/r \rangle = 1/a_0$. Hence

$$|\sigma_d| = \frac{(4.802 \times 10^{-10} \text{ esu})^2}{(4) \times (9.108 \times 10^{-28} \text{ gm})(3.0 \times 10^{10} \text{ cm sec}^{-1}} \frac{1}{a_0}$$

$$= 1.3 \times 10^{-6} \tag{3.45}$$

At this point it is helpful to note that the shielding tensor, which is a 3×3 matrix, is a representation of the value of a particular function in terms of the coordinates relative to a particular coordinate system (in this case, the coordinates defined by the magnetic field direction). The forms of these functions are typically bilinear and, provided that certain inequalities are satisfied, the values of the function can be represented pictorially by the length of a line parallel to one of the axes of the coordinate system (the magnetic field) through the center of an ellipsoid of revolution, like the one shown in Fig. 3.5.

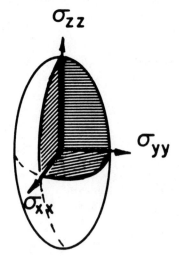

FIG. 3.5. The chemical shielding ellipsoid, which is used to indicate that different orientations of the magnetic field relative to the molecular framework result in different resonance positions for the same chemical species.

For an arbitrary orientation in which the coordinate system is oriented such that the magnetic field is not parallel to one of the semiaxes of the ellipsoid, the chemical shift will depend on the angles that specify the orientation of the field direction relative to the semiaxes of the ellipsoid, as well as the absolute magnitudes of the lengths of each semiaxis. These lengths are known as the *principal values*, and the coordinate system oriented such that the unit vectors determining the system are oriented along the semiaxes of the ellipsoid is known as the *principal-axis system*. By the relationship that describes the ellipsoid, one can determine the chemical shift at any arbitrary orientation of the magnetic field relative to the principal-axis system. This relationship is just a unitary transformation like that represented by the 3×3 transformation matrix found in uncoupling differential equations, as was done in Chapter 2.

Consider a nucleus in a molecule with some particular shielding environment. The effect of the environment is to produce a term in the Hamiltonian,

$$\mathscr{H} = \gamma \mathbf{I} \cdot \boldsymbol{\sigma} \cdot \mathbf{B} \tag{3.46}$$

We may choose to express this Hamiltonian in one of two coordinate systems: (1) The principal-axis system in which $\mathbf{B}$ has some arbitrary orientation relative to (x', y', z') axes, and (2) the coordinate system of the magnetic field $(\mathbf{B}_0 \| \mathbf{z})$.

In whichever way we choose to express this Hamiltonian, it must give the same result because it *represents the same physical system*. If we represent the spin vector by $\mathbf{I}'$ when it is expressed relative to the principal-axis system and by $\mathbf{I}$ when it is expressed relative to the coordinate system of the magnetic field (and similarly for $\mathbf{B}_0$ and $\boldsymbol{\sigma}$), one must have

$$\gamma \mathbf{I}' \cdot \boldsymbol{\sigma}' \cdot \mathbf{B}_0' = \gamma \mathbf{I} \cdot \boldsymbol{\sigma} \cdot \mathbf{B}_0 \tag{3.47a}$$

which is just a statement of the identity relation

$$\mathscr{H} = \mathscr{H} \tag{3.47b}$$

There is, however, a relationship between the forms of the vector $\mathbf{B}_0$ and $\mathbf{B}_0'$ (as well as $\mathbf{I}$ and $\mathbf{I}'$) from analytic geometry—the unitary transformation

$$\mathbf{B}_0' = \mathbf{R} \cdot \mathbf{B}_0, \qquad \mathbf{I}' = \mathbf{I} \cdot \mathbf{R}^{-1} \tag{3.48}$$

(Remember that $\mathbf{B}_0$ is a column vector and $\mathbf{I}$ is a row vector.) Hence, Eq. (3.46) may be rewritten as

$$\mathbf{I} \cdot \mathbf{R}^{-1} \cdot \boldsymbol{\sigma}' \cdot \mathbf{R} \cdot \mathbf{B}_0 = \mathbf{I} \cdot \boldsymbol{\sigma} \cdot \mathbf{B}_0 \tag{3.48'}$$

from which it is obvious that the relation we seek is

$$\boldsymbol{\sigma} = \mathbf{R}^{-1} \cdot \boldsymbol{\sigma}' \cdot \mathbf{R} \tag{3.49}$$

That is, the components in one frame (the magnetic field coordinates) have a functional relation to those in another frame (in this case, the principal-axis system) and the connection is through the transformation that expresses the components of a vector relative to one frame of reference in terms of its components relative to the second frame.

The general form of the algebra of rotations that is the connecting link between the two descriptions of the same vector in two different systems merits a closer examination. The determination of the transformation matrix **R** is found from analytic geometry. For example, consider the description of a vector in two frames that differ by having the x and y axes of one frame not coincident with the x' and y' axes of the second frame, as in Fig. 3.6. From geometry, the relation between the components of this vector in the two frames is:

$$x' = x\cos\theta + y\sin\theta, \qquad y' = -x\sin\theta + y\cos\theta, \qquad z' = z$$

In matrix form, this is stated as

$$\mathbf{r}' = \begin{bmatrix} \cos\theta & \sin\theta & 0 \\ -\sin\theta & \cos\theta & 0 \\ 0 & 0 & 1 \end{bmatrix} \begin{bmatrix} x \\ y \\ z \end{bmatrix} \tag{3.50}$$

or

$$\mathbf{r}' = \mathbf{R}_z \cdot \mathbf{r} \tag{3.51}$$

Similarly, a rotation about the y axis by angle θ is expressed in a rotation matrix $\mathbf{R}_y$, which relates the components of the vector in the rotated and un-rotated frames:

$$\mathbf{R}_y = \begin{bmatrix} \cos\theta & 0 & \sin\theta \\ 0 & 1 & 0 \\ -\sin\theta & 0 & \cos\theta \end{bmatrix}$$

The inverses of these matrices are given by the transposes of the matrices.

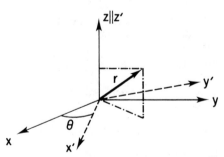

FIG. 3.6. Relationship of a rotated coordinate system to a fixed coordinate system.

Having obtained the appropriate rotation matrices, one may relate the elements of the chemical-shift tensor in one frame to those in the principal-axis frame by Eq. (3.49). The NMR experiment measures (for fields high enough that only the secular part of the chemical shift contributes to the resonance position) σ_{zz} in the frame with the z axis along the field. The usual NMR experiment consists of determining the chemical shift σ_{zz} for several different orientations of some axis system associated with the sample relative to the magnetic field. The experiment is begun with some known axis of the sample (for example, one of the axes of a single crystal) oriented along the field direction. For successive experiments, this axis is moved away from the field direction and the chemical shift is observed to change. Thus, one must relate σ_{zz} (in the magnetic field frame) for each one of these orientations to its value in some other frame, for example, the coordinate frame cooresponding to the initial orientation. One would prefer that this be the principal-axis system, but, life being what it is, it may not be. In this frame, then, the chemical-shift tensor is given as

$$
\sigma' = \begin{bmatrix} \sigma_{xx} & \sigma_{xy} & \sigma_{xz} \\ \sigma_{yx} & \sigma_{yy} & \sigma_{yz} \\ \sigma_{zx} & \sigma_{zy} & \sigma_{zz} \end{bmatrix}
$$

If the crystal has been rotated about the x axis by an angle θ, the chemical-shift tensor in the magnetic field frame is related to σ' by Eq (3.49) (see Problem 3.2).

$$
\sigma = \mathbf{R}_x^{-1}(\theta)\sigma'\mathbf{R}_x(\theta)
$$

$$
= \begin{bmatrix} \sigma_{xx} & \sigma_{xy}\cos\theta + \sigma_{xz}\sin\theta & -\sigma_{xy}\sin\theta + \sigma_{xz}\cos\theta \\ \sigma_{xy}\cos\theta + \sigma_{xz}\sin\theta & \sigma_{yy}\cos^2\theta + \sigma_{zz}\sin^2\theta & (\sigma_{zz} - \sigma_{yy})\sin\theta\cos\theta \\ & + 2\sigma_{yz}\sin\theta\cos\theta & + \sigma_{yz}(1 - 2\sin^2\theta) \\ -\sigma_{xy}\sin\theta + \sigma_{xz}\cos\theta & (\sigma_{zz} - \sigma_{yy})\sin\theta\cos\theta & \sigma_{yy}\sin^2\theta + \sigma_{zz}\cos^2\theta \\ & + \sigma_{yz}(1 - 2\sin^2\theta) & + 2\sigma_{yz}\sin\theta\cos\theta \end{bmatrix}
$$

We observe that the exact position of the line depends on the values of the components of σ' and on θ.

The Hamiltonian for the chemical-shift interaction $\mathbf{I} \cdot \sigma \cdot \mathbf{B}_0$ contains terms in all components of the angular momentum, as one may see from the matrix σ above. In static fields $\mathbf{B}_0$ of strengths commonly used by NMR spectroscopists, the Zeeman states are the zero-order description of the system. (This is not true for some quadrupolar nuclei.) The terms like the chemical shift are perturbation corrections. If one needs only the first-order correction, the only part of the chemical-shift interaction that needs to be considered is that part which commutes with the Zeeman Hamiltonian, the *secular* part (see Chapter 2). For the chemical shift, this is the term

$$\mathscr{H}_{cs}^{sec} = -\gamma B_0 \{\sigma_{yy} \sin^2 \theta + \sigma_{zz} \cos^2 \theta + 2\sigma_{yz} \sin \theta \cos \theta\} I_z \qquad (3.52a)$$

The shift, in parts per million, as a function of θ is

$$\sigma(\theta) = \sigma_{yy} \sin^2 \theta + 2\sigma_{yz} \sin \theta \cos \theta + \sigma_{zz} \cos^2 \theta \qquad (3.52b)$$

An example of the variation of the proton resonance of the waters of hydration of gypsum is given in Fig. 3.7. The angle gives the direction of the 001 axis of the single crystal relative to the magnetic field. From such a plot one may extract the parameters σ_{yy}, σ_{zz}, and σ_{yz} [see Eq. (3.52b)], for example, by a fit of the data by some nonlinear regression analysis with a computer.

One rotation can determine three parameters. In order to determine all of the components of the chemical-shift tensor, one performs experiments in which the chemical shift is measured for orientations of the magnetic field in three orthogonal planes. The simplest way is to perform the rotation experiments on a crystal first oriented with one of its crystal axes along the axis of rotation, then to repeat the experiment with a different axis of the crystal along the axis of rotation. The resultant variation can be analyzed to determine three more parameters. The parameters will be components of the chemical-shift tensor different from those determined in the first rotation experiment. A third rotation about the axis mutually orthogonal to the axes about which rotations were performed will give three more parameters. Of

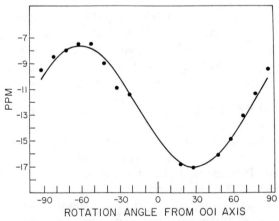

FIG. 3.7. Chemical shift for the protons in a single crystal of gypsum, $CaSO_4 \cdot 2H_2O$, as a function of the orientation of magnetic field relative to the 001 axis of the crystal. (From McKnett *et. al.* [4].)

the nine parameters, only six will be independent parameters. (In each rotation, one of the diagonal components will be equal to one of the diagonal components determined in another rotation experiment.)

These parameters will be the components of the chemical-shift tensor in the coordinate frame of the initial orientation. If this frame is the principal-axis frame, the off-diagonal elements will be zero. However, if it is not, the task is to find the orientation of the principal axes relative to the (known) coordinate frame of the initial orientation. This can be accomplished by finding the rotation matrix that diagonalizes the product in Eq. (3.49). This process is particularly easy to do with a digital computer that has standard matrix-diagonalization programs. The result of this diagonalization process gives the *principal values* of the chemical-shift tensor and the angles that transform the (known) initial coordinate system into the principal-axis system. Thus, one may gain information on the orientation of the principal axes and the crystalline axes.

A. The Powder Spectrum

We have just seen that NMR spectroscopy of single crystals can yield the six numbers characterizing a shielding tensor in a molecule (these are the three principal values and the three angles orienting these principal axes with respect to the molecular framework). It is more common to have a powder of a material available rather than a single crystal, however. Nuclear magnetic resonance measurements on a powder lose the advantage of yielding the tensor orientation with respect to the molecular frame, but the principal values may still be determined. To see how this is accomplished, we consider the resonance frequency of a shielding tensor having principal axes oriented with the laboratory frame as indicated in Fig. 3.8(a). The acute angle between z' and z is β. A rotation about the axis z by the angle α will place the original y axis in the xy plane of the laboratory system $[\sigma'_y$ in Fig. 3.8(b)]. A further rotation by β about y' will align z with the laboratory coordinate system (cs) [Fig. 3.8(c)]. The result of these rotations yields a transformed tensor, initially diagonal because it is described in its principal-axis system. The observed resonance frequency associated with this orientation of the tensor is given by

$$\omega = -\gamma B_0 \sigma_{33} \equiv -\gamma B_0 (\mathbf{R}_{y'_\beta}^{-1} \mathbf{R}_{z_\alpha}^{-1} \sigma_{PA} \mathbf{R}_{z_\alpha} \mathbf{R}_{y'_\beta})$$
$$= -\gamma B_0 (\sigma_{xx} \cos^2 \alpha \sin^2 \beta + \sigma_{yy} \sin^2 \alpha \sin^2 \beta + \sigma_{zz} \cos^2 \beta) \quad (3.53)$$

where σ_{xx}, σ_{yy}, and σ_{zz} are the principal-axis values of the chemical shift. When integrated over all angles, this line shape represents the absorption spectrum of a powder sample, in which all orientations of the principal axes with respect to the magnetic field are present with equal probability. We

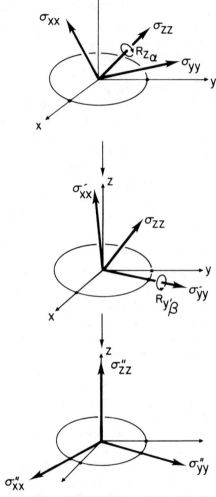

FIG. 3.8. The two-step process for bringing two arbitrarily oriented coordinate systems into a situation such that the two z axes are parallel.

derive the line-shape function $I(\omega)$ for a powder spectrum of a system with an axially symmetric tensor. The intensity is directly proportional to the number of molecules with chemical-shift tensor orientation (α, β), where

$$\int_{\Omega} P(\alpha, \beta)\, d\alpha \sin\beta\, d\beta = \int_{-\infty}^{\infty} I(\omega)\, d\omega = \int_{-\infty}^{\infty} F(\sigma)\, d\sigma$$

Here $P(\alpha, \beta)$ is the probability that a molecule's chemical-shift principal axes will have the orientation Ω, specified by α, β.

For an axially symmetric chemical-shift tensor, two of the three elements are equal. Letting σ_{zz} be the unique element ($\sigma_{||}$) and $\sigma_{xx} = \sigma_{yy} = \sigma_{\perp}$, the resonance frequency is given by

$$\omega - \omega_0 = \gamma B_0 \{\sigma_{\perp} \sin^2 \beta + \sigma_{||} \cos^2 \beta\} \tag{3.54}$$

Then

$$d\omega = \gamma B_0 \{\sigma_{\perp} 2 \sin \beta \cos \beta - \sigma_{||} 2 \sin \beta \cos \beta\} \, d\beta$$

or

$$\sin \beta \, d\beta = \frac{-1}{2\gamma B_0} \frac{1}{\sigma_{||} - \sigma_{\perp}} \frac{1}{\cos \beta} \, d\omega \tag{3.55}$$

From Eq. (3.54), however,

$$\cos \beta = \left\{ \frac{[(\omega - \omega_0)/\omega_0] - \sigma_{\perp}}{\sigma_{||} - \sigma_{\perp}} \right\}^{1/2} \tag{3.56}$$

or

$$\sin \beta \, d\beta = \frac{-1}{2\omega_0} \frac{1}{\{[(\omega - \omega_0)/\omega_0] - \sigma_{\perp}\}^{1/2} (\sigma_{||} - \sigma_{\perp})^{1/2}} \, d\omega$$

and making the identification

$$\int_{+\infty}^{-\infty} \frac{-1}{2\omega_0} \frac{1}{\{[(\omega - \omega_0)/\omega_0] - \sigma_{\perp}\}^{1/2} (\sigma_{||} - \sigma_{\perp})^{1/2}} \, d\omega = \int_{-\infty}^{+\infty} I(\omega) \, d\omega \tag{3.57}$$

We find

$$I(\omega) = \frac{1}{2\omega_0} \frac{1}{\{[(\omega - \omega_0)/\omega_0)] - \sigma_{\perp}\}^{1/2} (\sigma_{||} - \sigma_{\perp})^{1/2}} \tag{3.58}$$

A plot of this function is shown in Fig. 3.9, which is the *powder spectrum* for a nucleus subject to an axially symmetric shielding tensor. There is an infinity at $\sigma_{\perp}$ and a discontinuity at $\sigma_{||}$ that are not observed experimentally because of the existence of residual broadening. The theoretical spectrum that is smoothed by residual, e.g., lifetime, broadening will have a form obtained by convoluting the sharp spectrum with a broadening function in the frequency domain. This is equivalent to multiplication of the time domain decay by a monotonically decreasing function of time, e.g., an exponential decay.

For the case of a nonaxially symmetric tensor,

$$\sigma_{xx} \neq \sigma_{yy} \neq \sigma_{zz} \tag{3.59}$$

For a powder sample, the integral of the spectrum is given by

$$\int_{-\infty}^{+\infty} F(\sigma) \, d\sigma = \int_0^{2\pi} d\alpha \int_0^{\pi} \sin \beta \, d\beta \, P(\alpha, \beta) \tag{3.60}$$

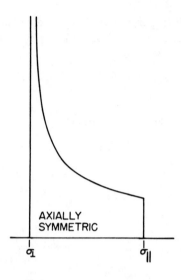

AXIALLY
SYMMETRIC

$\sigma_\perp$ $\sigma_\parallel$

FIG. 3.9. Theoretical chemical-shift spectrum of a randomly oriented powder of spins subject to an axially symmetric chemical-shift tensor.

For a random distribution, $P(\alpha, \beta) = (4\pi)^{-1}$ and the integral may be reduced by symmetry to four times an integral over the first quadrant:

$$\int_{-\infty}^{+\infty} F(\sigma)\, d\sigma = \frac{1}{\pi} \int_0^{\pi/2} d\alpha \int_0^\pi d\beta \sin\beta \qquad (3.61)$$

For nonaxial symmetry

$$\sigma = (\sigma_{xx}\cos^2\alpha + \sigma_{yy}\sin^2\alpha)\sin^2\beta + \sigma_{zz}\cos^2\beta \qquad (3.62)$$

The integral over any region of the spectrum (from σ_a to σ_b) for which $F(\sigma)$ is continuous is proportional to the number of spins with orientations in the ranges that give values of σ between σ_a and σ_b. The integration of this equation for a specific value of β must be restricted to a specific set of values of α. Let us call those limits α_l and α_u. Then, over that restricted region

$$\int_{\sigma_a}^{\sigma_b} F(\sigma)\, d\sigma = \frac{1}{\pi} \int_{\alpha_l}^{\alpha_u} d\alpha \sin\beta\, d\beta$$

We perform a transformation at constant α to show that (see Problem 3.9)

$$\sin\beta\, d\beta = -d(\cos\beta) = \left.\frac{\partial\cos\beta}{\partial\sigma}\right|_{d\sigma} \qquad (3.63)$$

and

$$\frac{\partial\cos\beta}{\partial\sigma} = \left(\frac{-1}{2}\right)(\sigma - \sigma_{xx}\cos^2\alpha - \sigma_{yy}\sin^2\alpha)^{-1/2}$$

$$\times (\sigma_{zz} - \sigma_{xx}\cos^2\alpha - \sigma_{yy}\sin^2\alpha)^{-1/2}$$

Thus we see that

$$\int_{\sigma_a}^{\sigma_b} F(\sigma)\,d\sigma = \frac{-1}{2\pi} \int_{\sigma_a}^{\sigma_b} \int_{\alpha_l}^{\alpha_u} (\sigma - \sigma_{xx}\cos^2\alpha - \sigma_{yy}\sin^2\alpha)^{-1/2}$$

$$\times\, (\sigma_{zz} - \sigma_{xx}\cos^2\alpha - \sigma_{yy}\sin^2\alpha)^{-1/2}\,d\alpha\,d\sigma$$

Making the identification of integrands, one obtains Eq. (3.64).

$$F(\sigma) = \frac{-1}{2\pi} \int_{\alpha_l}^{\alpha_u} \frac{d\alpha}{(\sigma - \sigma_{xx}\cos^2\alpha - \sigma_{yy}\sin^2\alpha)^{1/2}(\sigma_{zz} - \sigma_{xx}\cos^2\alpha - \sigma_{yy}\sin^2\alpha)^{1/2}}$$

$$\tag{3.64}$$

The form of the integral in Eq. (3.64) is that of an elliptic integral that can be put in the form [5]

$$F(\sigma) = C \int_0^{\pi/2} \frac{d\epsilon}{(1 - k^2\sin^2\epsilon)^{1/2}} = CK(k) \tag{3.65}$$

by a series of transformations. Here C and k are determined by the limits of integration, which in turn depend on the region of the spectrum in which σ lies. For $\sigma_{yy} < \sigma < \sigma_{zz}$, the line shape has the form of Eq. (3.65) with

$$\alpha_l = 0 \tag{3.66a}$$

$$\alpha_u = \pi/2 \tag{3.66b}$$

$$C = (1/\pi)(\sigma - \sigma_{xx})^{-1/2}(\sigma_{zz} - \sigma_{yy})^{-1/2} \tag{3.67a}$$

$$k = (\sigma_{yy} - \sigma_{xx})^{1/2}(\sigma_{zz} - \sigma)^{1/2}/(\sigma_{zz} - \sigma_{yy})^{1/2}(\sigma - \sigma_{xx})^{1/2} \tag{3.67b}$$

For the region $\sigma_{xx} < \sigma < \sigma_{yy}$, the line shape has the form of Eq. (3.65), but with the definitions

$$2\alpha_l = \arccos\left[\frac{(\sigma_{zz} + 2\sigma)(2\sigma_{zz} - \sigma_{yy} - \sigma_{xx})}{3\sigma_{zz}(\sigma_{yy} - \sigma_{xx})}\right] \tag{3.68a}$$

$$\alpha_u = \frac{\pi}{2} \tag{3.68b}$$

$$C = \pi^{-1}(\sigma_{zz} - \sigma)^{-1/2}(\sigma_{yy} - \sigma_{xx})^{-1/2} \tag{3.69a}$$

$$k = (\sigma - \sigma_{xx})^{1/2}(\sigma_{zz} - \sigma_{yy})^{1/2}/(\sigma_{zz} - \sigma)^{1/2}(\sigma_{yy} - \sigma_{xx})^{1/2} \tag{3.69b}$$

The graphical representation of this function is shown in Fig. 3.10, from which the values of the principal components of the nonaxially symmetryic tensor can be identified by the singularities.

In the liquid state, one does not see such broad, asymmetric lines. The resonances are sharp and symmetric. The distinction between the liquid and

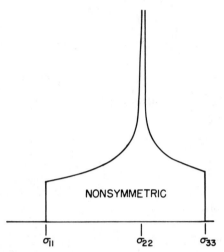

FIG. 3.10. Theoretical chemical-shift spectrum of a randomly oriented powder of spins subject to a nonaxially symmetric chemical-shift tensor.

a powder spectrum is the fact that, in a powder, there are a large number of orientations present in a sample, each evidenced by a single molecule. These orientations persist for a time long compared to the inverse of the anisotropy of the spectral absorption. In a liquid, however, all orientations are assumed by a single molecule in a time short compared to the inverse of the chemical-shift anisotropy. Thus the statistical averaging is different in the two cases. In a liquid, the time average over orientations results in a resonance line only at the average chemical shift, termed the *isotropic* value of $\bar{\sigma}$

$$\bar{\sigma} = \tfrac{1}{3}\{\sigma_{xx} + \sigma_{yy} + \sigma_{zz}\} \tag{3.70}$$

The ideal line shapes of Figs. 3.9 and 3.10 are limiting cases. In practice, the lifetimes of transverse magnetizations that lead to such patterns are finite. As we saw in Chapter 1, this finite lifetime leads to a damping of the time-dependent magnetization. For these broad lines a similar problem occurs in which the ideal line shape is mathematically folded with the broadening function characteristic of the decay. Hence, the observable is

$$I_{\mathrm{obs}}(\omega) = \int_{-\infty}^{+\infty} I_i(\omega')B(\omega' + \omega)\,d\omega' \tag{3.71}$$

where $I_{\mathrm{obs}}(\omega)$ is the actual absorption, $I_i(\omega)$ the ideal absorption already discussed, and $B(\omega)$ the broadening function for the line. The net result is a smearing of the ideal powder pattern, which causes some distortion of the shape. For example, Fig. 3.11 shows the effect of adding Lorentzian broadening to the powder pattern.

We have seen that the powder pattern line shape can be used to determine the three principal components of the shielding tensor, σ_{xx}, σ_{yy}, and

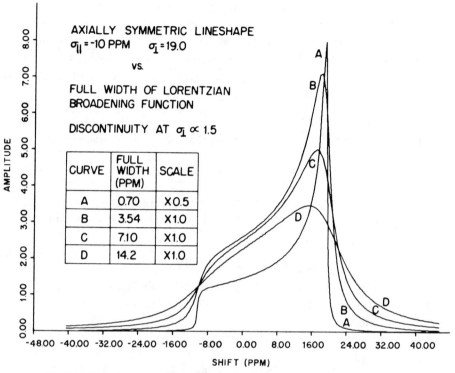

FIG. 3.11. The effect of broadening on the powder spectrum of spins subject to an axially symmetric chemical-shift tensor.

σ_{zz}. A more useful description of these components is combinations that describe the symmetry of the tensor.

$$\bar{\sigma} = (\sigma_{xx} + \sigma_{yy} + \sigma_{zz})/3 \tag{3.70a}$$

$$\delta = \bar{\sigma} - \sigma_{zz} \tag{3.70b}$$

$$\eta = (\sigma_{yy} - \sigma_{xx})/\delta \tag{3.70c}$$

For an axially symmetric tensor, the symmetry η is zero, and for a spherically symmetric tensor the anisotropy δ and asymmetry are zero.

V. Indirect Nuclear–Nuclear Interactions

We saw in the preceding section that an effective coupling of nuclear spins to the Zeeman field could occur through a direct coupling of the nuclear magnetic moment to the orbital angular momentum of the surrounding electrons, with simultaneous coupling of the electron orbital moment to the static field. Again we ask,"Is there a mechanism whereby the nuclear spins may be coupled, each to the others, by way of simultaneous coupling of the electrons to both nuclei?" Once again the answer is yes. There are two couplings that may contribute; the first is the electron–nuclear dipole–dipole interaction

$$\mathcal{H}_1 = g\beta \sum_k \sum_N \gamma_N \left\{ \frac{3(\mathbf{S}_k \cdot (\mathbf{r}_k - \mathbf{r}_N))(\mathbf{I}_N \cdot (\mathbf{r}_k - \mathbf{r}_N))}{|\mathbf{r}_k - \mathbf{r}_N|^5} - \frac{\mathbf{S}_k \cdot \mathbf{I}_N}{|\mathbf{r}_k - \mathbf{r}_N|^3} \right\} \qquad (3.72)$$

where the sums over k and N are over electrons and nuclei. The second interaction we have not discussed before. It is the Fermi contact interaction

$$\mathcal{H}_2 = \frac{16\pi}{3} \beta \sum_k \sum_N \gamma_N \mathbf{S}_k \cdot \mathbf{I}_N \, \delta(\mathbf{r}_k - \mathbf{r}_N) \qquad (3.73)$$

We may again take recourse to second-order perturbation theory to show that there exists a term that is an operator relative to the nuclear spins and has the form

$$\mathcal{H}_J = \sum_{NN'} \mathbf{I}_N \cdot \mathbf{J} \cdot \mathbf{I}_{N'} \qquad (3.74)$$

There are a number of parts to the tensor $\mathbf{J}$. The one due to cross coupling of the two electron–dipole nuclear–dipole interactions is

$$J^{(1)}_{NN'} = -\frac{8}{3} \beta^2 \hbar \gamma_N \gamma_{N'} \sum_{n \neq 0} \sum_{k,j} \left\langle 0 \left| \frac{\mathbf{S}_k}{|\mathbf{r}_k - \mathbf{r}_N|^3} - 3 \frac{(\mathbf{S}_k \cdot (\mathbf{r}_k - \mathbf{r}_N))(\mathbf{r}_k - \mathbf{r}_N)}{|\mathbf{r}_k - \mathbf{r}_N|^5} \right| n \right\rangle$$

$$\left\langle n \left| \frac{\mathbf{S}_j}{|\mathbf{r}_j - \mathbf{r}_{N'}|^3} - 3 \frac{(\mathbf{S}_j \cdot (\mathbf{r}_j - \mathbf{r}_{N'}))(\mathbf{r}_j - \mathbf{r}_{N'})}{|\mathbf{r}_j - \mathbf{r}_{N'}|^5} \right| 0 \right\rangle \cdot \frac{1}{E_n - E_0} \qquad (3.75)$$

The cross term between the Fermi contact interactions also leads to another component of $\mathbf{J}$:

$$J^{(2)}_{NN'} = -\frac{512}{27} \pi^2 \beta^2 \hbar \gamma_N \gamma_{N'} \sum_{n \neq 0} \sum_{k,j} \frac{\langle 0 | \delta(\mathbf{r}_k - \mathbf{r}_N) \mathbf{S}_k | n \rangle \langle n | \mathbf{S}_j \delta(\mathbf{r}_j - \mathbf{r}_{N'}) | 0 \rangle}{E_n - E_0} \qquad (3.76)$$

Finally, the coupling of each nucleus to the orbital angular momentum of the electrons leads to a contribution of the form

$$J^{(3)}_{NN'} = \frac{2}{3} \frac{e^2 \hbar}{2mc^2} \gamma_N \gamma_{N'} \left\langle 0 \left| \sum_k \frac{(\mathbf{r}_k - \mathbf{r}_N) \cdot (\mathbf{r}_k - \mathbf{r}_{N'})}{|\mathbf{r}_k - \mathbf{r}_N|^3 |\mathbf{r}_k - \mathbf{r}_{N'}|^3} \right| 0 \right\rangle$$

$$-\frac{8}{3} \beta^2 \hbar \gamma_N \gamma_{N'} \left\langle 0 \left| \sum_{k,j,n} \frac{[(\mathbf{r}_k - \mathbf{r}_N) \times \mathbf{V}_k]}{|\mathbf{r}_k - \mathbf{r}_N|^3 |\mathbf{r}_j - \mathbf{r}_{N'}|^3} \right| n \right\rangle \left\langle n \left| \frac{[(\mathbf{r}_j - \mathbf{r}_{N'}) \times \mathbf{V}_j]}{E_n - E_0} \right| 0 \right\rangle \qquad (3.77)$$

For the elements of the first row, the J couplings are usually small, and in the solid state, other interactions such as the direct nuclear–dipole–dipole interaction dominate spectra to the point that J couplings are often ignored.

Such is not the case for the elements of the second and higher rows [3]. In the liquid state, however, J coupling may persist as the trace of the tensor. These couplings along with the chemical shift make liquid-state NMR spectroscopy useful analytically, for they allow a distinction of modes of bonding not revealed by chemical shifts alone.

At the end of Chapter 1, we showed that the Carr–Purcell and Meiboom–Gill–Carr–Purcell sequences effectively eliminate the effects of the chemical shift and magnetic field inhomogeneity Hamiltonians upon stroboscopic observation. The J coupling, however, for homonuclear systems is invariant to the effects of nonselective 180° pulses used in this sequence, so stroboscopic observation during such a sequence allows measurement of such J couplings. The reason for this result is quite simple to picture. The 180° pulse inverts both I_N and $I_{N'}$. Hence, the Hamiltonian before and after the 180° pulse, being a bilinear function of I_N and $I_{N'}$, causes the system to evolve in time, undisturbed by the action of the pulse.

VI. The Quadrupolar Interaction

Of the roughly 110 NMR-active nuclear isotopes in the periodic table, 29 have spin-$\frac{1}{2}$. The rest are quadrupolar nuclei. The quadrupolar interaction is therefore, in principle (or statistically), of great interest to scientists whose work involves a broad range of elements. It has also been largely responsible, until recently, for avoidance of the use of non-spin-$\frac{1}{2}$ nuclei as labels in the solid state when high-resolution spectra are desired. We shall see that high-resolution spectra of most quadrupolar species are possible in the solid state. In addition, the dipolar coupling between quadrupolar nuclei and spin-$\frac{1}{2}$ nuclei (e.g., ^{14}N to ^{13}C) cause broadening of the solid-state spectra of the spin-$\frac{1}{2}$ species even under "high-resolution" conditions. This broadening is both a blessing and a curse. The blessing is that when the nature of the interaction under conditions of the experiment is understood, it can be used to measure the internuclear distance between the two species involved. The curse is that broad is not always beautiful in the presence of many overlapping lines such as might be found in biological molecules.

Under any circumstances, the materials scientist who wishes to use the full power of NMR spectroscopy to probe matter is forced to confront the quadrupolar interaction. We start by considering the classical electrostatic interaction between a charged nucleus with charge distribution $\rho(\mathbf{r})$ and a potential $V(\mathbf{r})$:

$$\hbar E_e = \int\limits_{\text{all space}} \rho(\mathbf{r})V(\mathbf{r})\,d^3r \qquad (3.78)$$

The potential may be expanded in a Maclaurin series to give

$$V(\mathbf{r}) = V(\mathbf{r} = 0) + \sum_{k=1}^{3} \left.\frac{\partial V}{\partial x_k}\right|_{r=0} x_k + \frac{1}{2!} \sum_{k,j=1}^{3} \left.\frac{\partial^2 V}{\partial x_k \partial x_j}\right|_{r=0} x_k x_j + \cdots \quad (3.79)$$

This potential may be used to obtain the electrostatic energy as an expansion. With $V_k = \partial V/\partial x_k$, evaluated at $r = 0$, and similarly for higher derivatives, we have

$$\hbar E_e = V(0) \int \rho(\mathbf{r}) d^3 r + \sum_{k=1}^{3} V_i \int x_k \rho(\mathbf{r}) d^3 r$$

$$+ \frac{1}{2} \sum_{k,j} V_{k,j} \int x_k x_j \rho(\mathbf{r}) d^3 r + \cdots \quad (3.80)$$

The first two integrals are identified as the overall charge of the distribution q and the components of the electric dipole moment of the charge distribution d. The remaining term is the nuclear quadrupole term. In discussing this term, we note that a coordinate system can be found in which V_{kj} is diagonal. This is the principal-axis system of the potential. In addition, when there is no electronic charge at the nucleus, V must satisfy Laplace's equation $\nabla^2 V = 0$, which leads to

$$\sum_{k=1}^{3} V_{kk} = 0 \quad (3.81)$$

when the derivatives are evaluated at the origin. Note that this condition leads to a vanishing quadrupolar coupling when all the V_{kk} are equal. This is the case for a nucleus in a site of cubic symmetry, e.g., ^{27}Al in a perfect tetrahedron of oxygens. It is also effectively true for a nucleus in a molecular framework that is tumbling isotropically with a correlation time short compared to the inverse of the magnitude of the quadrupolar coupling constant (*vide infra*), e.g., ^{14}N in NH_4^+ (soln) but not ^{33}S in benzthiophene in solution.

It is convenient to define the quantities Q_{kj}, defined by

$$Q_{kj} = \int (3x_k x_j - \delta_{kj} r^2) \rho(\mathbf{r}) d^3 r \quad (3.82)$$

so that, in terms of Q_{kj}, the integral in the quadrupolar term of the nuclear electrostatic energy is given by

$$\int x_k x_j \rho(\mathbf{r}) d^3 r = \frac{1}{3} \left[Q_{kj} + \delta_{kj} \int r^2 \rho(\mathbf{r}) d^3 r \right] \quad (3.83)$$

The quadrupolar contribution to the nuclear electrostatic energy then becomes

$$\hbar E_e^Q = \frac{1}{6} \sum_{k,j} \left[V_{kj} Q_{kj} + V_{kj} \delta_{kj} \int r^2 \rho(\mathbf{r}) d^3 r \right] \quad (3.84)$$

Since V satisfies Laplace's equation, the second terms in (3.84) do not contribute to the sum leaving

$$\hbar E_e^Q = \frac{1}{6} \sum_{kj} Q_{kj} V_{kj} \tag{3.85}$$

so that the introduction of the Q_{kj}'s amounts to subtracting an orientation-independent term from the left-hand side of (3.83). Since the Q_{kj}'s in (3.85) contain the term $\rho(r)$, the density of nuclear charge within the nucleus, for which we wish an operator form, we substitute the quantum-mechanical operator

$$\rho^{\mathrm{op}}(\mathbf{r}) = \sum_{\substack{k \\ \text{all nucleons} \\ \text{in the} \\ \text{nucleus}}} q_k \delta(\mathbf{r} - \mathbf{r}_k) \tag{3.86}$$

The sum over nucleons amounts to a sum over protons since the neutrons in the nucleus are uncharged. We then find

$$Q_{kj}^{\mathrm{op}} = e \sum_{\substack{k \\ \text{protons} \\ \text{in the} \\ \text{nucleus}}} (3x_k x_j - \delta_{kj} r_k^2) \tag{3.87}$$

and the quadrupole coupling Hamiltonian is

$$\mathcal{H}_Q = \frac{1}{6} \sum_{k,j} V_{kj} Q_{kj}^{(\mathrm{op})} \tag{3.88}$$

In order to relate $\mathcal{H}_Q$ to nuclear angular momentum operators, we realize that we are, in general, interested only in the ground state of the nucleus in question, which having a quadrupole moment, i.e., orbital angular momentum, is characterized by a total angular momentum, which is the sum of orbital and spin angular momentum:

$$\mathbf{I} = \sum_k (\boldsymbol{\ell}_k + \mathbf{s}_k) \tag{3.89}$$

With the use of the commutation relations such as

$$[\ell_{xk}, y_k] = i z_k, \qquad [s_{xk}, y_k] = 0, \qquad [\ell_{xk}, s_{yk}] = 0 \tag{3.90}$$

which result from the relations developed in Chapter 2, and from the fact that coordinates and orbital angular momenta belong to "real" space, which is separate from, and commutes with components of, "spin space," we find relations such as

$$[I_{zk}, x_k] = i y_k \tag{3.91}$$

These relations, inserted into Q_{kj}, Eq. (3.87), may be shown to yield the result

$$\mathcal{H}_Q = \frac{eQ}{6I(2I - 1)} \sum_{k,j} V_{kj} \left[\frac{3}{2} (I_k I_j + I_j I_k) - \delta_{kj} I^2 \right] \tag{3.92}$$

Here Q is the quadrupole moment of the nucleus, found from the expectation value of the *difference* of the nuclear charge distribution parallel and transverse to z, evaluated classically as

$$\int (z^2 - x^2)\rho(\mathbf{r})\, d^3r = \frac{1}{2}\int (2z^2 - x^2 - y^2)\rho(\mathbf{r})\, d^3r$$

$$= \frac{1}{2}\int (3z^2 - r^2)\rho(\mathbf{r})\, d^3r \qquad (3.93)$$

The quantum-mechanical analog is the expectation value of the analogous sum for all protons in the nucleus, evaluted for a nuclear ground state

$$|I, I_z, \eta\rangle = |I, I, \eta\rangle \qquad (3.94)$$

With units of square centimeters, Q is a constant that may be experimentally determined for each nucleus. A tabulation of these values may be obtained from Appendix 1 of Semin *et. al.* [6]. They are listed in Table 1.1.

We recognize that (3.92) is a quadratic form in V_{kj} and that such a form may always be expressed in its principal-axis system as being diagonal. In this system we find

$$\mathcal{H}_Q = \frac{eQ}{6I(2I - 1)\hbar}\sum_j [3I_j^2 - I^2] \qquad (3.95)$$

With the help of Laplace's equation $\sum V_{ii} = 0$, we may rewrite (3.95) in terms of the symmetry properties of the electric-field-gradient tensor:

$$\mathcal{H}_Q = (e^2 qQ/\hbar)[3I_z^2 - I^2 + \eta(I^{+2} - I^{-2})/2e]/4I(2I - 1)$$

$$\equiv \omega_Q[3I_z^2 - I^2 + e^{-1}\eta(I_x^2 - I_y^2)]/4I(2I - 1) \qquad (3.96)$$

where

$$\omega_Q = e^2 qQ/\hbar \qquad \text{(the quadrupole frequency)}$$

$$eq = V_{zz} = \partial^2 V/\partial z^2 \qquad \text{(the field gradient)} \qquad (3.97)$$

$$\eta = \frac{V_{xx} - V_{yy}}{V_{zz}} \qquad \text{(the asymmetry parameter)}$$

Equation (3.96) may be written in the form (Problem 3.12)

$$H_Q = \text{const} \times \mathbf{I} \cdot \mathbf{V} \cdot \mathbf{I}$$

where $\mathbf{V}$ is the electric field gradient tensor. To estimate the size of this energy for deuterium ($I = 1$),

$$|H_Q| = eQ|\mathbf{V}|/6$$

To estimate this quantity, one needs Q for ^{2}H. This value can be found from tables as

$$Q = 2.8 \times 10^{-27} \quad \text{cm}^2 \tag{3.98}$$

Of course, $|V|$ depends on the environment. Suppose that we model the situation as a ^{2}H nucleus sitting next to an electron that is sitting at a point 1 bohr radius away and opposite it a positively charged ion, also 1 bohr radius away. The electric field at a point r away from the positive charge along the vector connecting the positive and negative charges is

$$|E| = \left\{ \frac{e}{r^2} - \frac{e}{(2a_0 - r)^2} \right\} \tag{3.99}$$

The field gradient is

$$\frac{\partial E}{\partial r} = \frac{-2e}{r^3} - \frac{2e}{(2a_0 - r)^3} \tag{3.100}$$

At $r = a_0$, then,

$$\left. \frac{\partial E}{\partial r} \right|_{a_0} = \frac{-4e}{a_0^3} \quad \text{(in erg/cm}^2 \text{ esu)}$$

The units of this quantity may be inferred as follows: in electrostatic units, Coulomb's law reads

$$F = \text{dynes} = \text{esu}^2/\text{cm}^2 = q^2/r^2$$

so a dyne cm $\sim$ erg has the units of esu cm^{-1}. Therefore, $e/a_0^3 \sim$ esu cm$^{-3} \sim$ (esu^2/cm)(1/esu)(1/cm^2) $\sim$ erg esu^{-1} cm^{-2}. So, with $e = 4.08 \times 10^{-20}$ esu, $a_0 = 0.529 \times 10^{-8}$ cm, we find

$$|V| = \frac{4e}{\hbar a_0^3} = \frac{4 \times 4.802 \times 10^{-10}}{(1.05 \times 10^{-27})(0.59 \times 10^{-8})^3} = 8.9 \times 10^{42} \quad \text{sec}^{-1} \text{ cm}^{-2} \text{ esu}^{-1}$$

Hence

$$|\mathcal{H}_Q| = 4.802 \times 10^{-10} \text{ esu} \times 2.8 \times 10^{-27} \text{ cm}^2 \times 8.9 \times 10^{42} \text{ sec}^{-1} \text{ cm}^2 \text{ esu}^{-1}/6$$
$$= 1.996 \times 10^6 \text{ rad sec}^{-1}$$
$$\equiv 3.1 \times 10^5 \text{ Hz} \tag{3.101}$$

We see that, depending on the size of V, the quadrupole energy may be large for deuterium. If the electronic distribution about the nucleus is cubic in symmetry, the field gradient must be zero and the quadrupole energy vanishes.

As with the dipolar coupling tensor and the shielding tensor, we describe the possible absorption spectra associated with energies obtained from (3.96).

To do so, it is necessary to understand that the quadrupolar interaction is inherently different from the dipolar and shielding interactions. For the latter two, the Zeeman field determined the axes of quantization under normal "high-field" conditions. Depending upon the quadrupole moment Q and the local bonding situation, i.e., the field gradient, the values of ω_Q can range from zero to hundreds of megahertz. One can therefore distinguish three cases for purposes of calculation.

Case (1): $\omega_Z \gg \omega_Q$. This would be true for small quadrupole moments and/or near cubic electronic symmetry about the nucleus in question. In this case, the eigenstates of $\mathcal{H}_Z$ form zero-order basis functions from which to calculate first-order quadrupolar energies using standard perturbation theory. An important result in this case is that for nuclei with spin $n + \frac{1}{2}(n$ integer), the central $(\frac{1}{2}-\frac{1}{2})$ transition is unaffected to first order by ω_Q). An immediate effect of this result is that the *second order* quadrupolar broadening of this central transition for a powder sample may be averaged by sample spinning at a frequency slow compared to the spectral width of the inhomogeneously broadened line. This is not true for the six nuclei that have integral spin, of which two (^{2}H and ^{14}N), unfortunately, are common in molecules studied by a large fraction of the chemical community.

Case (2): $\omega_Q \gg \omega_Z$. In this situation, the splitting of the nuclear states is in the megahertz range in the absence of a static magnetic field. All states are doubly degenerate as per Kramers rule, and "pure quadrupole resonance" may be observed. In a static field the Zeeman splitting may be treated as a perturbation upon the quadrupole levels.

Case (3): $\omega_Q \simeq \omega_Z$. Here, a perturbation treatment is of no help, and the problem using the full Zeeman plus the quadrupolar Hamiltonian must be solved.

We discuss special examples of these three cases. We first discuss the strong field case, $\omega_Z \gg \omega_Q$. For simplicity, and because it appears to be the simplest case for many systems, we specify axial symmetry for the field gradient tensor, i.e., $\eta = 0$. The quadrupolar Hamiltonian is then

$$\mathcal{H}_Q = \omega_Q[3I_z^2 - I^2]/4I(2I - 1) \qquad (3.96)$$

This Hamiltonian was written in the principal-axis frame of the field-gradient tensor V, which is not in general aligned with the frame of quantization, where B_0 defines z' (but x and y may be arbitrarily chosen as long as they are perpendicular to z and each other). We are free to choose the y axis of V parallel to that of the frame of quantization. Therefore a rotation by θ about y will bring the z axis of V into coincidence with $\mathbf{B}_0$. The rotation of the operators I_k about y in (3.96) is effected by the exponential operators $\exp(\pm iI_y\theta)$ as discussed in Chapter 2. Here I^2 is a scalar and invariant to rotation, so only the term I_z^2 need be considered: the rotated value of I_z^2 is given by

$$I_{z'}^2 = e^{iI_y\theta}I_z e^{-iI_y\theta}e^{iI_y\theta}I_z e^{-iI_y\theta} = (I_{z'}\cos\theta - I_{x'}\sin\theta)^2 \tag{3.102}$$

which yields for $\mathscr{H}_Q$ in the frame with z' along $\mathbf{B}_0$,

$$\mathscr{H}_Q = \omega_Q \frac{[3I_{z'}^2\cos^2\theta + 3I_{x'}^2\sin^2\theta + 3(I_{z'}I_{x'} + I_{x'}I_{z'})\sin\theta\cos\theta - I^2]}{4I(2I-1)} \tag{3.103}$$

Before proceeding further, it is useful to remind ourselves of the results of perturbation theory. With a Hamiltonian

$$\mathscr{H} = \mathscr{H}_0 + \mathscr{H}_P \tag{3.104}$$

Where $\mathscr{H}_P \ll \mathscr{H}_0$, and zero-order wave function $|\Psi_k^0\rangle$, which are eigenfunctions of $\mathscr{H}_0$ with energies $E_k^{(0)}$, the first-order correction to the energies are

$$E_k^{(1)} = \langle\Psi_k^0|\mathscr{H}_P|\Psi_k^0\rangle \tag{3.105}$$

This means that only the secular part of $\mathscr{H}_P$, i.e., those terms commuting with $\mathscr{H}_0$, contribute to $E_k^{(1)}$. It may be seen (Problem 3.13) that the diagonal matrix elements of $I_{x'}^2$ and I_y^2 are identical and nonzero and that terms $I_{x'}I_{z'}$ and $I_{z'}I_{x'}$ are nondiagonal. This leads to the following result:

$$E_k^{(1)} = \omega_Q(3\cos^2\theta - 1)\{3k^2 - I(I+1)\}/8I(2I-1) \tag{3.106}$$

The second-order corrections to the zero-order energies are

$$E_k^{(2)} = \sum_{j\neq k} \frac{|\langle k|\mathscr{H}_P|j\rangle|^2}{E_k^0 - E_j^0} \tag{3.107}$$

This means that only the nonsecular terms in $\mathscr{H}_P$ contribute to $E_k^{(2)}$. The nonsecular portion of $\mathscr{H}_{Q'}$ is found to be [Eq. (3.96) and Problem (3.13)]

$$\mathscr{H}_Q^{ns} = 3\omega_Q\{\sin\theta\cos\theta[I_z(I_+ + I_-) + (I_+ + I_-)I_z] + \sin^2\theta[I_+^2 + I_-^2]/2\}/8I(2I-1) \tag{3.108}$$

from which the terms in 3.104 may be evaluated. With $\mathscr{H}_0 = \mathscr{H}_z = -\omega_0 I_z$, the values of $E_k^{(0)}$ and $E_k^{(1)}$ are indicated for spin-1 and spin-$\frac{3}{2}$ in Fig. 3.12. Note that, with the exception of the $(\frac{1}{2} \leftrightarrow -\frac{1}{2})$ transition of the spin-$\frac{3}{2}$ system, all levels perturbed by $\mathscr{H}_Q$ are functions of the orientation of the field-gradient tensor with respect to B_0. This means that, in a powder with all

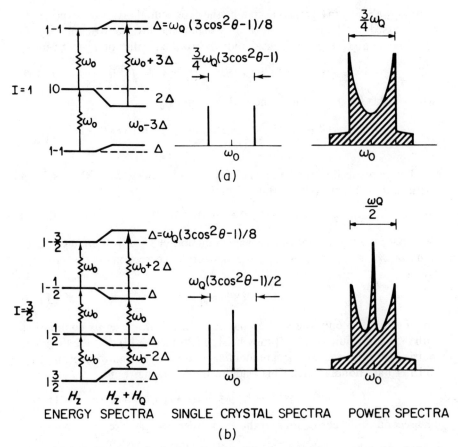

FIG. 3.12. The energy-level schemes and ideal powder absorption spectra for (a) spin-1 and (b) spin-$\frac{3}{2}$. $\omega_Q = e^2qQ/\hbar$. The spike in the second spectrum is the $\frac{1}{2} \to -\frac{1}{2}$ transition and is calculated to occur at ω_0 through the first order. (From Duncan and Dybowski [7].)

possible such orientations, the resulting line will be inhomogeneously broadened due to the superposition of sharp lines associated with each value of $\mathcal{H}_Q$. The powder spectra associated with this broadening are shown on the right of Fig. 3.12.

The sharp, central peak in the powder spectrum of the spin-$\frac{3}{2}$ case is associated with $(\frac{1}{2} \leftrightarrow -\frac{1}{2})$ transition.

For the powder spectrum of spin-$\frac{3}{2}$ in an axially symmetric field gradient, even this central transition is broadened by second-order quadrupolar terms. The second-order correction to each E_k, calculated by using (3.107) and

(3.108), is found to be (Problem 3.14)

$$E^{(2)}_{3/2, -3/2} = \Gamma(1 + 7\cos^2\theta)\sin^2\theta$$
$$E^{(2)}_{3/2, -1/2} = \Gamma(1 - 9\cos^2\theta)\sin^2\theta$$
$$E^{(2)}_{3/2, 1/2} = -\Gamma(1 - 9\cos^2\theta)\sin^2\theta \tag{3.109}$$
$$E^{(2)}_{3/2, 3/2} = -\Gamma(1 + 7\cos^2\theta)\sin^2\theta$$

where $\Gamma = \frac{3}{128}(\omega_Q^2/\omega_0)$. Note that the center of gravity of the states is invariant to addition of both $E_k^{(1)}$ and $E_k^{(2)}$, which is a useful check on such calculations.

To calculate the powder line shape of the central transition for the spin-$\frac{3}{2}$ case, we proceed in the same spirit with which we calculated the powder line shapes of shielding tensors. We seek the intensity of the absorption spectrum as a function of frequency $I(\omega)$. If $P(\theta)$ is the probability that the environment of a single spin will have the field-gradient orientation θ, then with N spins in the system and a random distribution $[P(\theta) = K(\text{a constant})]$, it must be true that

$$\int_{\omega_{min}}^{\omega_{max}} I(\omega)\,d\omega = N\int_0^\pi P(\theta)\sin\theta\,d\theta = NK\int_0^\pi \sin\theta\,d\theta \tag{3.110}$$

where we neglect the integration over ϕ. Since the probability of a given spin having *any* orientation is unity,

$$K\int_0^\pi \sin\theta\,d\theta = 1$$

which means $K = \frac{1}{2}$, so, with $dx = \sin\theta\,d\theta = -d(\cos\theta)$, we have

$$\int_{\omega_{min}}^{\omega_{max}} I(\omega)\,d\omega = -\frac{N}{2}\int_0^\pi d(\cos\theta) = -\frac{N}{2}\int_1^{-1} dx = \frac{N}{2}\int_{x=-1}^1 \frac{dx}{d\omega}\,d\omega \tag{3.111}$$

from which the line shape function $I(\omega) = N\,dx/d\omega$ is identified. For the central transition, Eq. (3.109) yields the relation between ω and $\cos\theta$:

$$\omega - \omega_0 = 2\Gamma\sin^2\theta(1 - 9\cos^2\theta) = 2\Gamma(1 - x^2)(1 - 9x^2) \tag{3.112}$$

We ask about the maximum and minimum values of ω on the range $-1 < x < 1$. By using $(d/dx)(\omega - \omega_0) = 0$, we find that

$$\omega_{max} = \omega_0 + \frac{3}{64}\frac{\omega_Q^2}{\omega_0}, \qquad x = 0 \tag{3.113a}$$

$$\omega_{min} = \omega_0 - \frac{\omega_Q^2}{12\omega_0}, \qquad x = \pm\frac{\sqrt{5}}{3} \tag{3.113b}$$

We transform the integrals of Eq. (3.111) to obtain the desired form for $I(\omega)$:

$$\int_{-\infty}^{+\infty} I(\omega)\, d\omega = \frac{N}{2}\int_{-1}^{1} dx$$

$$= \frac{N}{2}\left[\int_{-1}^{-\sqrt{5}/3} dx + \int_{-\sqrt{5}/3}^{0} dx + \int_{0}^{\sqrt{5}/3} dx + \int_{\sqrt{5}/3}^{1} dx\right]$$

$$= -\frac{N}{2}\int_{\omega_{min}}^{\omega_0} \frac{dx}{d\omega}\bigg|_{I}\, d\omega + \frac{N}{2}\int_{\omega_{min}}^{\omega_{max}} \frac{dx}{d\omega}\bigg|_{II}\, d\omega$$

$$- \frac{N}{2}\int_{\omega_{min}}^{\omega_{max}} \frac{dx}{d\omega}\bigg|_{III}\, d\omega + \frac{N}{2}\int_{\omega_{min}}^{\omega_0} \frac{dx}{d\omega}\bigg|_{IV}\, d\omega$$

$$= \frac{N}{2}\int_{\omega_{min}}^{\omega_0}\left[\frac{dx}{d\omega}\bigg|_{IV} - \frac{dx}{d\omega}\bigg|_{III} + \frac{dx}{d\omega}\bigg|_{II} - \frac{dx}{d\omega}\bigg|_{I}\right] d\omega$$

$$+ \frac{N}{2}\int_{\omega_0}^{\omega_{max}}\left[\frac{dx}{d\omega}\bigg|_{II} - \frac{dx}{d\omega}\bigg|_{III}\right] d\omega \tag{3.114}$$

From Eq. (3.114), we can make the identification

$$I(\omega) = \begin{cases} \dfrac{N}{2}\left[\dfrac{dx}{d\omega}\bigg|_{IV} - \dfrac{dx}{d\omega}\bigg|_{III} + \dfrac{dx}{d\omega}\bigg|_{II} - \dfrac{dx}{d\omega}\bigg|_{I}\right], & \omega_{min} \le \omega < \omega_0 \\[2ex] \dfrac{N}{2}\left[\dfrac{dx}{d\omega}\bigg|_{II} - \dfrac{dx}{d\omega}\bigg|_{III}\right], & \omega_0 < \omega \le \omega_{max} \end{cases} \tag{3.115}$$

From Eq. [112] we calculate the derivative

$$[d(\omega - \omega_0)/dx] = (d\omega/dx) = \tfrac{27}{16}(\omega_Q^2/\omega_0)x(x^2 - \tfrac{5}{9}) \tag{3.116}$$

Letting

$$y = (\omega - \omega_0)/8\Gamma = \tfrac{1}{4}(1 - x^2)(1 - 9x^2) \tag{3.117}$$

we can obtain the spectral line shape from Eq. (3.116) in the two regimes, it being zero for all other values:

$$I(\omega) = 3[(5 + \sqrt{16 + 9y})^{-1/2} + (5 - \sqrt{16 - 9y})^{-1/2}]/4\sqrt{16 + 9y}$$
$$\omega_0 - (\omega_Q^2/12\omega_0) \le \omega \le \omega_0 \tag{3.118}$$

and

$$I(\omega) = \tfrac{3}{4}(5 - \sqrt{16 + 9y})^{-1/2}(16 + 9y)^{-1/2},$$
$$\omega_0 \le \omega \le \omega_0 + (3\omega_Q^2/64\omega_0) \tag{3.119}$$

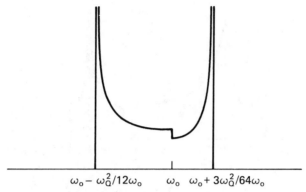

$$\omega_0 - \omega_Q^2/12\omega_0 \qquad \omega_0 \quad \omega_0 + 3\omega_Q^2/64\omega_0$$

FIG. 3.13. The powder spectrum for the central transition for a spin with $I = \frac{3}{2}$ and axial symmetry.

It is important to remember that the results apply *only* to a spin-$\frac{3}{2}$ nucleus subject to an axially symmetric field gradient. This function is plotted in Fig. 3.13.

We see that the central transition for the spin-$\frac{3}{2}$ case powder pattern is really a nonsymmetric, two-peak structure, the form of which could not be indicated on the scale of Fig. 3.9.

The central transition powder pattern for the *non*axially symmetric case of spin $n + \frac{1}{2}$ is a more complicated problem, so we simply state the result [8] here. The energy of the transition for an arbitrary orientation of the efg principal axes relative to the magnetic field is

$$\omega = \omega_0 + (\omega_Q^2/48I^2(2I - 1)^2\omega_0)\{\tfrac{3}{2}\sin^2\theta[(\tfrac{27}{8} - \tfrac{9}{2}I(I + 1))\cos^2\theta$$
$$+ \tfrac{3}{8} - (I(I + 1))/2] + (\eta^2/6)[3 - 4I(I + 1) - (\tfrac{3}{2} - 2I(I + 1))\cos^2\theta$$
$$- (\tfrac{27}{8} - \tfrac{9}{2}I(I + 1))(\cos^2 2\phi(\cos^2\theta - 1))^2]\} \qquad (3.120)$$

The powder line shape of the central transition is shown in Fig. 14. There are discontinuities that are functions of the asymmetry parameter η from which η may be determined.

In the case of strong electrostatic-field gradients and/or large quadrupole moments, i.e., the weak-field case, nuclear quadrupole interactions can be hundreds of megahertz in the absence of magnetic fields. The resonance spectra in this case are "pure quadrupolar resonances" with the energies of the levels given by

$$\omega_{NQR} = \frac{\omega_Q}{4}\left(\frac{3m^2 - I(I + 1)}{I(2I - 1)}\right) \qquad (3.121)$$

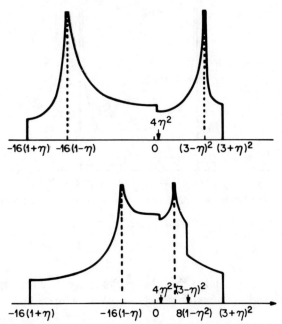

FIG. 3.14. The powder spectrum for the central transition for a spin with $I = \frac{3}{2}$ in which the spin is subject to an asymmetric interaction. The upper spectrum is the shape for $\eta < \frac{1}{3}$. The lower spectrum is the shape for $\eta > \frac{1}{3}$. (From Duncan and Dybowski [7], after Taylor *et al.* [8])

The addition of a Zeeman field results in a splitting of these Kramers degenerate levels for the spin $n + \frac{1}{2}$ case. Pure NQR spectroscopy comprises an entire branch of spectroscopy for which we refer the reader to other sources for extended discussions [6].

VII. T_1 for Randomly Modulated Hamiltonians

In Chapter 2 we saw that the spin–lattice relaxation time T_1 can be calculated in a second-order time-dependent perturbation treatment. This calculation assumes the existence of some randomly time-dependent Hamiltonian that affects the populations of the Zeeman energy levels in such a manner as to drive it toward thermal equilibrium with the other modes of the system. Any of the Hamiltonians we have discussed in the previous sections of this chapter can be time dependent because of the random movement of molecules. Thus, any of them can be the basis for relaxation of the Zeeman energy.

The motions that modulate the Hamiltonians may be quite complex. For example, the condition in which various components are modulated at different rates leads to anisotropic motion, which affects the spin–lattice relaxation in a manner that is different from that produced by random isotropic motion. The subtleties of these different models of motion can sometimes be distinguished by careful study of the dependence of the spin–lattice relaxation time on various experimental parameters such as the field, the temperature, or the concentration. Here we present the simplest models of motion and the effect of the motion on relaxation due to modulation of the Hamiltonians by isotropic random motion [9]. With the caveat that these isotropic models lead to simple results that may not be appropriately applied to complex motions, we present the results. These results do demonstrate the general appearance of the expressions for spin–lattice relaxation and are thus useful as a starting point for examining the behavior of systems.

A. Dipole–Dipole Interaction

There are two types of dipole–dipole interactions we must consider, homonuclear dipole–dipole interactions and heteronuclear dipole–dipole interactions. The result for each is useful, so we give them both. The model of the motion is one in which the nuclei interact pairwise and are rigidly held relative to each other. However, the angular position of one nucleus relative to another is randomly modulated, with an exponential correlation function. The correlation function decays with a time constant τ_c. In the case of homonuclear dipole–dipole interactions the form of T_1 is found from Eq. (3.122).

$$T_1^{-1} = \frac{2\gamma^4\hbar^2}{5r^6} I(I+1)\left\{\frac{\tau_c}{1+\omega_0^2\tau_c^2} + \frac{4\tau_c}{1+4\omega_0^2\tau_c^2}\right\} \qquad (3.122)$$

where γ is the gyromagnetic ratio of the nucleus, I spin quantum number, ω_0 the Larmor frequency of the nucleus, and r the separation of the pair of spins. In a similar manner, one may calculate the contribution to T_1 of the I spin due to the modulation of the heteronuclear dipole-dipole interaction from Eq. (3.123).

$$T_{1_I}^{-1} = \frac{2\gamma_I^2\gamma_S^2\hbar^2}{5r^6} S(S+1)\left\{\frac{1}{3}\frac{\tau_c}{1+(\omega_{0_I}-\omega_{0_S})^2\tau_c^2} + \frac{\tau_c}{1+\omega_{0_I}^2\tau_c^2}\right.$$
$$\left. + \frac{2\tau_c}{1+(\omega_{0_I}+\omega_{0_S})^2\tau_c^2}\right\} \qquad (3.123)$$

In Eq. (3.120), γ_I and γ_S are the gyromagnetic ratios of the two spins and ω_{0_I} and ω_{0_S} are the respective Larmor frequencies; S is the spin quantum number of the spin S. In each of these, we see the strong dependence of the relaxation time on the separation of the two spins.

B. Anisotropic Chemical-Shift Interaction

The rotation of the molecule in which a nucleus subject to an anisotropic chemical shift is present also affects the relaxation of the Zeeman energy of that nucleus. The simplest model of this interaction results from considering a nucleus subject to an axially symmetric chemical-shift interaction and which is undergoing random isotropic tumbling with correlation time τ_c. In this model, one would calculate T_1 according to Eq. (3.124).

$$T_1^{-1} = \tfrac{2}{15}\gamma^2 B_0^2 \delta^2 [\tau_c/(1 + \omega_0^2 \tau_c^2)] \tag{3.124}$$

In this equation, γ is the gyromagnetic ratio, B_0 the magnetic field intensity, δ the chemical-shift anisotropy, and ω_0 the Larmor frequency of the nucleus at the magnetic field strength. This mechanism is interesting in that in the fast-motion limit, T_1 depends on the square of the strength of the magnetic field.

C. Indirect Coupling

The indirect coupling of two nuclei can also be modulated by motions in the system and thus provide a pathway for spin–lattice relaxation. We quote the result first as

$$T_{1_I}^{-1} = \tfrac{2}{3} J^2 S(S + 1) [\tau_c/(1 + (\omega_{0I} - \omega_{0S})^2 \tau_c^2)] \tag{3.125}$$

where J is the magnitude of the scalar coupling, S is the spin quantum number of the one spin, and the Larmor frequencies are the same as in the preceding subsection. The nature of the relaxation depends on the model one chooses for the motion inducing relaxation. In this case, there are two models. In the first model, the scalar interaction is itself time dependent by virtue of the movement of the two spins relative to each other. This situation frequently happens if one of the spins is involved in an exchange process. This is called scalar relaxation of the first kind and the correlation time is measuring the exchange time.

The second model is subtle. If the spin S is involved in a more efficient relaxation process, then the spin itself is randomly changing state with a time constant T_{1S}, which is correlation time τ_c in Eq. (3.125). In effect, the coupling of the spin I to the spin S allows it to share in the efficient relaxation of spin S.

D. Quadrupolar Interaction

Because the quadrupolar coupling constant can be large, this mechanism can be very efficacious in relaxation of spins with $I > \tfrac{1}{2}$. The form of the relaxation of the Zeeman energy in the fast-motion limit is given as

$$T_1^{-1} = \frac{3}{40}\frac{2I + 3}{I^2(2I - 1)}\left(1 + \frac{\eta^2}{3}\right)\left(\frac{e^2qQ}{\hbar}\right)^2 \tau_c \tag{3.126}$$

Here, I is the spin quantum number, Q the quadrupolar coupling constant, eq the electric-field gradient at the site of the nucleus that is being modulated with the time constant τ_c. Here, η is the asymmetry parameter of the quadrupolar interaction.

Many more complicated models of the molecular motion can be used and the resulting forms of T_1 are more complicated than those given here. What we have attempted here is a cataloging of the forms under the simplest motions possible. These do show the general trends in the dependence of the relaxation time on various experimental parameters, such as temperature, field, coupling constant, and internuclear separation.

VIII. Concluding Remarks

A system of spins can have a variety of interactions that alter its energy. Each of these interactions is represented by a Hamiltonian, many of which have the form of a Zeeman interaction of a magnetic moment with a magnetic field. The total energy of the system is a sum of all the interaction energies. The interactions range from rather weak indirect couplings to strong direct couplings. The study of each of these interactions involves evaluating from NMR experiments the parameters such as the dipole–dipole coupling constant, the chemical shift, the spin–spin coupling constant or the electric-field gradient. The exact value of each of these parameters is useful for analyzing the chemical environment of the substance. In succeeding chapters, we shall see how one may examine various of these interactions by use of special spectroscopic experiments.

PROBLEMS

3.1 What are the units of $\mathcal{H}_D$ in Eq. (3.4)? What would the units be if the constant term were $\gamma^2\hbar^2$ instead of $\gamma^2\hbar$?

3.2 Derive Eq. (3.5) from Eq. (3.4) by using $\mathbf{r}_i - \mathbf{r}_k = \mathbf{r}$ and expressing the internuclear vector in Cartesian coordinates.

3.3 Derive each term of Eq. (3.7) by transforming the Cartesian coordinates of Eq. (3.5) into spherical polar coordinates, using stepping operators to represent transverse components of angular momentum.

3.4 At what value of θ will the secular contribution to the dipolar splitting of two spins [Eq. (3.7), terms A and B] be zero? (This value of θ is known as the "magic angle.")

3.5 Referring to Eq. (3.46), what are the dimensions of $\mathbf{I}$ and of $\mathbf{B}_0$ and why?

3.6 Given the form for $\mathbf{R}_z$ [Eq. (3.51a)] show that $\mathbf{R}_z^{-1}$ is the transpose of $\mathbf{R}_z$.

3.7 Given the forms of $\mathbf{R}_z$ and $\mathbf{R}_y$, write down the matrix for $\mathbf{R}_x$.

3.8 Show that $R_x^{-1} \cdot \sigma \cdot R_x$ yields Eq. (3.52).

3.9 Derive Eq. (3.53).

3.10 What is the value of ω_D, the dipolar frequency, in rad $\sec^{-1}$, for two ^{13}C nuclei separated by a distance of 1 Å? [See Eq. (3.7)]

3.11 Verify that (3.95) and (3.96) are equivalent.

3.12 Show that (3.96) may be written in the form

$$\mathscr{H}_Q = \mathbf{T} \cdot \mathbf{V} \cdot \mathbf{T}^\dagger$$

where $\mathbf{T}$ is a row vector with $T_k = I_k - iI/3$ and $\mathbf{T}^\dagger$ is the Hermitian conjugate of $\mathbf{T}$. What is the value of the constant?

3.13 The quadrupolar Hamiltonian in the case of axial symmetry and high field, in the frame of the Zeeman Hamiltonian, is given by Eq. (3.103). Using the usual stepping operators for I_x and I_y [Eqs. (2.17), (2.25), (2.26)] show that

(a)

$$\langle I, m | I_{x'}^2 | I, m \rangle = \langle I, m | I_{y'}^2 | I, m \rangle = \langle I, m | I^2 - I_{z'}^2 | I, m \rangle / 2$$
$$= \tfrac{1}{2}[I(I + 1) - m^2]$$

(b)

$$\langle I, m | I_{z'} I_{x'} | I, m \rangle = \langle I, m | I_{x'} I_{z'} | I, m \rangle = 0$$

(c) the nonvanishing diagonal parts of $\mathscr{H}_Q$ may be represented by

$$\mathscr{H}_Q^s = \omega_Q [3\cos^2 \theta - 1][3I_{z'}^2 - I^2]/8I(2I - 1)$$

(d) the portions of $\mathscr{H}_{Q'}$ that do not commute with the Zeeman Hamiltonian are given by Eq. (3.108).

3.14 Use (3.107) and (3.108) to verify (3.109).

REFERENCES

1. J. H. van Vleck, *Phys. Rev.* **74**, 1168 (1948).
2. I. Lowe and M. Engelsberg, *Phys. Rev.* **B10**, 822 (1974).
3. J. A. Pople, W. G. Schneider, and H. J. Bernstein, "High Resolution Nuclear Magnetic Resonance," McGraw-Hill, New York, 1959.
4. C. L. McKnett, C. R. Dybowski, and R. W. Vaughan, *J. Chem. Phys.* **63**, 4578 (1975).
5. F. Bowman, "Introduction to Elliptic Functions," Chapter 2, Wiley, New York, 1953.
6. G. K. Semin, T. A. Babushikina, and G. G. Yakobson, "Nuclear Quadrupole Resonance in Chemistry," Keter, Jerusalem, 1975.
7. T. M. Duncan and C. R. Dybowski, *Surf. Sci. Rep.* **1**, 157 (1981).
8. P. C. Taylor, J. F. Baugher, and H. M. Kriz, *Chem. Rev.* **75**, 203 (1975).
9. N. Bloembergen, E. M. Purcell, and R. V. Pound, *Phys. Rev.* **73**, 679 (1948).

CHAPTER 4

EXPONENTIAL APPROXIMATIONS
FOR EVOLUTION OPERATORS:
THE BCH FORMULA,
THE MAGNUS EXPANSION,
AND THE DYSON EXPRESSION[†]

I. Introduction

In Chapter 1 we saw how it is convenient to think of the time dependence of magnetization classically. Many of the aspects of magnetic resonance are fruitfully observed in this manner, e.g., sequences as complicated as the Carr–Purcell sequence. However, a fuller understanding of the inherent physical processes requires the introduction of a quantum-mechanical formalism and with it, the attendant mathematical complication resulting from non-commutativity of operators. In Chapter 2 we stated the problem in terms of the density matrix (operator) formalism so useful in magnetic resonance. The resultant problem that arose out of that exposition was solution of the equation of time evolution of the density operator. Only under favorable conditions could it be solved exactly. However, we found a perturbation solution that could, for example, yield a mechanism for spin–lattice relaxation consistent with the damping observed and that had been explained only classically by the *ad hoc* introduction of terms in the equation of motion. In addition, we pointed to the fact that the equation of motion could be solved by postulating the existence of a time development operator, the form of which must be found mathematically. It should therefore be obvious that the time development of the density operator is the pivot on which an understanding of transient techniques in NMR swings.

We have seen that the expression of the evolution operator as an exponential operator in certain limits leads directly to solution of the problem of time evolution of ρ. We have also seen that the effect of an exponential

---

[†] See R. M. Wilcox [1].

of spin operators on other spin operators has been relatively easy to characterize. We have already developed a physical feeling for exponentials of spin operators as rotation operators in N-dimensional space. Thus, the necessity of making approximations for the time dependence of ρ might best be ameliorated by couching the approximations in exponential form to maintain resemblance to familiar forms. In succeeding chapters we shall see that, with the idea of cancelling the effect of certain Hamiltonians (most particularly the dipole–dipole Hamiltonian) to allow observation of other Hamiltonians, one may use, among others, strong rf irradiation to counteract these effects. It will be convenient to approximate the results as being due to exponential evolution operators. To handle this formalism, one needs special tools, which we develop in this chapter.

The material in Section II of this chapter is rather mathematical in nature and is not read without a pencil in bed before retiring. The results, however, are quite terse and easy to apply, provided that one has some concept of time ordering. Rather than relegate the developement of this material to an appendix, the authors are keeping it in the main body of the text in the hope that the student will take the time to read it with understanding. There are at least two reasons for this approach in the current text. The first is that, in keeping with the idea of the manuscript as a text, the authors feel that a knowledge of how tools work the way they do is an important part of using them intelligently. The second is that there are some portions of this development that are almost certain to involve tools with which the audience at which this book is primarily aimed is unfamiliar and that will importantly enhance their operational analytical abilities. One such example, mentioned previously, is the use of time ordering to simplify the evaluation of integrals over a multidimensional time space. In plain fact, the authors have had fun learning this material themselves, and we feel that the present exposition is the clearest extant. We do not wish to deprive the student of this stimulating experience.

For those wishing to be deprived, the results are succinctly stated in Section III of this chapter, and some simple examples are given of their use.

To proceed with the development of the main relations with which the chapter deals, we develop certain ancillary formulae used in obtaining series solutions to the time-evolution equations.

II. The Baker–Campbell–Hausdorff Formula, the Magnus Expansion, and the Dyson Expression

In this section we develop three results important to the mechanics of calculating the results of applying a series of rf pulses to a resonant spin system. The first, the Baker–Campbell–Hausdorff (BCH) formula explicitly develops the result of multiplying the exponentials of two noncommuting operators A and B. The result is not the exponential of $A + B$. In fact we will find that the result is an infinite series in commutators of A and B:

$$e^A e^B = \exp(A + B + \tfrac{1}{2}[A, B] + \tfrac{1}{12}[A[A, B]] + \tfrac{1}{12}[[A, B], B] + \cdots) \quad (4.1)$$

The development involves an interesting bit of arithmetic, but it is not necessary to follow it in order to be able to use the result. Those with a wish to proceed with the use of the BCH formula may proceed directly to Section III of this chapter. The same remark applies to the development of the Magnus expansion, which will immediately follow that of the BCH expression.

A. Preliminary Derivations

We initially develop two important equations that are necessary for the derivations of the BCH formula and the Magnus expansion. The first is the expression for the derivative of an exponential operator. The result to be proved is

$$\frac{\partial e^{-\beta \mathscr{H}(\lambda)}}{\partial \lambda} = -e^{-\beta \mathscr{H}} \int_0^\beta du \, e^{u \mathscr{H}} \frac{\partial \mathscr{H}}{\partial \lambda} e^{-u \mathscr{H}} \quad (4.2)$$

Before proving this result, it is important to note that $\mathscr{H}$ is an operator that does not in general commute with $\partial \mathscr{H} / \partial \lambda$. Therefore, the left-hand side of (4.2) cannot be simply expressed as a product of the exponential and $\partial \mathscr{H} / \partial \lambda$, since one does not know whether to multiply by $\partial \mathscr{H} / \partial \lambda$ from the right or the left. The right-hand side of (4.2) gives us the recipe for evaluating this derivative. It states that the result is the exponential operator multiplied *on the right* by the derivative of $\mathscr{H}$ with respect to λ weighted as prescribed in the integral.

To prove the relation (4.2) we consider the exponential of the noncommuting operators A and B,

$$F(A, B) = e^{uA} e^{-u(A + B)} \quad (4.3)$$

As will be seen in the development of the BCH formula

$$e^A e^B \neq e^{(A + B)} \quad (4.4)$$

if A and B do not commute, so the right-hand side of (4.3) is not equal to e^{-uB}. Taking the differential of (4.3) with respect to u, we find

$$d[e^{uA} e^{-u(A + B)}] = -e^{uA} B e^{-u(A + B)} \, du \quad (4.5)$$

Integrating between zero and β yields

$$e^{\beta A} e^{-\beta(A + B)} = 1 - \int_0^\beta du \, e^{uA} B e^{-u(A + B)} \quad (4.6)$$

Multiplying on the left by $e^{-\beta A}$ gives

$$e^{-\beta(A+B)} = e^{-\beta A} - \int_0^\beta du\, e^{-(\beta-u)A} B e^{-u(A+B)} \qquad (4.7)$$

We now identify A and B in terms of $\mathscr{H}(\lambda)$ as

$$A = \mathscr{H}(\lambda_0), \qquad B = \int_{\lambda_0}^\lambda d\lambda'\, \frac{\partial \mathscr{H}(\lambda')}{\partial \lambda'}$$

so that

$$\mathscr{H}(\lambda) = A + B \qquad (4.8)$$

and

$$\partial \mathscr{H}(\lambda)/\partial \lambda = \partial B/\partial \lambda \qquad (4.9)$$

Equation (4.7) then becomes

$$e^{-\beta \mathscr{H}(\lambda)} = e^{-\beta \mathscr{H}(\lambda_0)} - \int_0^\beta du\, e^{-(\beta-u)\mathscr{H}(\lambda_0)} B e^{-u\mathscr{H}(\lambda)} \qquad (4.10)$$

Differentiating with respect to λ yields

$$\frac{\partial e^{-\beta\mathscr{H}(\lambda)}}{\partial \lambda} = -\int_0^\beta du\, e^{-(\beta-u)\mathscr{H}(\lambda_0)} \frac{\partial B}{\partial \lambda} e^{-u\mathscr{H}(\lambda)}$$

$$- \int_0^\beta du\, e^{-(\beta-u)\mathscr{H}(\lambda_0)} \left[\int_{\lambda_0}^\lambda d\lambda'\, \frac{\partial \mathscr{H}(\lambda')}{\partial \lambda'} \right] \frac{\partial}{\partial \lambda} e^{-u\mathscr{H}(\lambda)} \qquad (4.11)$$

Equation (4.11) must be true for any value of λ_0. In particular it must be true for $\lambda_0 = \lambda$, at which point the second term on the right-hand side of (4.11) vanishes, and the entire equation reduces to (4.2).

The second preliminary result to be developed is the expansion of the expression

$$e^{\lambda A} B e^{-\lambda A} \equiv C(\lambda) \qquad (4.12)$$

where A and B are operators and λ is some parameter. Taking the derivative of $C(\lambda)$ with respect to λ, we find

$$dC(\lambda)/d\lambda = A e^{\lambda A} B e^{-\lambda A} - e^{\lambda A} B e^{-\lambda A} A$$
$$= AC(\lambda) - C(\lambda)A \qquad (4.13)$$

(We note that A commutes with $e^{\lambda A}$, since $e^{\lambda A}$ is just a power series in A, so $A e^{-\lambda A} = e^{-\lambda A} A$.) Therefore,

$$[dC(\lambda)/d\lambda] + [C(\lambda), A] = 0 \qquad (4.14)$$

We now introduce the concept of the "superoperator," $[\,,A]$, which has the meaning

$$[\,,A]D = [D,A] \tag{4.15}$$

we may therefore recast (4.14) into the form

$$[dC(\lambda)/d\lambda] + [\,,A]C(\lambda) = 0 \tag{4.16}$$

or

$$\{d/d\lambda + [\,,A]\}C(\lambda) = 0$$

By analogy with the obvious solution for the case where $[\,,A]$ is a constant, we might guess a solution

$$C(\lambda) = C(0)\exp\{-\lambda[\,,A]\} \tag{4.17}$$

This solution turns out to be correct, as may be verified by direct differentiation. We also see, since $C(0) = B$, by (4.12), that

$$C(\lambda) = e^{\lambda A}Be^{-\lambda A} = e^{-\lambda[\,,A]}B \tag{4.18}$$

The exponential, as always, means the expansion

$$e^{-\lambda[\,,A]}B = B + \lambda[A,B] + \lambda^2[A,[A,B]]/2! + \cdots \tag{4.19}$$

For compactness, we denote this expansion as

$$e^{\lambda A}Be^{-\lambda A} = \sum_{j=0}^{\infty} \lambda^j(j!)^{-1}\{A^j,B\} \tag{4.20}$$

where

$$\{A^0,B\} = B \quad\text{and}\quad \{A^n,B\} = [A,\{A^{n-1},B\}] \tag{4.21}$$

B. The BCH Formula

In order to make this derivation, we recast (4.2) into the form

$$\frac{\partial e^Z}{\partial \lambda} = \int_0^1 dx\, e^{xZ}Z'(\lambda)e^{-xZ}e^Z \tag{4.22}$$

which is obtained from (4.2) making the identities $Z \equiv Z(\lambda)$ for $-\mathcal{H}$, 1 for β, $1-x$ for u, and $Z'(\lambda) = \partial Z/\partial\lambda$. We wish to express Z in a power series in λ such that

$$e^Z = e^{\lambda A}e^{\lambda B} \tag{4.23}$$

is identically satisfied. For future purposes, we remark that A, B, and Z are Hermitian operators. As a power series in λ, therefore, we have

$$Z = \sum_{n=1}^{\infty} \lambda^n Z_n \quad \text{and} \quad Z'(\lambda) = \sum_{n=1}^{\infty} n\lambda^{n-1} Z_n \tag{4.24}$$

where the Z_n are to be determined. We note that $Z = 0$ when $\lambda = 0$. Differentiating (4.23) with respect to λ yields

$$\frac{\partial e^Z}{\partial \lambda} \equiv \int_0^1 dx\, e^{xZ} Z' e^{-xZ} e^Z = A e^{\lambda A} e^{\lambda B} + e^{\lambda A} B e^{\lambda B} \tag{4.25}$$

which is obtained by utilizing (4.22) and (4.23). Multiplication on the right by $\exp\{-Z\} = \exp\{-\lambda B\} \exp\{-\lambda A\}$ yields

$$\int_0^1 dx\, e^{xZ} Z' e^{-xZ} = A + e^{\lambda A} B e^{-\lambda A} \tag{4.26}$$

By utilizing the result (4.20) for both $\exp\{xZ\} Z' \exp\{-xZ\}$ and $\exp\{\lambda A\} B \times \exp\{-\lambda A\}$, and by Eq. (4.24), we find

$$\int_0^1 dx\, e^{xZ} Z' e^{-xZ} = \sum_{k=0}^{\infty} \left\{ \frac{1}{(k+1)!} \left(\sum_{m=1}^{\infty} \lambda^m Z_m \right)^k, \sum_{n=1}^{\infty} n\lambda^{n-1} Z_n \right\}$$

$$= A + \sum_{j=0}^{\infty} \lambda^j (j!)^{-1} \{A^j, B\} \tag{4.27}$$

We require that (4.27) must be satisfied identically in λ, so we equate coefficients of λ^j on both sides of the equation. Expanding the first few terms of the left-hand sum of Eq. (4.27) gives

$$\sum_{k=0}^{\infty} \left\{ \frac{1}{(k+1)!} (\lambda Z_1 + \lambda^2 Z_2 + \cdots)^k, (Z_1 + 2\lambda Z_2 + \cdots) \right\}$$

$$= (Z_1 + 2\lambda Z_2 + 3\lambda^2 Z_3 + \cdots)$$

$$+ \frac{1}{2!} [(\lambda Z_1 + \lambda^2 Z_2 + \cdots), (Z_1 + 2\lambda Z_2 + 3\lambda^2 Z_3 + \cdots)]$$

$$+ \frac{1}{3!} [(\lambda Z_1 + \lambda^2 Z_2 + \cdots), [(\lambda Z_1 + \lambda^2 Z_2 + \cdots), (Z_1 + 2\lambda Z_2 + \cdots)]]$$

$$+ \cdots \tag{4.28}$$

Equating powers of λ from (4.28) with the right-hand side of (4.27) yields

$$j = 0, \quad Z_1 = A + B \tag{4.29}$$

$$j = 1, \quad Z_2 = \tfrac{1}{2}[A, B] \tag{4.30}$$

$$j = 2, \quad 3Z_3 + \tfrac{1}{2}[Z_2, Z_1] + \tfrac{1}{2}[Z_1, 2Z_2] = \tfrac{1}{2}[A, [A, B]] \tag{4.31}$$

Equation (4.31) is simplified by utilizing (4.30) and (4.29):

$$3Z_3 + \tfrac{1}{2}[\tfrac{1}{2}[A, B], (A + B)] + \tfrac{1}{2}[(A + B), 2 \cdot \tfrac{1}{2}[A, B]] = \tfrac{1}{2}[A, [A, B]] \quad (4.32)$$

yielding

$$Z_3 = \tfrac{1}{12}[A, [A, B]] + \tfrac{1}{12}[[A, B], B] \quad (4.33)$$

Setting $\lambda = 1$ in Eqs. (4.23) and (4.24), therefore, we have the BCH formula to second order:

$$e^A e^B = \exp\{A + B + \tfrac{1}{2}[A, B] + \tfrac{1}{12}[A, [A, B]] + \tfrac{1}{12}[[A, B], B] + \cdots\} \quad (4.34)$$

C. The Magnus Expansion

In Chapter 2, we noted that the trace on an oscilloscope screen, corresponding to the response of a nuclear spin system to a radio-frequency transient is proportional to the expectation value of the magnetization, as a function of time, at a phase in the xy plane of the rotating frame chosen by the experimenter. This expectation value is obtained by utilizing the time dependence of the density matrix.

The operator differential equation

$$(\partial/\partial t)U(t) = \mathscr{H}(t)U(t) \quad (4.35)$$

is particularly important in calculating the time dependence of the density matrix; i.e., in our case it must be solved to determine the "evolution operators" (*vide infra*) that cause the time dependence of the magnetization in a transient NMR experiment. We note in passing that since $\mathscr{H}$ is an operator whose forms at various times do not have to commute with each other, the obvious solution

$$U(t) = U(0)\exp \int_0^t \mathscr{H}(t')\, dt' \quad (4.36)$$

is not generally valid.

However, it seems reasonable to expect that the form of the exponential operator might be a form suitable for the solution, particularly since one knows that the above solution is the correct solution to (4.35) if $\mathscr{H}(t)$ is time independent. Thus we assume the solution of (4.35) is of the form

$$U(t) = Ke^{\Omega(t)} \quad (4.37)$$

where $\Omega(t)$ is some, as yet unspecified, function of the operators and of time. Here $U(t)$ will turn out to be an exponential "rotation" operator in a rather special sense. It will be an operator that "rotates" a system from one time to another. We call this type of operator a "time evolution" operator. For such an operator, of the form (4.37), it is appropriate to set the boundary condition $K = 1$, because at zero time, the evolution operator applied to the density operator yields just the density operator at zero time. Thus, in the following derivation, we take $K = 1$, and $\Omega(0) = 0$, so $U(0) = 1$.

Substituting (4.37) into (4.35) and taking note of (4.22) with the identifications

$$
Z \to \Omega, \qquad \lambda \to t, \qquad Z' \to \frac{\partial \Omega}{\partial t}, \qquad \frac{\partial e^{Z}}{\partial \lambda} \to \frac{\partial e^{\Omega}}{\partial t} \to \frac{\partial U}{\partial t}
$$

yields

$$
\mathscr{H}(t) = \int_0^1 dx \, e^{x\Omega} \frac{\partial \Omega}{\partial t} e^{-x\Omega} \tag{4.38}
$$

Equation (4.20) is used to obtain a form for the integrand of (4.38) by defining the function of η

$$
f(\eta) = \sum_{j=0}^{\infty} x^j \eta^j (j!)^{-1} \{\Omega^j, \dot{\Omega}\} \tag{4.39}
$$

which becomes $\exp\{x\Omega\}\dot{\Omega}\exp\{-x\Omega\}$ when η is unity. Integrating between limits yields

$$
\int_0^1 f(\eta) \, dx = \sum_{j=0}^{\infty} \frac{\eta^j}{(j+1)!} \{\Omega^j, \dot{\Omega}\}
$$

$$
= \{\Omega^0, \dot{\Omega}\} + \frac{\eta}{2!} \{\Omega, \dot{\Omega}\}
$$

$$
+ \frac{\eta^2}{3!} \{\Omega^2, \dot{\Omega}\} + \cdots \tag{4.40}
$$

We further expand Ω in terms of powers of the complete set of expansion parameters η:

$$
\Omega = \sum_{n=0}^{\infty} \eta^n \Delta_n(t) \tag{4.41}
$$

which yields

$$
\dot{\Omega}(t) = \sum_{n=0}^{\infty} \eta^n \dot{\Delta}_n \tag{4.42}
$$

Evaluating the terms in curly brackets in terms of commutators, one has

$$
\mathscr{H}(t) = \dot{\Omega} + \frac{\eta}{2!} [\Omega, \dot{\Omega}] + \frac{\eta^2}{3!} [\Omega, [\Omega, \dot{\Omega}]] + \cdots \tag{4.43}
$$

Finally, substitution for Ω in terms of powers of η [Eqs. (4.41) and (4.42)] yields

$$\int_0^1 f(\eta)\,dx = \sum_{n=0}^{\infty} \eta^n \dot\Delta_n + \frac{\eta}{2!}\left[\sum_{n=0}^{\infty} \eta^n \Delta_n, \sum_{k=0}^{\infty} \eta^k \dot\Delta_k\right]$$

$$+ \frac{\eta^2}{3!}\left[\sum_{n=0}^{\infty} \eta^n \Delta_n, \left[\sum_{k=0}^{\infty} \eta^k \Delta_k, \sum_{m=0}^{\infty} \eta^m \dot\Delta_m\right]\right] + \cdots$$

$$= \dot\Delta_0 + \eta \dot\Delta_1 + \eta^2 \dot\Delta_2 + \cdots + \frac{1}{2!}\sum_{n=0}^{\infty}\sum_{k=0}^{\infty} \eta^{n+k+1}[\Delta_n, \dot\Delta_k]$$

$$+ \frac{1}{3!}\sum_{n,m,k=0}^{\infty} \eta^{n+m+k+2}[\Delta_n,[\Delta_k,\dot\Delta_m]] + \cdots$$

$$= \dot\Delta_0 + \eta \dot\Delta_1 + \eta^2 \dot\Delta_2 + \cdots$$

$$+ \frac{\eta}{2!}[\Delta_0,\dot\Delta_0] + \frac{\eta^2}{2!}([\Delta_0,\dot\Delta_1] + [\Delta_1,\dot\Delta_0]) + \cdots$$

$$+ \frac{\eta^2}{3!}([\Delta_0,[\Delta_0,\dot\Delta_0]]) + \cdots \tag{4.44}$$

where we have carried the expansion out only to second order in η. Equating the coefficients of η^n on the left-hand side and right-hand side of (4.44) we obtain the following differential equations for the coefficients Δ_n, now setting $\eta = 1$:

$$\dot\Delta_0 = \mathscr{H}(t) \tag{4.45}$$

$$\dot\Delta_1 + \frac{1}{2}[\Delta_0,\dot\Delta_0] = 0 \tag{4.46}$$

$$\dot\Delta_2 + \frac{1}{2!}([\Delta_0,\dot\Delta_1] + [\Delta_1,\dot\Delta_0]) + \frac{1}{3!}[\Delta_0,[\Delta_0,\dot\Delta_0]] = 0 \tag{4.47}$$

For higher-order terms one must carry the expansion still further.

The solution to (4.45) is

$$\Delta_0(t_c) = \int_0^{t_c} \mathscr{H}(t)\,dt \tag{4.48}$$

where $\Delta_0(0)$ is required to be zero in order that

$$U(0) = \exp[\Delta_0(0) + \Delta_1(0) + \Delta_2(0) + \cdots] \equiv 1 \tag{4.49}$$

The solution of (4.46) is

$$\Delta_1(t_c) = \frac{1}{2}\int_0^{t_c} d\tau\left[\mathscr{H}(\tau), \int_0^{\tau} \mathscr{H}(\epsilon)\,d\epsilon\right] \tag{4.50}$$

A crucial point to be made with regard to (4.50) is that the integral is undefined for $\epsilon > \tau$. Therefore if the time t_c in question naturally divides itself into intervals $\delta t_1 = t_1 - 0$, $\delta t_2 = t_2 - t_1, \ldots, \delta t_n = t_n - t_{n-1}$, we may express (4.50) as:

$$2\Delta_1(t_c) = \int_0^{t_c} d\tau \left[\mathcal{H}(\tau), \int_0^\tau d\epsilon \, \mathcal{H}(\epsilon) \right]$$

$$= \int_0^{t_1} d\tau \left[\mathcal{H}(\tau), \int_0^\tau d\epsilon \, \mathcal{H}(\epsilon) \right]$$

$$+ \int_{t_1}^{t_2} d\tau \left[\mathcal{H}(\tau), \left\{ \int_0^{t_1} d\epsilon \, \mathcal{H}(\epsilon) + \int_{t_1}^\tau d\epsilon \, \mathcal{H}(\epsilon) \right\} \right] + \cdots$$

$$+ \int_{t_{n-1}}^{t_n} d\tau \left[\mathcal{H}(\tau), \left\{ \int_0^{t_1} d\epsilon \, \mathcal{H}(\epsilon) + \cdots + \int_{t_{n-1}}^\tau d\epsilon \, \mathcal{H}(\epsilon) \right\} \right] \quad (4.51)$$

The utility of this form of the integral will become clear once we start making calculations, in particular when we calculate the first-order coupling term between a phase error and an offset Hamiltonian for the flip-flop phase tuning sequence in Chapter 5. That the two sides of (4.51) are equal may be seen with reference to Fig. 4.1. Clearly, the space of the integration is that in which $\tau > \epsilon$. This is the area below the diagonal in this figure. The lower left area is

$$\int_0^{t_1} d\tau \left[\mathcal{H}(\tau), \int_0^\tau d\epsilon \, \mathcal{H}(\epsilon) \right]$$

The integral over the lower-right area is

$$\int_{t_1}^{t_2} d\tau \left[\mathcal{H}(\tau), \int_0^{t_1} d\epsilon \, \mathcal{H}(\epsilon) \right]$$

and that over the upper-right area is

$$\int_{t_1}^{t_2} d\tau \left[\mathcal{H}(\tau), \int_{t_1}^\tau d\epsilon \, \mathcal{H}(\epsilon) \right]$$

The formalism looks a bit oppressive at this point. It should be kept in mind that the $\mathcal{H}$ operators will correspond, e.g., to rf pulses, which in turn will be represented as angular momentum operators for which the commutation rules are well known.

The term in η^2 yields

$$\dot{\Delta}_2 = \tfrac{1}{2}[\dot{\Delta}_1, \Delta_0] + \tfrac{1}{2}[\dot{\Delta}_0, \Delta_1] + \tfrac{1}{6}[[\Delta_0, \dot{\Delta}_0], \Delta_0] \quad (4.52)$$

Since $\dot{\Delta}_1 = \tfrac{1}{2}[\dot{\Delta}_0, \Delta_0]$, the first terms in $\dot{\Delta}_2$ combine to yield

$$\dot{\Delta}_2 = \tfrac{1}{6}[\dot{\Delta}_1, \Delta_0] + \tfrac{1}{2}[\dot{\Delta}_0, \Delta_1] \quad (4.53)$$

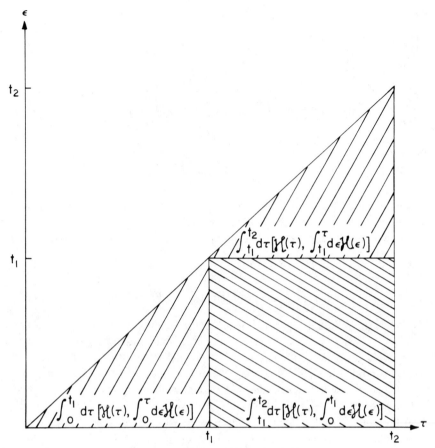

FIG. 4.1. The region of integration in time.

With careful attention to the limits of integration, we find

$$\Delta_2 = \frac{1}{12} \int_0^{t_c} d\tau \int_0^\tau d\epsilon \int_0^\tau d\phi [[\mathcal{H}(\tau), \mathcal{H}(\epsilon)], \mathcal{H}(\phi)]$$

$$+ \frac{1}{4} \int_0^{t_c} d\tau \int_0^\tau d\epsilon \int_0^\epsilon d\phi [\mathcal{H}(\tau), [\mathcal{H}(\epsilon), \mathcal{H}(\phi)]]$$ (4.54)

We note that the limits of the two integrals in (4.54) are not the same. In order to combine these two integrals into one term, it will be necessary to express the first integral as a sum of two integrals having the same limits as the first integral in (4.54). To do so, we consider the concept of "time ordering," which will also be useful in our discussion of the Dyson expression. Consider the

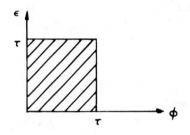

FIG. 4.2. The region of integration of the integral $\int_0^\tau d\epsilon \int_0^\tau d\phi \, F(\epsilon)F(\phi)$.

integral

$$\int_0^\tau d\epsilon \int_0^\tau d\phi \, F(\epsilon)F(\phi) \qquad (4.55)$$

This integral is clearly over the space $\epsilon = 0$ to τ, and $\phi = 0$ to τ, which is indicated as the cross-hatched square in Fig. 4.2.

Now consider expressing this integral as a sum of two integrals as

$$T \int_0^\tau d\epsilon \int_0^\tau d\phi \, F(\epsilon) \, F(\phi)$$
$$= \int_0^\tau d\epsilon \int_0^\epsilon d\phi \, F(\epsilon)F(\phi) + \int_0^\tau d\phi \int_0^\phi d\epsilon \, F(\phi)F(\epsilon) \qquad (4.56)$$

The operator T takes all possible $n!$ permutations of an n-fold product $F(\epsilon)F(\phi) \cdots F(\chi)$ and "time orders" each permutation such that the variables successively decrease from left to right. Therefore in the first integral on the right-hand side of (4.56) $\epsilon \geq \phi$, and in the second, $\phi \geq \epsilon$. The region of the integral on the left-hand side of (4.56) is the entire square of Fig. 4.3. The first integral on right-hand side of (4.56) is over the triangle of region I (white area) and the second is over the triangle of region II (cross-hatched area). Relabeling $\phi \to \epsilon$ and $\epsilon \to \phi$ in the second integral, we see this integral may be expressed as

$$\int_0^\tau d\epsilon \int_0^\epsilon d\phi \, F(\epsilon)F(\phi) \qquad (4.57)$$

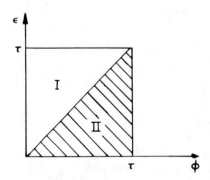

FIG. 4.3. The region of integration showing the two subregions in Eq. (4.56).

so that

$$T \int_0^\tau d\epsilon \int_0^\tau d\phi \; \mathcal{H}(\epsilon)\mathcal{H}(\phi) = 2! \int_0^\tau d\epsilon \int_0^\epsilon d\phi \; \mathcal{H}(\epsilon)\mathcal{H}(\phi) \tag{4.58}$$

For an n-fold product, the result is

$$T \int_0^\tau dt_1 \int_0^\tau dt_2 \cdots \int_0^\tau dt_n \; \mathcal{H}(t_1)\mathcal{H}(t_2) \cdots \mathcal{H}(t_n)$$
$$= n! \int_0^\tau dt_1 \int_0^{t_1} dt_2 \cdots \int_0^{t_{n-1}} dt_n \; \mathcal{H}(t_1) \cdots \mathcal{H}(t_n) \tag{4.59}$$

We now use these ideas to obtain a more compact form for Δ_2, Eq. (4.54). The first integral in (4.54) may be rewritten as

$$\int_0^{t_c} d\tau \int_0^\tau d\epsilon \int_0^\tau d\phi [[\mathcal{H}(\tau), \mathcal{H}(\epsilon)], \mathcal{H}(\phi)]$$
$$= \int_0^{t_c} d\tau \int_0^\tau d\epsilon \int_0^\epsilon d\phi [[\mathcal{H}(\tau), \mathcal{H}(\epsilon)], \mathcal{H}(\phi)]$$
$$+ \int_0^{t_c} d\tau \int_0^\tau d\phi \int_0^\phi d\epsilon [[\mathcal{H}(\tau), \mathcal{H}(\epsilon)], \mathcal{H}(\phi)]$$
$$\equiv \int_0^{t_c} d\tau \int_0^\tau d\epsilon \int_0^\epsilon d\phi([[\mathcal{H}(\tau), \mathcal{H}(\epsilon)], \mathcal{H}(\phi)]$$
$$+ [[\mathcal{H}(\tau), \mathcal{H}(\phi)], \mathcal{H}(\epsilon)]) \tag{4.60}$$

Utilizing the result

$$[[\mathcal{H}(\tau), \mathcal{H}(\phi)], \mathcal{H}(\epsilon)] + [[\mathcal{H}(\epsilon), \mathcal{H}(\tau)], \mathcal{H}(\phi)]$$
$$+ [[\mathcal{H}(\phi), \mathcal{H}(\epsilon)], \mathcal{H}(\tau)] = 0$$

to evaluate the second set of commutators on the right-hand side of (4.60), we find that (4.54) becomes

$$\Delta_2 = \frac{1}{6} \int_0^{t_c} d\tau \int_0^\tau d\epsilon \int_0^\epsilon d\phi([[\mathcal{H}(\tau), \mathcal{H}(\epsilon)], \mathcal{H}(\phi)]$$
$$+ [[\mathcal{H}(\phi), \mathcal{H}(\epsilon)], \mathcal{H}(\tau)]) \tag{4.61}$$

One therefore finds, with $\eta = 1$,

$$U(t) = \exp(\Delta_0(t) + \Delta_1(t) + \Delta_2(t) + \cdots)$$
$$= \exp\left[\int_0^{t_c} \mathcal{H}(t)\,dt + \frac{1}{2}\int_0^{t_c} d\tau \int_0^\tau d\epsilon [\mathcal{H}(\tau), \mathcal{H}(\epsilon)] \right.$$
$$+ \frac{1}{6}\int_0^{t_c} d\tau \int_0^\tau d\epsilon \int_0^\epsilon d\phi\{[[\mathcal{H}(\tau), \mathcal{H}(\epsilon)], \mathcal{H}(\phi)]$$
$$\left. + [[\mathcal{H}(\phi), \mathcal{H}(\epsilon)], \mathcal{H}(\tau)]\} + \cdots \right] \tag{4.62}$$

which is the Magnus expansion for the solution of the operator differential equation

$$dU(t)/dt = \mathscr{H}(t)U(t) \qquad (4.35)$$

For the present work, we will encounter this equation in the form

$$idU/dt = \mathscr{H}(t)U(t) \qquad (4.63)$$

where $U(t)$ will be a unitary exponential operator that explicitly gives the form of the time development of the density matrix in terms of the Hamiltonian $\mathscr{H}(t)$ responsible for this behavior. Thus, the Magnus expansion allows one to retain the form of exponential operators, but the argument of the exponential becomes a complicated function of the Hamiltonian at various times. The advantage of such a scheme is that it is possible, by limiting the time of attention, to make the first-order term completely dominant for the system. Under these conditions, the form of the operator becomes relatively simple, and the transformations can be performed with relative ease.

D. The Dyson Expression

One further solution for Eq. (4.35) is the Dyson expression

$$U(t_c) = T \exp \int_0^{t_c} \mathscr{H}(\tau)\, d\tau \qquad (4.64)$$

where T is the previously discussed time ordering operator. We require $U(t)$ to be a solution of (4.35), with the usual boundary condition $U(0) = 1$, We try an iterative solution

$$U(t) \simeq U(0) = 1 \qquad (4.65)$$

which yields

$$\frac{dU}{dt} = \mathscr{H}(t) \qquad (4.66)$$

or

$$U(t_c) - 1 = \int_0^{t_c} d\tau\, \mathscr{H}(\tau) \qquad (4.67)$$

Substituting this result back into (4.35) yields

$$\frac{dU}{dt} = \mathscr{H}(t)\left[1 + \int_0^{t_c} d\tau\, \mathscr{H}(\tau) \right] \qquad (4.68)$$

We again integrate and obtain

$$U(t_c) = 1 + \int_0^{t_c} d\tau\, \mathscr{H}(\tau) + \int_0^{t_c} d\tau \int_0^{\tau} d\epsilon\, \mathscr{H}(\tau)\mathscr{H}(\epsilon) \qquad (4.69)$$

We continue in this manner to obtain

$$U(t_c) = 1 + \sum_{n=1}^{\infty} \int_0^{t_c} d\tau \int_0^{\tau} d\epsilon \cdots \int d\chi \, \mathcal{H}(\tau)\mathcal{H}(\epsilon) \cdots \mathcal{H}(\chi) \qquad (4.70a)$$

We note that in order for the integral to make sense, $\tau > \epsilon > \cdots > \chi$; i.e., each integral implies a definite sense of time ordering.

Dyson first recognized that Eq. (4.70a), as discussed in Subsection B, could be expressed as an integral over the entire time from 0 to t_c, with the restriction $\tau > \epsilon > \cdots > \chi$.

If one rewrites Eq. (4.70a) utilizing the Dyson time ordering operator, which takes any product of time-dependent operators and places them in order of decreasing time from left to right, the result is

$$U(t_c) = T\left[1 + \int_0^{t_c} d\tau \, \mathcal{H}(\tau) + \frac{1}{2!} \int_0^{t_c} d\tau \int_0^{t_c} d\epsilon \, \mathcal{H}(\tau)\mathcal{H}(\epsilon) \right.$$
$$\left. + \frac{1}{3!} \int_0^{t_c} d\tau \int_0^{t_c} d\epsilon \int_0^{t_c} d\phi \, \mathcal{H}(\tau)\mathcal{H}(\epsilon)\mathcal{H}(\phi) + \cdots \right] \qquad (4.70b)$$

which we formally identify as

$$U(t_c) = T\left[1 + \int_0^{t_c} d\tau \, \mathcal{H}(\tau) + \frac{1}{2!} \left(\int_0^{t_c} d\tau \, \mathcal{H}(\tau) \right)^2 \right.$$
$$\left. + \frac{1}{3!} \left(\int_0^{t_c} d\tau \, \mathcal{H}(\tau) \right)^3 + \cdots \right] \qquad (4.71)$$

But (4.71) is just the expansion of the exponential

$$U(t_c) = T \exp \int_0^{t_c} d\tau \, \mathcal{H}(\tau) \qquad (4.72)$$

We are always to understand that this integral means (4.70). Thus, the Dyson expression gives a means of determining the time evolution operator at any time t_c. In order to perform the calculation, however, it is necessary to know the precise time dependence of the operator $\mathcal{H}(t)$.

III. Interaction Frames Revisited: Examples of the Use of the Magnus Expansion and the Dyson Expression

The rather involved results of Section II as applied to time development of an ensemble of spins may be summarized as follows:

(a) The time development of the ensemble, under the Hamiltonian

$$\mathcal{H} = \mathcal{H}_0 + \mathcal{H}_1 + \mathcal{H}_2 + \cdots$$

is given by the solution of the equation

$$id\rho/dt = [(\mathcal{H}_0 + \mathcal{H}_1 + \cdots), \rho] \tag{4.73}$$

where the $\mathcal{H}_i$ have been ordered in magnitude such that $\mathcal{H}_i > \mathcal{H}_{i+1}$.

(b) If the unitary transformation

$$\rho(t) = U_0\tilde{\rho}(t)U_0^{-1} \tag{4.74}$$

is made, where U_0 is specified by

$$idU_0/dt = \mathcal{H}_0 U_0, \qquad U_0(0,0) = 1 \tag{4.75}$$

then in the interaction frame of $\mathcal{H}_0$, the time development of ρ is given by

$$id\tilde{\rho}/dt = [(\tilde{\mathcal{H}}_1 + \tilde{\mathcal{H}}_2 + \cdots), \tilde{\rho}] \tag{4.76}$$

In this frame $\tilde{\mathcal{H}}_i$ is given by

$$\tilde{\mathcal{H}}_i = U_0^{-1}\mathcal{H}_i U_0 \tag{4.77}$$

and U_0 is given by the solutions of (4.35), but with $\mathcal{H}(t)$ now replaced by $-i\mathcal{H}_0$, i.e., by the solution of

$$idU_0/dt = \mathcal{H}_0 U_0 \tag{4.78a}$$

which, as we have seen in Section II, is given by either the Magnus expansion or by the Dyson expression. In terms of the Dyson expression, U_0 is given by

$$U_0 = T\exp\left\{-i\int_0^t \mathcal{H}_0(t')\,dt'\right\} \tag{4.78b}$$

Alternatively, an equally valid solution is given by the Magnus expression

$$U_0 = \exp\{-it[\bar{H}_0^{(0)} + \bar{H}_0^{(1)} + \bar{H}_0^{(2)} + \cdots]\} \tag{4.79}$$

where

$$\bar{H}_0^{(0)} = \frac{1}{t}\int_0^t \mathcal{H}_0(t')\,dt' \tag{4.80}$$

$$\bar{H}_0^{(1)} = -\frac{i}{2t}\int_0^t dt'\left[\mathcal{H}_0(t'), \int_0^{t'} dt''\,\mathcal{H}_0(t'')\right] \tag{4.81}$$

$$\bar{H}_0^{(2)} = \frac{1}{6t}\int_0^t dt'\int_0^{t'} dt''\int_0^{t''} dt'''\{[[\mathcal{H}_0(t'), \mathcal{H}_0(t'')], \mathcal{H}_0(t''')]$$

$$+ [[\mathcal{H}_0(t'''), \mathcal{H}_0(t'')], \mathcal{H}_0(t')]\} \tag{4.82}$$

The process of transforming to the interaction frame of $\mathcal{H}_0$ results in a description of the system in which $\mathcal{H}_0$ is absent. If $\mathcal{H}_0$ was the largest interaction determining the behavior of the system, then the transformation to the interaction frame of $\mathcal{H}_0$ results in the system being described in terms of the eigenfunctions of $\mathcal{H}_0$, as specified by the inversion of (4.74); i.e., $\tilde{\rho}(t) = U_0^{-1}\rho(t)U_0$. With $\mathcal{H}_0 \gg \mathcal{H}_1, \mathcal{H}_2, \ldots$, the system may then be described in terms of a perturbation by $\tilde{\mathcal{H}}_1, \tilde{\mathcal{H}}_2, \ldots$ on the eigenstates of $\mathcal{H}_0$. This is a valid description of the system *if* the exponential series representing U_0, i.e., Eq. (4.78) or (4.79), rapidly converges. If $\mathcal{H}_0$ is not a function of time, then only the first term in the Magnus expansion does not vanish, and the exact expression for U_0 is

$$U_0 = \exp(-i\mathcal{H}_0 t) \tag{4.83}$$

With $\mathcal{H}_0$ the Zeeman interaction, which is time-independent, the transformation to the frame of $\mathcal{H}_0$ is that to the observation frame, i.e., to the rotating frame. Subsequent transformations to the frames of $\tilde{\mathcal{H}}_1, \tilde{\mathcal{H}}_2$, etc., may be performed to remove $\tilde{\mathcal{H}}_1, \tilde{\mathcal{H}}_2, \ldots$ from a description of the system in the interaction frame in question, until ultimately a frame is attained in which $^n\tilde{\rho}$ is time-independent or dependent on only the interaction one wishes to observe, plus small perturbations. If it is then possible to relate this final interaction frame to the measurement frame, then only the effect of the interaction one wishes to observe will be measured. It is the job of $\mathcal{H}_1 = \mathcal{H}_{rf}$, the time-dependent radio-frequency Hamiltonian set by the experimenter, to effect this result. The details of this operation will be developed in Chapters 5 and 6. For the present, we perform some simple, but important, calculations exhibiting the use of the Magnus and Dyson solutions to (4.75) as an introduction to the power of the expressions in helping to determine the results of pulse NMR experiments.

Consider a spin system placed in a static magnetic field in the z direction. How does ρ describe the motion in the laboratory frame? The basic formula is that ρ develops in time as

$$\rho(t) = U(t, 0)\rho(0)U^+(t, 0)$$

Here U is determined by (4.75) with $\mathcal{H}(t) = -\gamma \hbar B_0 I_z \equiv V(t)$. First consider use of the Magnus expansion

$$U(t, 0) = \exp\left\{ -i \int_0^t -\gamma \hbar B_0 I_z \, dt + \frac{1}{2} \int_0^t d\tau \int_0^\tau d\epsilon [\mathcal{H}(\tau), \mathcal{H}(\epsilon)] + \cdots \right\} \tag{4.84}$$

but

$$[\mathcal{H}(\tau), \mathcal{H}(\epsilon)] \sim [I_z, I_z] \equiv 0 \tag{4.85}$$

so only the first term in the series is nonzero. Therefore,

$$U(t, 0) = \exp i\gamma B_0 I_z t = \exp i\omega_0 I_z t \tag{4.86}$$

We have seen, however, that this is just the form of an operator for rotation about the z axis by $\theta = \omega_0 t$, so we have the by-now-familiar result that in a static field, the spin system precesses about the z axis, with precession frequency $\omega_0 = -\gamma B_0$.

Next consider use of the Dyson expression to calculate $U(t, 0)$:

$$U(t,0) = 1 + \frac{1}{i} \int_0^t d\tau \, V(\tau) + \frac{1}{i^2} \int_0^t d\tau \int_0^\tau d\epsilon \, V(\tau)V(\epsilon) + \cdots$$

$$= 1 + \frac{1}{i}(-\gamma B_0 I_z t) + \frac{1}{2!(i)^2}(-\gamma B_0 I_z t)^2 + \cdots \tag{4.87}$$

This is just the expansion of $\exp(i\omega_0 I_z t)$ as before, and again the result is precession of the spin system about the z axis. For example, suppose that the ensemble of spins were equilibrated in a magnetic field in the z direction, as originally postulated. Then the operator form of the density matrix would linearly contain I_z as the operator, and one would find that the density operator develops in time as

$$\rho(t) \sim \exp(iI_z\theta)I_z \exp(-iI_z\theta) = I_z \tag{4.88}$$

to no great surprise.

On the other hand, suppose at $t = 0$ an intense pulse of rf power is directed with the B_1 field along the y axis of the rotating frame. The density operator at this point transforms as

$$\rho(t = 0) \rightarrow \exp[(i(\pi/2)I_y]I_z \exp[-i(\pi/2)I_z] = -I_x \tag{4.89}$$

(We will more fully discuss the meaning of the arrow in this transformation later.) Following this pulse, the system is allowed to develop in time under a static field δB corresponding to the offset frequency $\delta\omega = -\gamma\delta B$. Then the density operator develops according to

$$\rho(t) \sim \exp(iI_z\delta\omega t)(-I_x)\exp(-iI_z\delta\omega t) \tag{4.90}$$

Use of Table 2.1 yields

$$\rho(t) = -I_x \cos(\delta\omega t) + I_y \sin(\delta\omega t) \tag{4.91}$$

As previously discussed, the observable is the expectation value of the magnetization in the xy plane of the rotating frame, or, $\langle \mathbf{M} \rangle \equiv \gamma\hbar\langle \mathbf{I} \rangle$. As seen in the discussion of the utility of the density operation,

$$\langle \mathbf{I} \rangle = \text{Tr}(\rho \mathbf{I})$$

More specifically,

$$\langle I(t) \rangle = \text{Tr}(\rho(t)\mathbf{I}) \tag{4.92}$$

Note that if $\rho(t) \equiv \rho(0)$, i.e., if the density operator feels no interactions causing it to evolve in time, then $\langle I(t) \rangle \equiv \langle I(0) \rangle$. A specification of $\rho(t)$ is

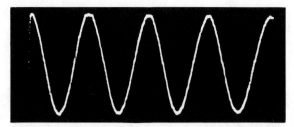

FIG. 4.4. The free-induction decay of the protons in H_2O. The carrier frequency is offset from the water resonance by 1 kHz.

therefore frequently all that is needed to inform us about $\langle I(t) \rangle$; i.e., it is frequently not necessary to calculate $\text{Tr}(\rho I)$ explicitly, and since the calculation involves time and effort, it is made only when needed. In the present case, we *are* interested in the expectation value of **I** at time t, so we must evaluate $\text{Tr}(\rho(t) \cdot \mathbf{I})$. Now $\mathbf{I} = I_0(\mathbf{i}I_x + \mathbf{j}I_y + \mathbf{k}I_z)$ with I_0 being the magnitude of the magnetization caused by the dc field. The x, y, and z components are evaluated separately. The angular momentum operator for a spin-$\frac{1}{2}$ system is related to the Pauli spin matrix vector by

$$\mathbf{I} = \tfrac{1}{2}\boldsymbol{\sigma} \tag{4.93}$$

so the components of **I** are therefore easily seen to obey the relations

$$\text{Tr}\,I_i = 0, \quad i = x, y, z; \quad \text{Tr}\,I_i I_j = \delta_{ij}/2 \tag{4.94}$$

Therefore

$$\langle I_z(t) \rangle = \text{Tr}(\{-I_x \cos(\delta\omega t) + I_y \sin(\delta\omega t)\}I_z) = 0$$

$$\langle I_x(t) \rangle = \text{Tr}(\{-I_x \cos(\delta\omega t) + I_y \sin(\delta\omega t)\}I_x) = -I_0 \cos\delta\omega t \tag{4.95}$$

$$\langle I_y(t) \rangle = I_0 \sin\delta\omega t$$

The expectation value of **I** is then given by

$$\langle \mathbf{I}(t) \rangle = I_0[-\mathbf{i}\cos(\delta\omega t) + \mathbf{j}\sin\delta\omega t] \tag{4.95a}$$

Therefore, if we "look" at the component of angular momentum along the x axis in the rotating frame, a signal that is proportional to $\cos\delta\omega t$, is detected as shown in Fig. 4.4, taken from an actual FID on ^1H in water at an offset of 1 kHz. The motion of **I** is simply a rotation in the x–y plane at angular velocity $\delta\omega$. Note that Fig. 4.4 is a bit uncharacteristic of what is expected for a FID because there is no apparent exponential decay. There is an excellent physical reason for this apparent violation of what is expected to be physical reality. The total time scan of Fig. 4.4 is 4 msec, and T_2 for the water sample on which the data in Fig. 4.4 was taken is 50 msec. Over the time scale of 4 msec, therefore, there is no appreciable decay; i.e., on the time scale τ, of the experiment,

$$\tau |\mathscr{H}_{\text{int}}| \ll 1 \qquad (4.96)$$

where $|\mathscr{H}_{\text{int}}|$ are the magnitude of the internal interactions responsible for spin–spin and spin–lattice relaxation. The inequality is critical to an understanding of multiple pulse experiments to remove internal interactions, as will be seen in Chapter 5.

As another example, we calculate the response of a spin system to a radio-frequency pulse as a function of the offset, where $\delta\omega = \omega_0 - \omega_z$, and of the pulse width t_p. The desired result will be that the initial amplitude (the amplitude immediately following the pulse) of the observed signal will decrease in an oscillatory fashion as the offset increases at fixed t. In order to derive this result it is necessary to realize that a pulse along x leads to an observable signal along y as discussed in Problem 4.5 at the end of this chapter. The amplitude of this signal will be proportional to $\langle I_y \rangle$, given by

$$\langle I_y \rangle = \text{Tr}\, \rho I_y \qquad (4.97)$$

One therefore wishes to evaluate $\langle I_y \rangle$ at a time corresponding to the end of the pulse. (Further examination will show why we do not choose $t_p = \pi/2\omega$.) The desired calculation, therefore, is that of $\rho(t_p)$.

In the rotating frame of the radio-frequency interaction having magnitude $\gamma B_1 = \omega_1$ and frequency ω_z, the Hamiltonian is given by equation

$$\mathscr{H} = -(\delta\omega I_z + \omega_1 I_x) \qquad (4.98)$$

The program is to determine ρ through using Eq. (2.62) and then $\langle I_y \rangle$ by utilizing Eq. (2.84) since the Hamiltonian in Eq. (4.98) is constant in the frame of observation and the evolution of ρ may be evaluated directly:

$$\rho(t) = \exp[i(\delta\omega I_z + \omega_1 I_x)t_p]\rho(0)\exp[-i(\delta\omega I_z + \omega_1 I_x)t_p] \qquad (4.99)$$

Since I_x and I_z do not commute, however, the exponential operator may not be simply expressed as a product. One way to solve this problem is to rotate the coordinate system such that the exponential operator in the rotated frame becomes a function of a single angular momentum operator, e.g., I_z'. Then, $\rho'(t)$ is evaluated in the rotated frame. The inverse rotation then yields the result in the frame of observation. This is to say that we wish to transform into a frame in which $\delta\omega I_z + \omega_1 I_x \to C I_z'$, where C is some constant. There are a number of ways of viewing the situation to determine which transformation is appropriate. One is to realize that the Hamiltonian $-(\delta\omega I_z + \omega_1 I_x)$ corresponds to an effective field

$$\mathbf{B}_{\text{eff}} = \delta\omega\mathbf{k} + \omega_1\mathbf{i} \qquad (4.100)$$

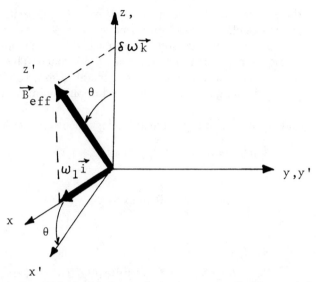

FIG. 4.5. The effective field during a radio-frequency pulse off resonance. $\sin\theta = \omega_1/(\omega_1^2 + \delta\omega^2)^{1/2}$ and $\cos\theta = \delta\omega_1/(\omega_1^2 + \delta\omega^2)^{1/2}$.

as shown in Fig. 4.5. A clockwise rotation of the coordinate system about the y axis by $\sin\theta = \omega_1/(\omega_1^2 + \delta\omega^2)^{1/2}$ (indicated in Fig. 4.5) will place the z' axis along the effective field. In the rotated frame, the Hamiltonian will be proportional to I_z'. The angle θ is given by

$$\theta = \text{arc sin } \omega_1/(\omega_1^2 + \delta\omega^2)^{1/2} \tag{4.101}$$

The operator for this rotation is given by the exponential form

$$U = \exp(-i\theta I_y) \tag{4.102}$$

The Hamiltonian in the rotated frame becomes (see Section I)

$$\mathscr{H}' = U^+ \mathscr{H} U = -\exp(i\theta I_y)(\delta\omega I_z + \omega_1 I_x)\exp(-i\theta I_y) \tag{4.103}$$

Use of Table 2.1 yields

$$\mathscr{H}' = -(\delta\omega^2 + \omega_1^2)^{1/2} I_z \equiv -\omega_e I_z \tag{4.104}$$

with ω_e being the effective frequency in the rotating frame. The density matrix in the rotated frame is

$$\rho'(0) = U^+ \rho(0) U = \exp(i\theta I_y)\rho(0)\exp(-i\theta I_y) \tag{4.105}$$

At this point there are at least two options in continuing the calulation. The first is to adopt an operator form for $\rho(0)$ immediately and to proceed as in the preceding example. The second is to leave $\rho(0)$ unspecified and to carry through the calculation in a more general form. We choose to exercise the second option for two reasons. The first is that some valuable calculational techniques are illustrated by such an approach. The second is that the answer is obtained in a much more general form involving no assumptions such as the magnitude of $\mathcal{H}/kT$ and about thermal equilibrium at $t = 0$.

In the rotated frame, since $\mathcal{H}'$ is again independent of time,

$$\rho'(t_p) = \exp(-i\mathcal{H}'t_p)\rho'(0)\exp(i\mathcal{H}'t_p) \tag{4.106}$$

But since

$$\rho = \exp(-i\theta I_y)\rho'\exp(i\theta I_y) \tag{4.107}$$

we find

$$
\begin{aligned}
\langle I_y(t_p)\rangle &= \mathrm{Tr}\,\rho(t_p)I_y \\
&= \mathrm{Tr}\exp(-i\theta I_y)\exp(-i\mathcal{H}'t_p)\rho'(0)\exp(i\mathcal{H}'t_p)\exp(i\theta I_y)I_y \quad (4.108)
\end{aligned}
$$

A useful property of the trace of a product of matrices is that cyclic permutation does not affect the result; i.e.,

$$\mathrm{Tr}\,ABC = \mathrm{Tr}\,CAB = \cdots$$

With this in mind, (4.108) can be arranged in the form

$$\langle I_y(t_p)\rangle = \mathrm{Tr}\exp(i\omega_e I_z t_p)I_y\exp(-i\omega_e I_z t_p)\exp(i\theta I_y)\rho(0)\exp(-i\theta I_y) \tag{4.109}$$

In obtaining (4.109), the relations

$$\exp(i\theta I_y)I_y\exp(-i\theta I_y) = I_y \tag{4.110}$$

and

$$\rho'(0) = \exp(i\theta I_y)\rho(0)\exp(-i\theta I_y) \tag{4.111}$$

were utilized.

Use of Table 2.1 then gives

$$
\begin{aligned}
\langle I_y(t_p)\rangle &= \mathrm{Tr}[I_y\cos(\omega_e t_p) + I_x\sin(\omega_e t_p)]\exp(-i\theta I_y)\rho(0)\exp(i\theta I_y) \\
&= \mathrm{Tr}\exp(i\theta I_y)[I_y\cos(\omega_e t_p) + I_x\sin(\omega_e t_p)]\exp(-i\theta I_y)\rho(0)
\end{aligned}
$$
$$\tag{4.112}$$

Again, use of Table 2.1 results in

$$\langle I_y(t_p)\rangle = \mathrm{Tr}[I_y\cos(\omega_e t_p) + I_x\cos\theta\sin(\omega_e t_p) + I_z\sin\theta\sin(\omega_e t_p)]\rho(0) \tag{4.113}$$

The generality of the treatment becomes apparent when it is seen that (4.113) is just

$$\langle I_y(t_p)\rangle = \cos(\omega_e t_p)\langle I_y(0)\rangle$$
$$+ \langle I_x(0)\rangle \cos\theta \sin(\omega_e t_p) + \langle I_z(0)\rangle \sin\theta \sin(\omega_e t_p) \quad (4.114)$$

This result allows us to obtain the result of the excitation corresponding to $\mathscr{H}$ for *any* initial values of $\langle I_x\rangle$, $\langle I_y\rangle$, and $\langle I_z\rangle$. In particular, if the system is prepared by soaking in a static dc field along z, then $\langle I_x(0)\rangle = \langle I_y(0)\rangle = 0$, and we call $\langle I_z(0)\rangle$ by the name I_0 (with operator form *still* unspecified!)

$$\langle I_z(t_p)\rangle = I_0 \sin\theta \sin\omega_e t_p$$
$$= \frac{\omega_1 I_0}{(\omega_1^2 + \delta\omega^2)^{1/2}} \sin(\omega_1^2 + \delta\omega^2)^{1/2} t_p \quad (4.115)$$

It is now convenient to look at $\langle I_y(t_p)\rangle$ as a function of the ratio of the radio frequency to the offset field, $\mathscr{A} = \omega_1/\delta\omega$. Equation (4.115) then becomes

$$\langle I_y(t_p)\rangle = \frac{I_0\mathscr{A}}{(1 + \mathscr{A}^2)^{1/2}} \sin(1 + \mathscr{A}^2)^{1/2}\delta\omega t_p \quad (4.116)$$

Very close to resonance, $\mathscr{A}^2 \gg 1$, and

$$\langle I_y(t_p)\rangle = I_0 \sin\omega_1 t_p \quad (4.117)$$

Under this condition, the value of I_y at the end of the pulse is maximized if t_p is chosen to satisfy

$$\omega_1 t_p = \pi/2 \quad (4.118)$$

the usual condition for a 90° pulse. However, as $\delta\omega$ becomes large compared to ω_1, we find

$$\langle I_y(t_p)\rangle = I_0(\omega_1/\delta\omega) \sin\delta\omega t \quad (4.119)$$

The amplitude of the signal immediately following the pulse now decreases with increasing offset and oscillates at a frequency $\delta\omega$. The maximum amplitude is no longer found by the condition $\pi/2 = \omega_1 t_p$, but by the condition

$$\delta\omega t_p = \pi/2 \quad (4.120)$$

A "90° pulse" therefore depends upon offset!

In order for a pulse to be precise, it must be short compared to the time scale of other interactions. This has its origin in the fact that the pulse only has finite intensity in a band around ω_0, the transmitter frequency. The band is roughly of the order of $2/t_p\omega_0$. For microsecond pulses this means the radio frequency affects only spins that resonate within about ± 1 MHz of the center frequency ω_0. If it is not large, substantial effects occur that are determined by other interactions, as in the preceding calculations of the effect of offset on the "90° pulse width."

IV. Secular Perturbations

In perturbation treatments, whatever form the operators may take, one dominant theme is that the time evolution is determined successively by the largest perturbation, the next largest, and so on down the line. By choosing the appropriate frame of reference, one may be able to ignore those large parts and focus on small parts. However, large interactions are still present. For example, the fact that $\mathcal{H}_Z$ is much larger than any other Hamiltonian leads one to describe the state of a system in terms of the eigenstates of $\mathcal{H}_Z$ despite the fact that at certain times (after a pulse) it is obvious that the system is nowhere near an eigenstate of $\mathcal{H}_Z$. If the state of the system after the pulse had been either α or β, the expectation value of both I_x and I_y would have been zero, and no signal would have been observed. There is a completely remarkable aspect to the behavior of the system perturbed by an interaction *very small* compared to the Zeeman interaction. This is that *in the rotating frame, on resonance*, this small interaction completely dominates the observed time evolution. Another way of stating this fact is that the eigenstates of the system are no longer determined by the largest interaction, but by the small perturbing Hamiltonian. The reason this is at first thought surprising is that physical intuition would lead us to infer that the motion of, e.g., a subway train would be dominated by the action of its driving engines and very little affected by the action of passengers walking in the train. The driving of a nuclear spin system from one state to another by an rf field small compared to the static Zeeman field would appear to be a bit like passengers in a moving train causing the train to jump from one track to another just by rocking the train. It seems reasonable that passengers could slightly change the speed of the train by walking either with, or against the motion caused by the engine, but a bit farfetched to expect that they could cause any displacement perpendicular to the train's velocity that would alter this vector. In a quantum-mechanical system, the analogy is that the eigenstates are predominantly determined by the largest interaction, in our case $\mathcal{H}_Z$. If there are other, relatively small interactions, they may be classified into groups that commute with the largest interaction (and thus possess the same eigenstates) and into terms that do not commute with $\mathcal{H}_Z$. Terms commuting with $\mathcal{H}_Z$, i.e., terms diagonal in the representation used to classify

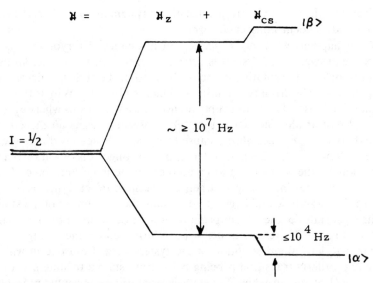

FIG. 4.6. The effect of the chemical shift on the energy levels of a spin in a magnetic field. The actual shifts are much smaller than indicated in the diagram.

the unperturbed system, are called "secular." Others are called "nonsecular." The secular terms obviously can only cause energy level shifts, i.e., changes in a *given* eigenvalue. The nonsecular terms, on the other hand, are the interactions responsible for transitions from one eigenstate to another. For example, a nuclear spin system in a static field has Zeeman Hamiltonian $\mathscr{H}_Z = -\omega_0 I_z$. The chemical-shift Hamiltonian $\mathscr{H}_{cs}$ also turns out to be proportional to I_z. We have seen that the form of $\mathscr{H}_{cs}$ is $\sigma_{zz}\omega_0 I_z$. Under normal conditions, $\omega_0 \geq 10^7$ Hz, and $\sigma_{zz}\omega_0 \leq 10^4$ Hz. Since $[\mathscr{H}_Z, \mathscr{H}_{cs}] = 0$, the states of the system experiencing both $\mathscr{H}_Z$ and $\mathscr{H}_{cs}$ are still α and β for a single spin. The energies, however, will be those of $\mathscr{H}_Z$ slightly changed by $\mathscr{H}_{cs}$, as shown in Fig. 4.6.

Turning on the rf interaction, however,

$$\mathscr{H}_{rf} = -\omega_1 \exp(-iI_z\omega_Z t)I_x \exp(iI_z\omega_Z t)$$

with magnitude of $|\omega_1|/\omega_0 = 10^{-4}$ causes transitions from $|\alpha\rangle$ to $|\beta\rangle$, as we saw in the previous section! The *nonsecular* terms are responsible for quantum jumps, i.e., for resonance.

A feature not really explored in the discussion of a quantum-mechanical description of a $\pi/2$ pulse was that $\omega_Z = \omega_0$. Under this condition, the Zeeman operator disappeared from the description of the system in the interaction frame in question, and the nonsecular term was time-independent.

In this frame, therefore, the eigenstates of the system are *only* determined by the time-independent nonsecular term.

What happens when $\omega_z \neq \omega_0$, i.e., off resonance? Everyone physically knows the answer. If the system is driven "slightly" off resonance, an oscillating response in the rotating frame is observed. If the system is driven with a frequency *very* far from resonance, no signal is observed. Why is this true? Is it that the nonsecular terms simply do not cause transitions when ω_z is not near ω_0? Not at all! The fact is that the nonsecular terms *always* connect different states of $\mathcal{H}_z$, i.e., *always* cause transitions. When ω_z differs greatly from ω_0, however, the Zeeman interaction no longer disappears from the description of the system's motion in the interaction frame of $\omega_z I_z$. As $|\omega_0 - \omega_z|$ becomes increasingly larger, the Zeeman offset term $(\omega_0 - \omega_z)I_z$ increases in magnitude and begins to dominate the description of the system; i.e., the eigenstates of the system begin to now be determined by the Zeeman offset interaction. Therefore, as $|\delta\omega| = |\omega_c - \omega_z|$ becomes increasingly larger compared to $|\omega_1|$, the description of the system turns from one in which a sedate and orderly transition is being made from state α to state β with frequency $\omega_1/2$ to one in which the state is primarily an eigenfunction of I_z, with a relatively small but very-high-frequency oscillation. In fact, as was demonstrated in Section III the amplitude of the observed signal decreases with increasing offset $\delta\omega$ and oscillates with $\delta\omega$ and the pulse width t_p, according to the relation

$$\text{observed amplitude} \propto \left[\omega_1/(\omega_1^2 + \delta\omega^2)^{1/2}\right] \sin(\omega_1^2 + \delta\omega^2)^{1/2} t_p$$

In fact, what is happening is that when the Hamiltonian in the observation frame contains a large secular term and a small nonsecular term, the eigenstates will be determined by the secular term. The effect of the nonsecular term will be to cause the eigenstates to oscillate about the mean energies associated with the secular Hamiltonian. The further ω_z is from ω_0, the more rapid is this oscillation. When the oscillation becomes sufficiently rapid compared to the time of observation, then one can only observe differences in the average values of the energies, i.e., differences in the eigenstates from those of the secular portion of the Hamiltonian.

Alternatively, what if the secular portions of $\mathcal{H}$ are small compared to the nonsecular terms in our frame of observation? The obvious thing to do is to transform to a frame in which the largest terms are secular and the smallest terms are nonsecular and then arrange to observe from this frame. In this case the eigenstates will again be determined by the large, secular interaction, and the nonsecular terms will cause oscillations that will not be so observable under appropriate conditions of offset. An example of such a situation will be given in Chapter 6 when continuous-wave decoupling is discussed.

PROBLEMS

4.1 Show that $\exp[i(I^2 - 3I_z^2)\theta] = \exp(-iI^2\theta)\exp(3iI_z^2\theta)$ (θ represents some angle).

4.2 Consider the integral $\int_0^a dx \int_0^x dy\, xy$.
  (a) Sketch the area in the xy plane corresponding to this integral.
  (b) Evaluate this integral.

4.3 Consider the pulse sequence $[\tau/2, \pi_y, \tau, \pi_y, \tau/2]$, where π_y represents a delta-function pulse along y in the frame of the Zeeman interaction (the rotating frame).

  (a) If $\omega_{1y} = \gamma B_{1y}$ measures the magnitude of the B_1 field in the y direction, evaluate $\mathcal{H}_{rf}$ at times (i) $0 \le t < \tau/2$, (ii) $t = \tau/2$, (iii) $\tau/2 < t < 3\tau/2$, (iv) $t = 3\tau/2$, (v) $3\tau/2 < t \le 2\tau$.

  (b) If U_{rf} is determined by the differential equation

$$idU_{rf}/dt = \mathcal{H}_{rf}U_{rf}$$

(i) calculate $\bar{H}_{rf}^{(0)}$, at time $t = 2\tau$; (ii) evaluate $[\mathcal{H}_{rf}(t), \mathcal{H}_{rf}(t')]$ for all times t, t' between $0 \le t \le 2\tau$ and $0 \le t' \le 2\tau$; (iii) show that $\bar{H}_{rf}^{(1)}(2\tau) = 0$; (iv) show that $\bar{H}_{rf}^{(k)} = 0$, all k; (v) show what the value of $U_{rf}(2\tau, 0)$ is.

4.4 Evaluate $U_{rf}(2\tau, 0)$ for the pulse sequence given in Problem 4.3 by using the Dyson expression.

4.5 (a) Show that $\exp[i(I_x + I_y)\theta] \ne \exp(iI_x\theta)\exp(iI_y\theta)$.

  (b) Show that $R_k \exp(iI_j\theta) = \exp(iR_kI_j\theta)$, where R_k is a rotation about some axis k. (*Hint*: Expand the exponential and realize $R_kI_j^2 = R_kI_zR_k^{-1}R_kI_z$, etc.)

  (c) Show that $\exp(iI_x\pi/2)\exp(iI_y\theta)\exp(-iI_x\pi/2) \equiv R(\pi/2)_x\exp(iI_y\theta) = \exp(iI_z\theta)$.

  (d) Show that $\exp(iI_z\pi/2)\exp[i(I_x + I_y)\theta]\exp(-iI_z\pi/2) = \exp(iI_x\theta)$.

  (e) Evaluate $\exp[i(I_x + I_y)\theta]I_z\exp[-i(I_x + I_y)\theta]$ by using the ideas from a–d.

REFERENCE

1. R. M. Wilcox, Exponential operators and parameter differentiation in quantum physics, *J. Math. Phys.* **8**, 962 (1967).

CHAPTER 5

HOMONUCLEAR PULSE
NMR EXPERIMENTS

I. Introduction

In previous chapters we have developed a mechanism for turning the crank to calculate the results of pulse NMR experiments. We have also, it is hoped, developed a physical feeling for how the machine to which the crank is attached is constructed, and why the individual gears work as they do. This chapter is involved primarily with cranking to illustrate how the machine works for a number of experiments in which the emphasis is irradiation and observation of a single type of spin, e.g., ^1H and the selective removal of interactions that contribute to the proton linewidth.

We first discuss the concept of the "average Hamiltonian," which is of critical importance in the understanding of a broad range of pulse experiments. The basic idea involved is that the time development of the density matrix is described by Eq. (2.95)

$$\rho(t) = U\rho(0)U^+$$

If the rf Hamiltonian responsible for the form of U (via either the Dyson or Magnus expansion) satisfies appropriate symmetry conditions in time, *and if the time interval over which this Hamiltonian acts between observations is short compared to T_2^**, then at selected observation times the spin system acts as if it were exposed to an internal Hamiltonian averaged over time for a given sequence of rf pulses.

After developing the idea of the average Hamiltonian, we discuss the use of the concept in a number of multiple-pulse experiments. We start with some sample calculations appropriate to pulse tuning sequences. Hamiltonians for pulse imperfections are considered, and the calculation of the effect of some of these interactions on the experimentally observed tuning sequences is given. In particular, the effect of an offset Hamiltonian on a pulse phase-tuning sequence is calculated to zeroth order. Also, the effect of the cross term between an offset and a phase error on the same sequence is calculated to first order. The calculations are carried through in fairly complete detail so that the reader may have an easy time in following the arithmetic. No apology is made for this treatment. The calculations are messy, and it is easy to lose sight of the use of the theory when the symbolism becomes tedious. This text is aimed at those who are not comfortable yet with these types of calculations, and we feel that at least one example of a completely worked problem is necessary for an unambiguous exposition of the theory. Any operation at all subtle, therefore, has been illustrated in detail. An

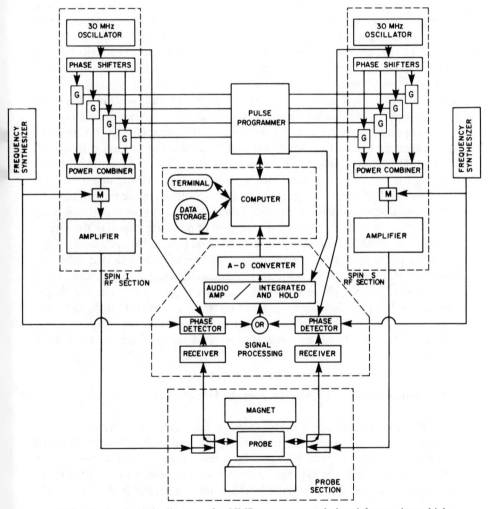

FIG. 5.1. A schematic diagram of a NMR spectrometer designed for use in multiple-pulse experiments.

extremely valuable connection is made between average Hamiltonians that are linear in angular momentum operators and a physical picture that allows one to visualize the result of these interactions on a pulse experiment. The reader is urged to read this material carefully, and, if necessary, as many times as is needed to obtain a picture of the physics of the situation. Problem 5.12 should help in obtaining expertise in this type of visualization.

We then proceed to describe a number of pulse experiments, some of which are now standard and others that are just coming into use, in the average Hamiltonian formulation. Before proceeding with this description, it might be useful to examine an example of an apparatus used for multiple-pulse experiments. A schematic of one such piece of equipment is given in Fig. 5.1. This unit is just an extension of the apparatus shown in Fig. 1.8. There are four important differences. The first is that instead of one frequency channel, there are two, for simultaneous irradiation of nuclei I and S, allowing the type of heteronuclear spin experiments discussed in Chapter 6. The second is that each frequency is produced by mixing the signal of a 30-MHz oscillator with that of a frequency synthesizer. The third is that the 30-MHz signal, prior to sidebanding, is divided into four (or more, if desired) channels, each of which may have an adjustable phase. The manner of producing the phase adjustment depends upon the type of experiment in question. For multiple-quantum experiments, not discussed in the present context, the phase shifting may be performed digitally. For the experiments described in this chapter, where precise, stable control of phases is required, the phases may be adjusted by using varying lengths of coaxial cable and constant-impedance phase trimmers or with some other type of analog device. Switches, or gates (G), activated by a pulse programmer, control the sequencing of pulses with a given phase used in a particular experiment. The fourth difference is that the NMR signal proceeding from the receiver, or video amplifier, may be stroboscopically observed at time periods, or windows, fixed by the experimenter via an integrate-and-hold circuit flagged by the pulse programmer.

II. Multiple-Pulse NMR and the Average Hamiltonian

We now specifically use the ideas of Chapter 4 to develop a useful method of viewing the effects of multiple-pulse experiments on the observed response of a spin system. We consider a spin system in the rotating frame (the frame in which the experimenter observes the results of a pulse experiment, as indicated in Chapter 1). In this frame, the Hamiltonian is

$$\mathscr{H} = \mathscr{H}_{rf}(t) + \mathscr{H}_{int} \tag{5.1}$$

where $\mathcal{H}_{rf}$ is the interaction associated with the sequence of rf pulses, and the internal Hamiltonian $\mathcal{H}_{int}$ is given by

$$\mathcal{H}_{int} = \mathcal{H}_O + \mathcal{H}_{cs} + \mathcal{H}_D \tag{5.2}$$

Here $\mathcal{H}_O$ is the resonance offset Hamiltonian, with operator $\Delta\omega I_z$, $\mathcal{H}_{cs}$ is the chemical-shift Hamiltonian, with operator proportional to I_z, and $\mathcal{H}_D$ is the dipolar Hamiltonian, as discussed in Chapter 3.

In the rotating frame, the equation of motion of ρ is

$$i\frac{d\rho}{dt} = [(\mathcal{H}_{rf} + \mathcal{H}_{int}), \rho] \tag{5.3}$$

where we may identify $\mathcal{H}_{rf}$ and $\mathcal{H}_{int}$ with $\mathcal{H}_O$ and $\mathcal{H}_1$ of (4.73). We define an operator U_{rf}, which obeys the Liouville equation

$$i\frac{dU_{rf}}{dt} = \mathcal{H}_{rf} U_{rf} \tag{5.4}$$

subject to $U_{rf}(0,0) = 1$. We make the transformation

$$\rho(t) = U_{rf}\tilde{\rho}(t)U_{rf}^+ \tag{5.5}$$

The motion of $\tilde{\rho}$ in the new frame is governed by

$$i\frac{d\tilde{\rho}}{dt} = [\tilde{\mathcal{H}}_{int}, \tilde{\rho}] \tag{5.6}$$

with

$$\tilde{\mathcal{H}}_{int}(t) = U_{rf}(t)^+ \mathcal{H}_{int} U_{rf}(t) \tag{5.7}$$

where U_{rf} is the solution of (5.4), given by

$$U_{rf} = T\exp\left\{-i\int_0^t d\tau\,\mathcal{H}_{rf}(\tau)\right\} \tag{5.8}$$

or by the Magnus expansion.

We now inquire, "What simplifications result in the description of $\rho(t)$ by imposing symmetry restrictions upon $\mathcal{H}_{rf}$?" Specifically, we insist that $\mathcal{H}_{rf}$ be a *periodic series of groups* of a *cyclic sequence* of a small number n (usually less than 10) of pulses. A *cyclic sequence* is defined by the requirement that the propagation operator $U_{rf}(t_c,0)$ defined by this sequence at a cycle time t_c, through (5.8), be unity;

$$U_{rf}(t_c,0) = T\exp\left[-i\int_0^{t_c}\mathcal{H}_{rf}(t)\,dt\right] = \pm 1. \tag{5.9}$$

Using the Magnus expansion to express the solution of (5.4), we see that a sufficient condition for designing a cyclic sequence is by the requirement that

$$\bar{H}_{\text{rf}}^{(i)} = 0 \tag{5.10}$$

be met for all i. It is therefore useful to explicitly examine the first two terms of the Magnus expansion to decide how to construct cyclic sequences of rf pulses;

$$\bar{H}_{\text{rf}}^{(0)} = \frac{1}{t_c} \int_0^{t_c} H_{\text{rf}}(t)\, dt \tag{5.11a}$$

$$\bar{H}_{\text{rf}}^{(1)} = \frac{-i}{2t_c} \int_0^{t_c} dt \int_0^{t} dt' \left[\mathcal{H}_{\text{rf}}(t),\ \mathcal{H}_{\text{rf}}(t') \right] \tag{5.11b}$$

Clearly if $\mathcal{H}_{\text{rf}}$ is *phase alternated*, such that for each pulse with rotation angle θ about the direction k corresponding to angular momentum operator I_k, there is a subsequent pulse about k with rotation angle *minus* θ, then $\bar{H}_{\text{rf}}^{(0)} = 0$ and to zeroth order, $U_{\text{rf}}(t_c, 0) = 1$. An example of such a sequence might be a 90° pulse along x, followed by a 90° pulse along $\bar{x}$, which we designate as $[\tau, P_x, 2\tau, P_{\bar{x}}, \tau]$. In this case the cycle time is $t_c = 4\tau$, and for ideal delta-function pulses it may be seen by direct integration that $\bar{H}_{\text{rf}}^{(0)} = (\pi/2)[I_x - I_x]/4\tau = 0$. It is also easy to see that since H_{rf} contains only I_x operators, $\bar{H}_{\text{rf}}^{(1)}$ vanishes, since I_x commutes with itself. Similarly, $\bar{H}_{\text{rf}}^{(i)}$ vanishes. The requirement that a sequence be *periodic* is

$$\mathcal{H}_{\text{rf}}(t) = \mathcal{H}_{\text{rf}}(t + Nt_c) \tag{5.12}$$

which is to say that the sequence is periodically repeated with a cycle time t_c. For a sequence both cyclic and periodic, we see that

$$U_{\text{rf}}(Nt_c, 0) = 1$$

where $N = 0, 1, 2, \ldots$.

It is important to note at this point that the famous sequence shown in Fig. 5.2, the WAHUHA sequence [1], with a cycle time of 6τ (which is a combination of "dipolar echo" sequences [2] $[\tau, P_x, \tau, P_y, \tau]$) is a phase-alternated sequence in which clearly $\bar{H}_{\text{rf}}^{(0)} = 0$. Because of the lack of commutation between I_x and I_y, however, $\bar{H}_{\text{rf}}^{(i)}$ may not vanish for higher-order terms. The student will benefit from proving that for such a phase-alternated sequence of ideal delta-function pulses $\bar{H}_{\text{rf}}^{(1)} = 0$, but $\bar{H}_{\text{rf}}^{(2)} \neq 0$. In this chapter, we shall nevertheless utilize such phase-alternated sequences in our discussions of homonuclear dipolar decoupling. We realize that these sequences are only cyclic at best to first order, but we allow the experimental results to speak for themselves.

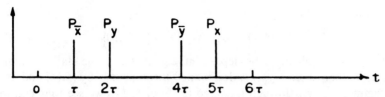

FIG. 5.2. The WAHUHA four-pulse experiment for suppressing homonuclear dipole–dipole interactions. Ideal delta-function pulses are assumed.

Now to see how the cyclic and periodic symmetries imposed upon our radio-frequency pulse sequences simplify the description of ρ, we name a second transformation,

$$\tilde{\rho} = U_{int}\tilde{\tilde{\rho}}\, U_{int}^{-1} \tag{5.13}$$

We specify that $U_{int}(t, 0)$ is controlled by our rf sequences through $\tilde{H}_{int}$ by insisting that U_{int} satisfy the Liouville equation

$$i\frac{dU_{int}}{dt} = \tilde{\mathscr{H}}_{int} U_{int} \tag{5.14}$$

with boundary value

$$U_{int}(0, 0) = 1 \tag{5.15}$$

Then $U_{int}(t, 0)$ is given by the Dyson expression

$$U_{int}(t, 0) = T \exp\left\{-i \int_0^t dt\, \tilde{\mathscr{H}}_{int}(t)\right\} \tag{5.16}$$

or by the Magnus expansion

$$U_{int}(t, 0) = \exp\{-it[\bar{H}_{int}^{(0)} + \bar{H}_{int}^{(1)} + \cdots]\} \tag{5.17}$$

Clearly, this last transformation removes $\mathscr{H}_{int}$ from the description of the density matrix, and in this frame, $\tilde{\tilde{\rho}}$ does not vary with time. The frame in which we take measurments is that described by $\rho(t)$. We have

$$\rho(t) = U_{rf}\tilde{\rho}(t)U_{rf}^+ = U_{rf}U_{int}\tilde{\tilde{\rho}}U_{int}^+U_{rf}^+ \tag{5.18}$$

Now let us set the "window" in our pulse experiment, the time we choose to take measurements. We restrict our "looking" to time $t = Nt_c$, i.e., to integral multiples of our cycle time. If our sequences are cyclic, Eq. (5.9) is valid and by (5.5) we find

$$\rho(Nt_c) = \tilde{\rho}(Nt_c) \tag{5.19}$$

Therefore, by (5.9),

$$\rho(Nt_c) = U_{int}(Nt_c, 0)\tilde{\bar{\rho}}U_{int}^+(Nt_c, 0) \tag{5.20}$$

If $\mathscr{H}_{int}$ is not explicitly time-dependent; e.g., if $\mathscr{H}_{int}$ is the static dipolar interaction, then since $U_{int}(t)$ is determined through $\mathscr{H}_{int}$ by (5.14) and U_{int} is determined by $\mathscr{H}_{rf}$ through (5.7) and (5.16) or (5.17), we see that U_{int} must similarly be periodic. Therefore, for example,

$$\begin{aligned}
\rho(2t_c) &= U_{int}(2t_c, t_c)U_{int}(t_c, 0)\tilde{\bar{\rho}}U_{int}^+(t_c, 0)U_{int}^+(2t_c, t_c) \\
&= U_{int}^2(t_c, 0)\tilde{\bar{\rho}}U_{int}^{-2}(t_c, 0)
\end{aligned} \tag{5.21}$$

with the general result

$$U_{int}(Nt_c, 0) = U_{int}^N(t_c, 0) \tag{5.22}$$

Therefore,

$$\rho(Nt_c) = U_{int}^N(t_c, 0)\tilde{\bar{\rho}}U_{int}^{-N}(t_c, 0) \tag{5.23}$$

Utilizing the Magnus expansion to evaluate $U_{int}(t_c, 0)$ yields

$$U_{int}(t_c, 0) = \exp\{-it_c(\bar{H}_{int}^{(0)} + \bar{H}_{int}^{(1)} + \cdots)\} \tag{5.24}$$

with

$$\bar{H}_{int}^{(0)} = \frac{1}{t_c}\int_0^{t_c} \tilde{\mathscr{H}}_{int}(\tau)\, dt$$

$$\bar{H}_{int}^{(1)} = \frac{-i}{2t_c}\int_0^{t_c} d\tau \left[\tilde{\mathscr{H}}_{int}(\tau), \int_0^{\tau} d\phi\, \tilde{\mathscr{H}}_{int}(\phi)\right] \tag{5.25}$$

$$\bar{H}_{int}^{(2)} = \cdots$$

The convergence of the series in (5.24) has been discussed by Maricq [3]. If this series is to be strongly convergent, the condition

$$t_c|\tilde{\mathscr{H}}_{int}(\tau)| \simeq t_c\omega_{int} \ll 1 \tag{5.26}$$

where $|\tilde{\mathscr{H}}_{int}|$ is the magnitude of $\tilde{\mathscr{H}}_{int}$, must be satisfied. For example, if $|\tilde{\mathscr{H}}_{int}|$ is 20 kHz, t_c must be less than 50 μsec.

Under this condition, we may express $\rho(Nt_c)$ to zero order as

$$\begin{aligned}
\rho(Nt_c) &= U_{int}^N(t_c, 0)\tilde{\bar{\rho}}U_{int}^{-N}(t_c, 0) \\
&\cong [\exp\{-i\bar{H}_{int}^{(0)}t_c\}]^N\tilde{\bar{\rho}}[\exp\{i\bar{H}_{int}^{(0)}t_c\}]^N
\end{aligned}$$

or

$$\rho(Nt_c) \cong \exp\{-i\bar{H}_{int}^{(0)}Nt_c\}\tilde{\bar{\rho}}\exp\{i\bar{H}_{int}^{(0)}Nt_c\} \tag{5.27}$$

When $U_{rf}(Nt_c, 0)$ and $U_{int}(Nt_c, 0)$ are both unity, from Eq. (5.15), $\tilde{\bar{\rho}}$ is equal to $\rho(0)$. By comparison with the *time-development* relation for a spin system evolving under a constant Hamiltonian $\mathscr{H}_0$,

$$\rho(t) = \exp(-i\mathscr{H}_0 t)\rho(0)\exp(i\mathscr{H}_0 t) \tag{5.28}$$

we see that (5.27) represents the time evolution of the spin system *in the frame of our measurement*, at "window" times $t = Nt_c$, as being caused by a *constant Hamiltonian* $\bar{H}^{(0)}$, *acting over the time period* $t = Nt_c$. Here we have identified $\tilde{\bar{\rho}}$ in (5.27) with $\rho(0)$ in (5.28). Clearly, however, $\bar{H}^{(0)}$ is due to a time-*dependent* interaction $\tilde{\mathscr{H}}_{\text{int}}(t)$, so in (5.27) we have replaced a time-dependent interaction with a formally time-independent interaction. We may say that at window times $t = Nt_c$, the spin system evolves *as if* it were being acted upon by a Hamiltonian averaged over the cycle time in question or by an "average Hamiltonian." We shall see that, for purposes of digging chemical-shift information out of NMR experiments in solids, one job of the multiple-pulse sequence is to arrange for the dipolar contribution to $\bar{H}_{\text{int}}^{(i)}$ to be zero.

The concept of the average Hamiltonian allows a particularly simple physical picture from which one may easily infer the signal expected as the response of a spin system to a cyclic, periodic rf perturbation under a given internal interaction. This picture will be illustrated in Section V.B when the response of a spin system to the "flip-flop" phase-tuning cycle, where the internal Hamiltonian is a dc offset, is considered.

In summary, there are four characteristics of multiple-pulse experiments used to obtain averaging of internal interactions:

1. The sequence of rf pulses must be cyclic;

$$U_{\text{rf}}(t_c) = \pm 1 \tag{5.9}$$

2. Groups of rf sequences must be periodic;

$$\mathscr{H}_{\text{rf}}(t) = \mathscr{H}_{\text{rf}}(t + Nt_c) \tag{5.12}$$

3. The cycle time of the pulse sequence used for averaging must be short compared to T_2 characterizing the homogenous broadening due to the interaction being averaged

$$t_c \omega_{\text{int}} \ll 1 \tag{5.26}$$

4. The response of the spin system must be observed at "windows" which are integral multiples of the cycle time.

III. The Symmetry of $\mathscr{H}_{\text{int}}$ in the Frame of the Radio-Frequency Excitation

We have seen that the recipe for calculating the time development of a spin system in which there are both time-dependent radio-frequency perturbations and internal perturbations that may or may not be time-dependent

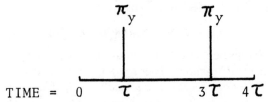

FIG. 5.3. A refocusing sequence for Hamiltonians depending on I_z, the Carr–Purcell sequence.

is provided by (5.18); identifying $\tilde{\tilde{\rho}}$ as $\rho(0)$, we find

$$\rho(t) = U_{rf}(t,0)U_{int}(t,0)\rho(0)U_{int}^{-1}(t,0)U_{rf}^{-1}(t,0) \tag{5.18}$$

When $\mathscr{H}_{rf}(t)$, which determines $U_{rf}(t,0)$, is appropriately cyclic and periodic,

$$\rho(Nt_c) = U_{int}(Nt_c,0)\rho(0)U^{-1}(Nt_c,0)$$

which is a re-writing of Eq. (5.27). Here $U_{int}(t_c,0)$ is determined by $\tilde{\mathscr{H}}_{int}(t)$, the internal Hamiltonian in the frame of the rf, where

$$\tilde{\mathscr{H}}_{int}(t) = U_{rf}^{-1}(t,0)\mathscr{H}_{int}U_{rf}(t,0) \tag{5.7}$$

and

$$U_{int}(t_c,0) = \exp\{-it_c[\bar{H}_{int}^{(0)} + \bar{H}_{int}^{(1)} + \cdots]\} \tag{5.17}$$

If $\mathscr{H}_{rf}(t)$ is so arranged that all of the terms in the exponent of (5.17) vanish, then $U_{int} = 1$, $\rho(t) = \rho(0)$, and the spin system *observed at the window times* $t = t_c$ behaves *as if* nothing were affecting it, i.e., *as if* $\mathscr{H}_{int}$ *were zero.* It is therefore useful to develop a scheme in which it becomes easily possible to visualize $\tilde{\mathscr{H}}_{int}$ as a function of time, so that the integrals implicit in the exponent of (5.17) may be evaluated by inspection. Such a scheme is the subject of this section. As a simple example, we consider the cycle illustrated in Fig. 5.3 that is common to a number of pulse sequences utilized for measurement of relaxation times and for removal of certain specified internal interactions. The pulses are taken to be ideal delta-function π pulses along y in the rotating frame. It should be clear by inspection that the rf sequence satisfies both cyclic and periodic boundary conditions at $t = Nt_c$. Let us investigate the values of an internal Hamiltonian that is proportional to I_z. Such an operator, for example, would be involved in dc-field inhomogeneities, in a portion of the homonuclear dipolar interaction, in the heteronuclear dipolar interaction, and in the offset interaction. As an initial portion of the scheme, a table of $\tilde{I}_z$ (I_z in the frame of the radio frequency) as a function of time is constructed. The entries in Table 5.1 are obtained as follows: In the interval

TABLE 5.1

Transformations of I_z under the Carr–Purcell (CP) Sequence

Interval	Time	$\mathscr{H}_{rf}$	U_{rf}^{-1}	$\tilde{I}_z$
δ_1	$0 \leq t < \tau$	0	1	I_z
Pulse 1	τ	I_y	$\exp(+i\pi I_y)$	Undefined
δ_2	$\tau < t < 3\tau$	0	1	$-I_z$
Pulse 2	3τ	I_y	$\exp(+i\pi I_y)$	Undefined
δ_3	$3\tau < t < 4\tau$	0	1	I_z

$0 \leq t < \tau$, the radio frequency is off, so $\mathscr{H}_{rf}$ is zero and U_{rf} is $\exp(0) = 1$. $\tilde{I}_z$ is $U_{rf}^{-1}I_zU_{rf}$, which is I_z. At τ there is an "instantaneous" (we shall amplify the meaning of this term further) pulse along y in the rotating frame by the angle $\omega_1 t = \gamma B_{1y}t = \pi$. In this case, as discussed in Chapter 3, $\mathscr{H}_{rf} = -\mu B = -\gamma B_{1y}I_y = -\omega_{1y}I_y$. Note that $\mathscr{H}_{rf}$ is constant in the frame of the Zeeman Hamiltonian. Also U_{rf} is given by Eq. (4.72), which for constant $\mathscr{H}_{rf}$ yields

$$U_{rf} = \exp\left\{-i\mathscr{H}_{rf}\int_0^{t_p} dt\right\} = \exp\left\{i\omega_{1y}I_y\int_0^{t_p} dt\right\} = \exp\{i\pi I_y\} \quad (5.29)$$

Since the pulse is "instantaneous," $\tilde{I}_z$ is undefined during the pulse. However, during the pulse, the instantaneous transformation takes place,

$$\tilde{I}_z = U_{rf}^{-1}I_zU_{rf} = \exp\{-i\pi I_y\}I_z\exp\{i\pi I_y\} = -I_z$$

so that immediately following the pulse, i.e., at the beginning of the period δ_2, $\tilde{I}_z = -I_z$. Since $\mathscr{H}_{rf} = 0$ during δ_2, $\tilde{I}_z$ remains $-I_z$ until $t = 3\tau$. At $t = 3\tau$, there is a second pulse. To evaluate $\tilde{I}_z$ resulting from a pulse P_1 at τ and a second pulse P_2 at 3τ, it is necessary to realize that

$$U_{rf}(3\tau, 0) = U_{rf}(3\tau, \tau)U_{rf}(\tau, 0) \quad (5.30a)$$

and therefore that

$$U_{rf}^{-1}(3\tau, 0) = U_{rf}^{-1}(\tau, 0)U_{rf}^{-1}(3\tau, \tau) \quad (5.30b)$$

or

$$U_{rf}^{-1}(\delta_2 + \delta_1, 0) = U_{rf}^{-1}(\delta_1)U_{rf}^{-1}(\delta_2) \quad (5.30c)$$

where $U_{rf}(\delta_1)$ is evaluated by utilizing P_1 and $U_{rf}(\delta_2)$ by utilizing P_2. In the present case the operators associated with P_1 and P_2 commute (both are I_y), so the order of operators in (5.30b) and (5.30c) is not important. In general, however, such will not be the case, and it will be necessary to exert care in ordering U_{rf}^{-1} to appropriately evaluate $\mathscr{H}_{rf}$. In the present case, we find

$$\tilde{I}_z(\text{at } t \geq 3\tau) = \exp\{-i\pi I_y\}\exp\{-i\pi I_y\}I_z\exp\{i\pi I_y\}\exp\{i\pi I_y\} = I_z \quad (5.31)$$

since $\exp\{2\pi i I_y\}$ is unity. We now combine Table 5.1 and Fig. 5.3 into one figure. From Fig. 5.4 it is immediately apparent by inspection that $\bar{H}_{int}^{(i)}$ is zero for all i, since

$$\bar{H}_{int}^{(0)} = \frac{\omega_1}{t_c} \int_0^{t_c} I_z(t')\,dt' = \frac{I_z\omega_1}{t_c}(\tau - 2\tau + \tau) = 0 \qquad (5.32)$$

and since all higher terms involve commutators of I_z with itself. Therefore, $U_{int}(t_c,0)$ is unity, and $\rho(t_c) = \rho(0)$, which means that, when observed at the window times $t = Nt_c$ under the pulse sequence $[\tau, \pi_y, 2\tau, \pi_y, \tau]$, the spin system appears to be unaffected by the operator I_z. This fact is fundamental to an understanding of a number of multiple-pulse experiments, which are, by now, becoming commonly used, e.g., the Carr–Purcell–Meiboom–Gill method of measuring T_2, and heteronuclear decoupling (Chapter 6). In experiments where $\mathcal{H}_{int}(t)$ does not commute at all times, it will be necessary to evaluate terms of the form $[\mathcal{H}_{int}(t'), \mathcal{H}_{int}(t'')]$ in order to obtain $\bar{H}_{int}^{(i)}$ for $i > 0$. In this case it will be necessary to construct two-dimensional tables of $\mathcal{H}_{int}(t)$ to evaluate the commutators inherent in $\bar{H}_{int}^{(1)}$, and we shall do so with examples later in this chapter. As one further example of the symmetry of $\mathcal{H}_{int}$, consider the internal Hamiltonians I_z and $(I_{z_1}I_{z_2})$ under the pulse sequence shown in Fig. 5.5. In Fig. 5.5, for simplicity $I_{x_1}I_{x_2}$ is represented by the symbol X and similarly for Y and Z. Inspection reveals that when $\mathcal{H}_{int} = aI_z$,

$$\bar{H}_{int}^{(0)} = (\tau a/t_c)(2I_x - 2I_y + 2I_z) = a(I_x - I_y + I_z)/3 \qquad (5.33)$$

and when

$$\mathcal{H}_{int} = \mathbf{I}_1 \cdot \mathbf{I}_2 - 3I_{z1}I_{z2}, \qquad \bar{H}_{int}^{(0)} = 0 \qquad (5.34)$$

(Keep in mind the fact that a scalar product of two vectors is invariant to rotation.) A physical interpretation of (5.33) is that, to zeroth-order, operators associated with I_z are not averaged to zero under this sequence. On the other hand, homonuclear dipolar interactions are averaged to zero, since $\mathbf{I}_1 \cdot \mathbf{I}_2$ is

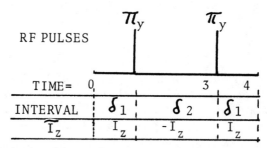

FIG. 5.4. Evaluation of the transformations of I_z at various points in the Carr–Purcell sequence.

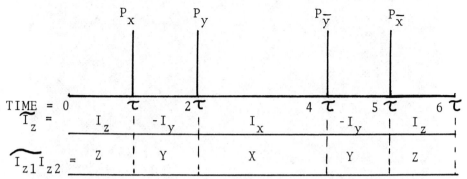

FIG. 5.5. A refocusing sequence for the dipole–dipole interaction, the WAHUHA sequence.

invariant to rotation. More will be said about this result later in the discussion of sequences used to remove homonuclear dipolar interactions and of the concept of "scaling."

IV. Treatment of Nonideal rf Perturbations

In the discussion of pulse experiments thus far, we have assumed that all rf pulses are "ideal delta-function" pulses. This means that when we turn on a $\pi/2$ pulse along the x direction in the rotating frame, for example, there are no errors in timing that might lead $\gamma B_1 t_p$ to be $\pi/2 \pm \delta$, where δ is due to a small error in pulse width t_p. It is also assumed that the pulse width is infinitely small compared to *any* characteristic time in which the system might decay, e.g., compared to T_2^*. This is the sense in which we have used "instantaneous transformation" to describe time development during the pulse. Further, it is assumed that $\mathbf{B}_1$ is *always* phased *exactly* in the rotating frame and that it is uniform over the entire sample. These are *all* assumptions that are *always* violated to some extent, the degree of violation being a function of experimental design and the care exercised in tuning the spectrometer. For example, suppose that we expose a spin system to a sequence of the form $[(\pi/2)_x, 2\tau]^n$, and we observe along y. For ideal delta-function pulses, we would expect the magnetization after the first pulse to lie along y for a time 2τ during which time we observe a positive signal. After the second pulse, we expect the magnetization to be along $-z$ for a time 2τ, during which time we observe a zero magnetization. After the third pulse, we expect to see a negative signal ($\mathbf{M}$ along $-y$), and after the fourth pulse, we again expect a zero response ($\mathbf{M}$ back along z). Such a response in an undoped water sample to this four-pulse experiment is shown in Fig. 5.6. The total scan shown is 4 msec. The times τ are 240 μsec, and the pulses

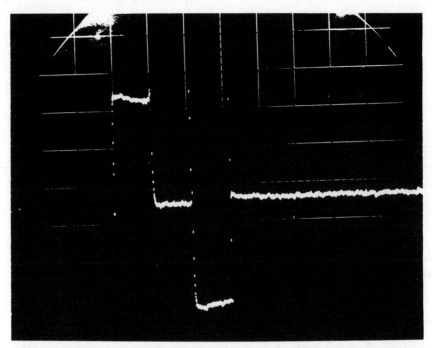

FIG. 5.6. The response of protons in water to a train of four $\pi/2$ pulses.

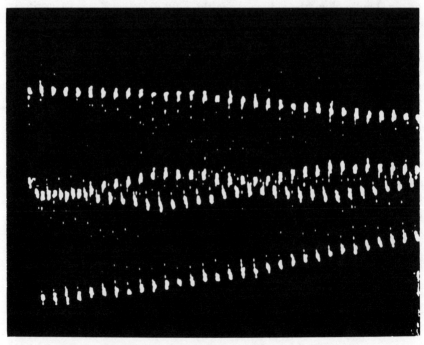

FIG. 5.7. The response of protons in water to a train of 128 $\pi/2$ pulses.

are 1.25 μsec in width. The result of 128 such pulses over a total time scan of 20 msec is shown in Fig. 5.7 for a spherical water sample with dimensions small compared to the inductance coil, i.e., for a sample in a relatively uniform $\mathbf{B}_1$. The decay of the positive and negative signals toward the zero line is simply that due to T_2^*, in this case dominated by dc field inhomogeneities. Note that there is a modulation of the amplitude of the decays that is particularly noticeable along the zero line, with period roughly 17 msec. This is simply a result of 60-Hz modulation of the power output of the transmitter. Also note that aside from this modulation, the response is quite close to the "zero-positive-zero-negative-etc." response expected for a string of $\pi/2)_x$ pulses detected along y. On the other hand, the top trace of Fig. 5.8 illustrates the result of the sample size not being small relative to the coil dimensions. No amount of adjustment of the pulse width is able to produce a better pattern, since the *inhomogeneity of the* $\mathbf{B}_1$ *field* leads to large portions of the sample being exposed to a field *intensity* far from that required to satisfy $B_1 \gamma t_w = \pi/2$ for any given pulse width t_w. This is effectively saying that there are pulse-width errors that vary throughout the region of the sample and that rapidly

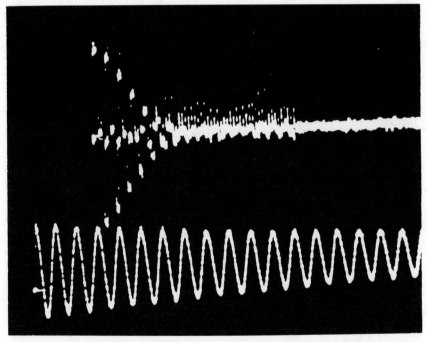

FIG. 5.8. The response of the nuclei in a sample of dimensions larger than the coil to a chain of nominally $\pi/2$ pulses. The lower trace indicates the free-induction decay of the same sample, which demonstrates that the short decay is not due to T_2^*.

accumulate to dephase the bulk magnetization in the rotating frame. It should be noted that this decay is much shorter than that associated with T_2^*, as can be seen by comparison with the lower trace in Fig. 5.8, which is an off-resonance FID from a single-pulse experiment with the same time scale.

Phase errors in pulses are of two types. There are those associated with the fact that a circuit driven into oscillation by a pulse of rf power is not capable of instantaneous response to the driving force. In general, there will be time constants associated with attainment of both phase and amplitude equilibrium after both the initiation and the termination of the pulse. This effect is illustrated in Fig. 5.9, in which the phase and amplitude of an ideal driving pulse, indicated by a solid line, is compared with the phase and amplitude of the pulse as detected by an antenna near the inductance of a resonant NMR probe. Omitted from this picture are transients associated with switching the rf gate open and closed. The time constant for the amplitude to achieve $1 - e^{-1}$ of its full value depends upon the details of the probe design, but for commonly used probes is roughly equal to $\tau \cong Q/3f$, where Q is the quality factor of the resonant circuit and f the resonant frequency in hertz. Similarly, the time constant for the signal to "ring down" after the rf gate controlling the signal is closed is $\tau \cong Q/3f$. There is clearly a phase lag between the ideal square-wave pulse and the real pulse "turn-on" region and in the turn-off region of the pulse, the magnitude of which is dependent on the circuit Q. This effect is more graphically illustrated in Fig. 5.10, which is a picture of the dc amplitude of an rf pulse detected both in phase (top scan) and 90° out of phase (bottom scan) with the driving pulse. The amplitude of the quadrature (90°-out-of-phase) component has been magnified by a factor of 10 relative to the in-phase component in this plate. In language used earlier in the text, we put a pulse along x and observe components of the pulse along x in the top scan. In the bottom scan, we are observing components of the pulse along y. If the pulse along x were a perfect square-wave pulse, we would expect a square wave for the dc envelope observed in phase and *nothing* when we observe along y. There is clearly a nonzero

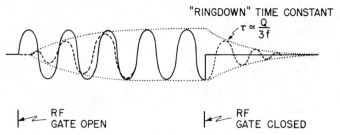

FIG. 5.9. The response of a tuned rf circuit to a pulse of rf radiation. (——, driving pulse; ————, response of probe circuit; ······, envelope of response.)

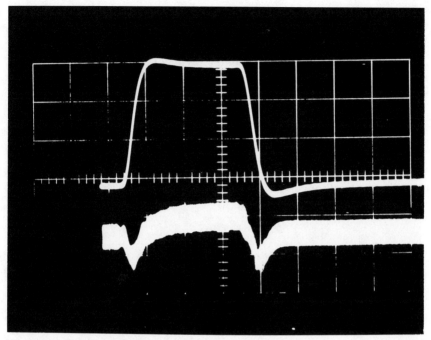

FIG. 5.10. The phase-detected rf pulse envelope: (upper trace) the in-phase component and (lower trace) the out-of-phase component.

contribution along y both during the turn-on and the turn-off periods of this pulse. This is to say that the effective rf field seen by a spin system contains an initial component along y in the rotating frame. The rf field then swings to the x axis for a portion of the pulse as indicated in Fig. 5.11. When the pulse is turned off, the field again swings back along y during the turn-off

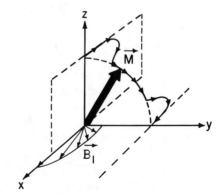

FIG. 5.11. The time evolution of the classical magnetization under the action of a nonideal pulse. (Arrows indicate vectors in the figures; boldface indicates vectors in the text.)

transient. The motion of a spin exposed to such a field is indicated in Fig. 5.11. Note that the transient in Fig. 5.10 approximately satisfies the symmetry conditions of (5.37) (*vide infra*).

In order to understand in detail the results of a pulse NMR experiment, it will be necessary to calculate the effects of these types of pulse errors. To do so, we realize that some of these errors meet both cyclic and periodic boundary conditions [Eqs. (5.9) and (5.12)] so average Hamiltonian theory may be used in a straightforward manner. On the other hand, some errors can lead to a break down of the cyclic conditions (e.g., consider the sequence $[\tau, \pi/2 + \delta)_x, 2\tau, \pi/2)_{\bar{x}}, \tau]$, where δ is a pulse width error). If the errors are sufficiently small, they may be treated by quite a straightforward extension of average Hamiltonian theory. In this treatment, we separate the error, $\mathscr{H}^1_{rf}(t)$ from the "ideal" rf pulse, $\mathscr{H}^0_{rf}(t)$:

$$\mathscr{H}_{rf}(t) = \mathscr{H}^0_{rf}(t) + \mathscr{H}^1_{rf}(t) \tag{5.35}$$

With the condition $|\mathscr{H}^1_{rf}(t)| \ll |\mathscr{H}^0_{rf}(t)|$, we simply combine $\mathscr{H}^1_{rf}(t)$ with $\mathscr{H}_{int}(t)$ in Eq. (5.2) and repeat the steps described by Eqs. (5.3)–(5.34). In general, $\mathscr{H}^1_{rf}(t)$ will be a sum of terms, so we have

$$\mathscr{H}^1_{rf}(t) = \sum_k \mathscr{H}_k(t)$$

For example, we shall use P for phase misadjustments, T for phase transients, δ for pulse-width errors, and ϵ for rf inhomogeneity. For spin-$\frac{1}{2}$ systems, the contributions we wish to consider from the internal Hamiltonian are the dipolar $\mathscr{H}_D$, and the sum of the dc-field inhomogeneity, the chemical shift, and the offset terms. Since the last three terms are all proportional to the operator I_z, we only consider $\mathscr{H}_O$, the offset term in the present discussion. With $\mathscr{H}_{int} = \mathscr{H}_O + \mathscr{H}_D$, the average Hamiltonians will be of the form

$$\bar{H}^{(0)}_{int} = \bar{H}^{(0)}_O + \bar{H}^{(0)}_D + \sum_k \bar{H}^{(0)}_k \tag{5.36a}$$

$$\bar{H}^{(1)}_{int} = \bar{H}^{(1)}_O + \bar{H}^{(1)}_D + \bar{H}^{(1)}_{OD} + \sum_k [\bar{H}^{(1)}_{Ok} + \bar{H}^{(1)}_{Dk}] + \cdots \tag{5.36b}$$

$$\bar{H}^{(2)}_{int} = \bar{H}^{(2)}_O + \bar{H}^{(2)}_D + \bar{H}^{(2)}_{OD} + \sum_k [\bar{H}^{(2)}_{Ok} + \bar{H}^{(2)}_{Dk}] + \cdots \tag{5.36c}$$

Here, $\bar{H}^{(1)}_{OT}$ represents, for example, the first-order coupling between the offset and a phase transient.

The forms of the Hamiltonians for pulse imperfections are, with a bit of reflection, self-obvious. They are listed in Table 5.2, for an x pulse.

The pulse-width-error Hamiltonian for an x pulse $(\delta_x/t_w)I_x$ simply expresses the fact that the total rotation for a pulse of nominal rotation angle θ is $\theta(1 + \delta/t_w)$, where t_w is the pulse width in seconds. The term δ/t_w is thus the fractional error associated with a pulse-width misadjustment.

TABLE 5.2

Hamiltonians Corresponding to Nonidealities in a Radio-frequency Pulse

Radio-frequency imperfection	Hamiltonian
Pulse-width error	$-(\delta_x/t_x)I_x$
B_1 inhomogeneity for ith nucleus	$-(\epsilon_1/t_w)I_{xi}$
Phase error by angle ϕ from the x direction of the rotating frame	$-\delta_1 I_y \sin\phi_x \simeq -\delta_1 I_y \phi_x$
Pulse-transient phase error for a pulse along x in rotating frame	$\omega_T(t)I_y$
Deviation from ideal square-wave pulse	$-\omega_I(t)I_x$

The B_1-inhomogeneity term expresses the fact that the rotation angle θ_i experienced by the ith nucleus due to a pulse of nominal width θ is $\theta_i = \theta(1 + \epsilon_i/t_w)$. The error term ϵ_i/t_w expresses the B_1 inhomogeneity in terms of a fraction of the pulse width.

The phase-error Hamiltonian associated with a phase misadjustment between channels stems from the fact that a pulse in the xy plane of the rotating frame has components $I_y \sin\phi$ along y, and $I_x \cos\phi$ along x. Here ϕ is the error in phase adjustment of an x pulse so if $\phi = 0$, the Hamiltonian for an x pulse for rotation is simply $-\gamma B_1 I_x \equiv -\omega_1 I_x$. We assume ϕ is sufficiently small that $\cos\phi \cong 1$, and the phase-*error* component is then $-\omega_1 I_y \sin\phi$. For sufficiently small, ϕ, this becomes $-\omega_1 I_y \phi_x$. We describe the phase error as positive if the error corresponds to a clockwise deviation from the axis in question, when viewed from above the xy plane. Therefore, an x pulse with a positive phase error would have phase-error Hamiltonian $\omega_1 I_y \phi_x$, corresponding to an imperfection operator along $\bar{y}$. Similarly, an $\bar{x}$ pulse with a positive phase error would have a phase-error Hamiltonian $-\omega_1 I_y \phi_{\bar{x}}$ corresponding to an imperfection along y.

The phase error associated with the phase transient of an x pulse, as we saw from Figs. 5.8–5.10, is a function of time and along $\pm y$. A straightforward representation of this interaction is $\omega_T(t)I_y$.

The term $\omega_I(t)I_x$ is the *difference* between the ideal square-wave pulse and the real pulse, the dc envelope of which is shown as the top trace in Fig. 5.10. We shall be particularly interested in the symmetry, in time, of the last two pulse imperfections. In particular, we impose the condition

$$\omega_T(t_w/2 - t) = \omega_T(t_w/2 + t) \tag{5.37}$$

for the amplitude of the transient associated with all pulses. We note that when we take into account the condition (5.37), a unique pulse width t_w is undefined, since the width necessary to attain a given rotation, e.g., $\pi/2$, will be dependent on the pulse shape. In general, we attempt to arrange our apparatus such that the conditions

$$|\epsilon_i/t_w|, |\delta_x/t_w|, |\phi_x\omega_1|, |\omega_I(t)|, |\omega_T(t)| \ll |\omega_1|$$

where the Hamiltonian for an ideal square-wave pulse in the x direction is $-\omega_1 I_x$ during the period of the pulse. Further, we always attempt to have sufficient power that $t_w/t_c \ll 1$, where t_c is the cycle time. Under these conditions, pulses are treated as delta functions for the purposes of calculating higher-order correction terms to the average Hamiltonian for internal interactions. If this condition is not met, calculation of the effects of finite pulse widths on higher-order terms may be necessary.

V. Sample Calculations

A. Calculation of $H_0^{(0)}$ for the Flip-Flop Phase-Tuning Cycle with Finite Pulse Widths

There are two pulse cycles that are pertinent to spectrometer tuning. These are

(a) a string of $\pi/2$ pulses in a given direction in the rotating frame, e.g., $\pi/2)_x, 2\tau, \pi/2)_x, 2\tau, \ldots$, and
(b) the phase-alternated, flip-flop cycle $\pi/2)_x, 2\tau, \pi/2)_{\bar{x}}, 2\tau, \pi/2)_x, \ldots$

Cycle (b) is used to set the phase of the $\bar{x}$ channel of the spectrometer relative to the x channel. An example of the response to cycle (a) is shown in Figs. 5.6 and 5.7. The timing of these two pulse sequences is shown in Fig. 5.12, in which P_x represents a $\pi/2)_x$ pulse. In order to satisfy periodic and cyclic boundary conditions for both cycles, it is necessary to consider a cycle time of $t_c = 4\tau$ for the flip-flop cycle, as indicated in Fig. 5.12, and a cycle time of $t_c = 8\tau$ for the pulse-width tuning cycle.

An instructive calculation is that of the zeroth-order term in the Magnus expansion associated with the offset Hamiltonian $\bar{H}_O^{(0)}$. To evaluate this term, we have the option of considering finite rf pulses or delta-function pulses. As discussed in the preceding section, and as we illustrate more clearly in the next section, this choice depends on the physics of the situation. In the case in which the Hamiltonian in question is acting over time periods long compared to the pulse width, a consideration of delta-function pulses is sufficient. However, if the Hamiltonian is *only* on during the pulse, e.g., such as would be the case for any Hamiltonian associated with a pulse imperfection, it is absolutely necessary to take into account finite pulse widths.

In the present case, the offset Hamiltonian is always on, and if the time between pulses is long compared to the pulse width, we may neglect the time

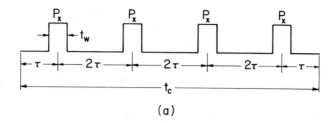

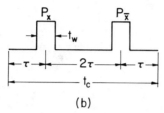

FIG. 5.12. (a) A train of $\pi/2$ pulses of finite width and (b) the flip-flop cycle, with pulses of finite width.

development of the system due to this Hamiltonian during the pulses. To make this point clear, we perform the calculation taking into account finite pulse widths for the phase-tuning flip-flop cycle.

To calculate $\bar{H}_0^{(0)}$ for the flip-flop cycle, we construct Table 5.3 and calculate $\tilde{I}_z$. In this table, α is used to represent half the pulse width; i.e., $2\alpha = t_w$. The entries in Table 5.3 are constructed as follows: for the first time interval $0 \leq t < \tau - \alpha$, the rf is off, so $U_{rf}^{+} = \exp\{-i \cdot 0\}t = 1$. For subsequent calculations of U_{rf}^{+}, it is important to keep in mind

$$U(t_n, 0) = U(t_n, t_{n-1})U(t_{n-1}, t_{n-2}) \dots U(t_1, 0) \qquad (5.38)$$

Therefore, for example,

$$U(\tau + \alpha, 0) = U(\delta_2)U(\delta_1) \qquad (5.39)$$

TABLE 5.3

Transformations of the Operator I_z during a Flip-Flop Cycle

Interval	Time	$\mathscr{H}_{rf}$	U_{rf}^{+}	$\tilde{I}_z$
δ_1	$0 \leq t < \tau - \alpha$	0	1	I_z
δ_2	$\tau - \alpha \leq t \leq \tau + \alpha$	$-\omega_1 I_x$	$\exp[-iI_x\theta(t)]$	$I_z \cos\theta(t) - I_y \sin\theta(t)$
δ_3	$\tau + \alpha < t < 2\tau - \alpha$	0	1	$-I_y$
δ_4	$2\tau - \alpha \leq t \leq 2\tau + \alpha$	$\omega_1 I_x$	$\exp[iI_x\theta(t)]$	$I_z \sin\theta(t) - I_y \cos\theta(t)$
δ_5	$2\tau + \alpha < t \leq 3\tau$	0	1	I_z

and

$$U^+(\tau + \alpha, 0) = U^+(\delta_1)U^+(\delta_2) \qquad (5.40)$$

For the interval δ_2, $\omega_1 t$ will be some $\theta(t)$ between 0 and $\pi/2$. We assume that $\theta(t)$ is linear in time. With $\theta = 0$ at $t = \tau - \alpha$, and $\theta = \pi/2$ at $t = \tau + \alpha$ we find

$$\theta(t) = (\pi/4)[1 + (t - \tau)/\alpha] \qquad (5.41)$$

Since $U_{rf}^+(\tau + \alpha, \tau - \alpha) \equiv \exp[-iI_x\theta(t)]$ corresponds to a clockwise rotation about x (when viewed along the direction of *positive* x, i.e., a rotation in which I_z is rotated into $-I_y$), we can immediately evaluate the result of the transformation of I_z as

$$\begin{aligned}
\tilde{I}_z(\delta_2) &= U_{rf}^+(\tau + \alpha, 0)I_z U_{rf}(\tau + \alpha, 0) \\
&= U_{rf}^+(\delta_1)U_{rf}^+(\delta_2)I_z U_{rf}(\delta_2)U_{rf}(\delta_1) \\
&= 1\exp[-i\theta(t)I_x]I_z\exp[i\theta(t)I_x] \cdot 1 \\
&= I_z\cos\theta(t) - I_y\sin\theta(t) \qquad (5.42)
\end{aligned}$$

If the physics of the rotation is not clear, one can always consult Table 2.1. At this stage of development, however, it should be clear that an expression such as

$$\exp(iI_x\theta)I_y\exp(-iI_x\theta) \qquad (5.43)$$

represents a rotation in which I_y is rotated counterclockwise (when viewed in the direction of positive x) toward the $-z$ axis. The result of this rotation for $0 \le \theta \le \pi/2$ will be

$$I_y\cos\theta - I_z\sin\theta \qquad (5.44)$$

All entries in Table 2.1 may thus be evaluated without recourse to the table if this simple picture is kept in mind. To reinforce the physical idea of the exponential operation as a rotation, we rewrite Eq. (5.42) in a "rotation" form; letting $R_{\theta(t)\bar{x}}$ symbolize a rotation by $\theta(t)$ about x in a sense clockwise when viewed along positive x, Eq. (5.42) reads

$$\tilde{I}_z(\delta_2) = 1 \cdot R_{\theta(t)\bar{x}} \cdot 1I_z \qquad (5.45)$$

Proceeding as before, for $\tau + \alpha \le t \le 3\tau - \alpha$, we observe that at $t = \tau + \alpha$, $\theta(t) = \pi/2$, so

$$\tilde{I}_z = R_{90\bar{x}} \cdot 1 \cdot I_z = -I_y \qquad (5.46)$$

and since $U_{rf} = 1$ in this time interval, the value of $\tilde{I}_z$ does not change.

For the time interval $3\tau - \alpha \le t < 3\tau + \alpha$, we have

$$\tilde{I}_z = 1 \cdot R_{90_{\bar{x}}} \cdot 1 \cdot R_{\theta(t)_x} \cdot I_z$$
$$= R_{90_{\bar{x}}}[I_z \cos\theta + I_y \sin\theta]$$
$$= -I_y \cos\theta + I_z \sin\theta \qquad (5.47)$$

As before, we find

$$\theta(t) = \pi/4[(1 - 3t_c/4\alpha) + t/\alpha] \equiv a' + bt \qquad (5.48)$$

At $t = 3\tau + \alpha$, $\theta(t) = \pi/2$, so $\tilde{I}_z = I_z$. The calculation of the last row in the table is left as a trivial exercise.

We calculate $\bar{H}_0^{(0)}$ by utilizing (5.25);

$$\bar{H}_0^{(0)} = \frac{1}{t_c} \int_0^{t_c} \tilde{\mathcal{H}}_0(t)\, dt$$

$$= \frac{\Delta\omega}{t_c}\left[I_z \int_0^{\tau-\alpha} dt - I_y \int_{\tau-\alpha}^{\tau+\alpha} \sin\theta(t)\, dt + I_z \int_{\tau-\alpha}^{\tau+\alpha} \cos\theta(t)\, dt \right.$$

$$- I_y \int_{\tau+\alpha}^{3\tau-\alpha} dt - I_y \int_{3\tau-\alpha}^{3\tau+\alpha} \cos\theta(t)\, dt + I_z \int_{3\tau-\alpha}^{3\tau+\alpha} \sin\theta(t)\, dt$$

$$\left. + I_z \int_{3\tau+\alpha}^{4\tau} dt \right] = \frac{\Delta\omega}{2}(I_z - I_y)\left(1 + 2\frac{t_w}{t_c}\left[\frac{4}{\pi} - 1\right]\right) \qquad (5.49)$$

In evaluating (5.49), we utilize the relations

$$4\tau = t_c, \qquad b = \pi/2t_w \equiv \pi/4\alpha$$

$$\int \cos(a + bt)\, dt = b^{-1} \sin(a + bt), \qquad \int \sin(a + bt)\, dt = -b^{-1} \cos(a + bt)$$

We similarly calculate $\bar{H}_0^{(0)} = 0$ for the pulse-width tuning sequence P_x, 2τ, P_x, 2τ, P_x, ...

B. A Simple Physical Picture of the Average Hamiltonian

We see that $\bar{H}_0^{(0)}$ is linear in the components of $\mathbf{I}$. To the extent that this linearity holds for terms in average-Hamiltonian operators, we may write them in the form

$$\bar{H}_{int} = \mathbf{\Omega} \cdot \mathbf{I} \qquad (5.50)$$

where $\bar{H}_{int}$ is the sum of all internal Hamiltonians that are linear in components of $\mathbf{I}$ and thus transform as vector operators. The indentification of Eq. (5.50) as the effective Hamiltonian in the frame of observation leads to a powerful method of determining the results of a given pulse experiment without the necessity of a detailed calculation of $\rho(t)$. The basis for this method is the fact that the Hamiltonian for a moment $\mathbf{M} = \gamma\mathbf{I}$ in a field $\mathbf{B}$ is

$$-\mathbf{B} \cdot \mathbf{M} = -\gamma\mathbf{B} \cdot \mathbf{I}$$

accompanied by the fact that the natural motion of a moment in this field is a precession with frequency $\omega = -\gamma\mathbf{B}$. We may therefore formally identify $\mathbf{\Omega}$ in Eq. (5.50) as a vector about which the magnetization precesses, with frequency $|\mathbf{\Omega}|$. This identification can be used to determine directly the effect of a given rf Hamiltonian on the time development of the magnetization, i.e., on the results of a given pulse experiment. For example, consider the phase-tuning flip-flop sequence P_x, 2τ, P_x, 2τ, P_x, ... for which we have just calculated $\bar{H}_O^{(0)}$, the average Hamiltonian for the offset operator. The result of this calculation was

$$\bar{H}_O^{(0)} = (\Delta\omega/2)(1 + 2a)(I_z - I_y) \tag{5.51}$$

where $a = (t_w/t_c)(4/\pi - 1)$. If we identify this Hamiltonian as $\mathbf{\Omega} \cdot \mathbf{I}$, where $\mathbf{\Omega}$ is a vector about which the magnetization precesses in the rotating frame, we have

$$\mathbf{\Omega} = (\Delta\omega/2)(1 + 2a)(\mathbf{k} - \mathbf{j}) \tag{5.52}$$

and

$$|\mathbf{\Omega}| = \sqrt{2}(\Delta\omega/2)(1 + 2a) \tag{5.53}$$

Here $\mathbf{k}$ and $\mathbf{j}$ are unit vectors in the z and y directions in the rotating frame. If the magnetization is observed in the windows after the P_x pulse of the flip-flop sequence, i.e., at cycle times $Nt_c = 4N\tau$, the initial magnetization will be along z. The motion of $\mathbf{M}$ will be a precession about $\mathbf{\Omega}$ as shown in Fig. 5.13. If we detect along the positive y axis, as indicated in Fig. 5.13, the motion we observe is the projection of this precession along y, starting at zero, since $\mathbf{M}_0$ is along $\mathbf{k}$. This projection is just the amplitude-versus-time signal we observe as the result of such an experiment and is shown at the bottom of Fig. 5.13. Note that the signal is negative relative to zero, since we observe along y, and the projection of $\mathbf{M}$ either is zero or is along $\bar{y}$.

On the other hand, what do we expect from the offset Hamiltonian if we observe the magnetization after the P_x pulses in the flip-flop sequence? Again, to make this determination, we calculate the average Hamiltonian for $\mathcal{H}_O$ under this sequence, but now, in order that the sequence considered be cyclic, we must redefine the cycle to be τ, $\pi/2)_x$, τ, $[\tau, \pi/2)_{\bar{x}}, 2\tau, \pi/2)_x, \tau]^{N-1}$ for N pulses. Under this sequence, we find

$$\bar{H}_O^{(0)} = \tfrac{1}{2}\Delta\omega(1 + 2a)(I_z + I_y) \tag{5.54}$$

which corresponds to a precession of the magnetization about the vector

$$\mathbf{\Omega} = \tfrac{1}{2}\Delta\omega(1 + 2a)(\mathbf{k} + \mathbf{j}) \tag{5.55}$$

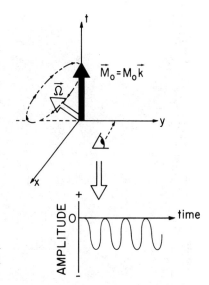

FIG. 5.13. The motion of a magnetization in the effective field and the observed response. [$\Omega = \Delta\omega/2(1 + 2/3a)(-\mathbf{j} + \mathbf{k})$.]

with magnitude $\Delta\omega(1 + 2a)(\sqrt{2})^{-1}$ in the rotating frame. We note that the *frequency* of the signal due to the offset is $(\sqrt{2})^{-1}\Delta\omega(1 + 2a)$, which is the same frequency as the signal observed after the $P_{\bar{x}}$ pulse. With $a = (t_w/t_c)\cdot(4/\pi - 1)$, and with $t_w \ll t_c$, we find that the frequency is $2^{-1/2}\Delta\omega$; that is, the observed frequency of the response is the offset *scaled* by the factor $2^{-1/2} = 0.71$. For an offset of 1000 Hz, for example, we would expect a frequency of 710 Hz for the signals observed after both the P_x and the $P_{\bar{x}}$ pulses. The sign of the signal observed after the P_x pulse, on the other hand, would be expected to be positive if observation is along the $+y$ axis in the rotating frame, since the vector about which $\mathbf{M}$ precesses is in the positive y–z quadrant. To illustrate more graphically the point made regarding the visualization of the response associated with a given average Hamiltonian that is linear in components of $\mathbf{I}$, we show Fig. 5.14. In this tracing, taken from the results of the flip-flop cycle on protons in water, the left third of the signal is due to observation of signals after both the x and $\bar{x}$ pulses. The right two-thirds of the scan is the FID following termination of the sequence. Note that the frequency of the FID on the right side of the screen is indeed higher than that of the signals on the left. The decay in amplitude with time is, in this case, a result of the dc-field inhomogeneity, which was not included in the calculation. At this point, it should be clear that if t_w/t_c is small compared to unity, then the "a" term in Eqs. (5.54) and (5.55) may be neglected, and the treatment could have been carried through by using ideal delta-function pulses.

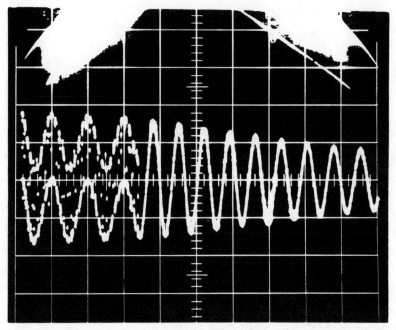

FIG. 5.14. The response to the flip-flop cycle, followed by a free-induction decay, showing the effect of retarded time evolution.

C. Calculation of the Effect of a Phase Error on the Flip-Flop Cycle

In the preceding section we saw that the effect of an offset on the flip-flop cycle was independent of pulse width, provided that the cycle time was large compared to the pulse width. Another way of viewing this result is that the evolution of the spin system during the cycle is negligibly dependent upon the evolution during the pulses if the pulses form only a small portion of the total time taken for the system to evolve. This is because the offset Hamiltonian is affecting the system for the entire cycle time, and its effect during the pulses is small relative to that during the remainder of the cycle.

We now inquire into the effect of a pulse phase-tuning error on the flip-flop cycle. Such a tuning error would be a result of, e.g., a misadjustment of one of the phase trimmers in the 30-MHz portion of the multiple-pulse spectrometer shown in Fig. 5.1. Since the flip-flop cycle is specifically designed to tune the phase of $\bar{x}$ relative to x, and $\bar{y}$ relative to y, an understanding of the effect of a phase error on this cycle is an important part of the tuning process.

A characteristic of the phase error on the time evolution of a system under the flip-flop sequence is that, unlike the offset Hamiltonian, this term *only* affects the system during the pulse; i.e., $\mathscr{H}_P(t) = 0$ for the time *between* pulses. Therefore, it is *always* necessary to consider the effect of finite pulse width on this term. With the timing of the sequence as shown in Fig. 5.12b, we again construct a table to evaluate the phase-error Hamiltonian in the interaction frame of the radio frequency. As indicated in our discussion of imperfection Hamiltonians,

$$\tilde{\mathscr{H}}_P = -\omega_1 \phi_x I_y \tag{5.56a}$$

for a pulse along x, and

$$\tilde{\mathscr{H}}_P = \omega_1 \phi_{\bar{x}} I_y \tag{5.56b}$$

for a pulse along $\bar{x}$. With $\mathscr{H}_P = 0$ in the time interval between pulses, we find for the flip-flop sequence the values given in Table 5.4. Straightforward integration yields, with $\omega_1 \alpha = \pi/4$,

$$\bar{H}_P^{(0)} = \frac{1}{t_c} \int_0^{t_c} dt \; \tilde{\mathscr{H}}_P(t) = \frac{1}{t_c} (\phi_{\bar{x}} - \phi_x)(I_y + I_z) \tag{5.57}$$

for the signal observed after the $\bar{x}$ pulse. The value of $\bar{H}_P^{(0)}$ for a signal observed after the x pulse is left as an exercise (Problem 5.12). Note that in evaluating $\tilde{\mathscr{H}}_P(t)$, we use the fact that

$$\tilde{\mathscr{H}}_P(t) = U_{rf}^+(t,0)\mathscr{H}_P(t)U_{rf}(t,0) \tag{5.58}$$

and that

$$U_{rf}^+(t,0) = U_{rf}^+(t_1,0)U_{rf}^+(t_2,t_1)\dots U_{rf}^+(t_{n-1},t_n) \tag{5.59}$$

as was explicitly pointed out in the preceding discussion. Therefore, to evaluate $\tilde{\mathscr{H}}_P(t)$ during the interval $3\tau - \alpha \le t \le 3\tau + \alpha$, we use the formula

$$\begin{aligned}
\tilde{\mathscr{H}}_P(3\tau - \alpha \le t \le 3\tau + \alpha) &= \exp(-iI_x\pi/2) \cdot 1 \cdot \exp[iI_x\theta(t)][\omega_1\phi_{\bar{x}}I_y] \\
&\quad \cdot \exp[-iI_x\theta(t)] \cdot 1 \cdot \exp(iI_x\pi/2) \\
&\equiv R_{90_{\bar{x}}} \cdot R_{\theta(t)_x}(\omega_1\phi_{\bar{x}}I_y) \tag{5.60}
\end{aligned}$$

TABLE 5.4

Transformations of the Hamiltonian for the Phase Error during a Flip-Flop Cycle

Interval	Time	U_{rf}^+	$\tilde{\mathscr{H}}_P(t)$
δ_1	$0 \le t < \tau - \alpha$	1	0
δ_2	$\tau - \alpha \le t \le \tau + \alpha$	$\exp[-iI_x\theta(t)]$	$-\omega_1\phi_{\bar{x}}(I_y\cos\theta + I_z\sin\theta)$
δ_3	$\tau + \alpha < t < 3\tau - \alpha$	1	0
δ_4	$3\tau - \alpha \le t \le 3\tau + \alpha$	$\exp[iI_x\theta(t)]$	$\omega_1\phi_{\bar{x}}(I_y\sin\theta + I_z\cos\theta)$
δ_5	$3\tau + \alpha < t \le 4\tau$	1	0

In this case, we have explicitly recognized that only for the time interval in question, i.e., for $3\tau - \alpha \leq t \leq 3\tau + \alpha$, will the time-evolution operator involve $\theta(t)$. For other times, e.g., for $\tau - \alpha \leq t \leq \tau + \alpha$, the time-evolution operator is simply that due to a $\pi/2$ pulse. Therefore, we find that

$$\tilde{\mathscr{H}}_{\mathrm{P}}(3\tau - \alpha \leq t \leq 3\tau + \alpha) = \omega_1 \phi_{\bar{x}} \cdot R_{90_{\bar{x}}}[I_y \cos\theta - I_z \sin\theta]$$

$$= \omega_1 \phi_{\bar{x}}(I_z \cos\theta + I_y \sin\theta) \qquad (5.61)$$

the fourth entry in the last column of Table 5.4.

D. Calculation of the First-Order Phase-Error-Offset Cross Term $\bar{H}_{\mathrm{PO}}^{(1)}$ for the Flip-Flop Cycle

As a final illustration of average-Hamiltonian theory applied to pulse errors, we calculate the operator for the cross terms between a phase error and an offset Hamiltonian. We wish to evaluate all of the *cross terms* in the expression

$$\bar{H}_{\mathrm{PO}}^{(1)} = \frac{-i}{2t_{\mathrm{c}}} \int_0^{t_c} dt \left[(\tilde{\mathscr{H}}_{\mathrm{P}}(t) + \tilde{\mathscr{H}}_{\mathrm{O}}(t)), \right.$$

$$\left. \int_0^t dt'(\tilde{\mathscr{H}}_{\mathrm{P}}(t') + \tilde{\mathscr{H}}_{\mathrm{O}}(t')) \right] \qquad (5.62a)$$

These are the terms

$$\bar{H}_{\mathrm{PO}}^{(1)} = \frac{-i}{2t_{\mathrm{c}}} \int_0^{t_c} dt \left[\tilde{\mathscr{H}}_{\mathrm{P}}(t), \int_0^t dt' \tilde{\mathscr{H}}_{\mathrm{O}}(t') \right]$$

$$+ \int_0^{t_c} dt \left[\tilde{\mathscr{H}}_{\mathrm{O}}(t), \int_0^t dt' \tilde{\mathscr{H}}_{\mathrm{P}}(t') \right] \qquad (5.62b)$$

This calculation will involve a wrinkle already implied but not specifically illustrated. Since $\tilde{\mathscr{H}}_{\mathrm{P}}(t)$ contributes to the evolution of the system only during the pulses, its value between pulses is zero. On the other hand, since $\tilde{\mathscr{H}}_{\mathrm{O}}(t)$ contributes to the evolution of the spin system primarily between pulses (if $t_{\mathrm{w}} \ll t_{\mathrm{c}}$), we take its values to be zero during the pulses; i.e., $\tilde{\mathscr{H}}_{\mathrm{O}}$ (during pulses) $= 0$. To evaluate the commutators implied in (5.62), we therefore construct a table of $\tilde{\mathscr{H}}_{\mathrm{P}}(t)$ and $\tilde{\mathscr{H}}_{\mathrm{O}}(t)$ [Table 5.5]. To evaluate commutators in the double integrals of (5.62b) we use (4.51) as illustrated in Fig. 4.1. It is helpful at this point to construct a two-dimensional table of the values of $[\tilde{\mathscr{H}}_{\mathrm{O}}(t), \tilde{\mathscr{H}}_{\mathrm{P}}(t')]$, as evaluated by inspection from Table 5.5. The values of these commutators are shown in Table 5.6. For example, the value in row δ_3 and column δ_2 is found from the commutator $\tilde{\mathscr{H}}_{\mathrm{O}} = a_3 = -\Delta\omega I_y$ with $\tilde{\mathscr{H}}_{\mathrm{P}} = b_2 = -\omega_1 \phi_x(I_y \cos\theta + I_z \sin\theta)$. This value is $\omega_1 \Delta\omega \phi_x I_x \sin\theta$. The value of the commutator $[\tilde{\mathscr{H}}_{\mathrm{P}}, \tilde{\mathscr{H}}_{\mathrm{O}}]$ with $\tilde{\mathscr{H}}_{\mathrm{P}}$ being evaluated in the interval δ_2

TABLE 5.5
Hamiltonians in the Interaction Frame for the Flip-Flop Cycle

Interval	Time	$\mathscr{H}_{rf}$	U_{rf}^+	$\tilde{\mathscr{H}}_O$	$\tilde{\mathscr{H}}_P$
δ_1	$0 \le t < \tau - \alpha$	0	1	$\Delta\omega I_z = a_1$	0
δ_2	$\tau - \alpha \le t \le \tau + \alpha$	$-\omega_1 I_x$	$\exp[-iI_x\theta]$	0	$-\omega_1\phi_x(I_y\cos\theta + I_z\sin\theta) = b_2$
δ_3	$\tau + \alpha < t \le 3\tau - \alpha$	0	1	$-\Delta\omega I_y = a_3$	0
δ_4	$3\tau - \alpha \le t \le 3\tau + \alpha$	$\omega_1 I_x$	$\exp[iI_x\theta]$	0	$\omega_1\phi_{\bar{x}}(I_y\sin\theta + I_z\cos\theta) = b_4$
δ_5	$3\tau + \alpha < t \le 4\tau$	0	1	$\Delta\omega I_z = a_5$	0

TABLE 5.6
Cross Terms for Calculation of the Correction due to the Presence of Phase Errors in the Pulses of a Flip-Flop Cycle

$\tilde{\mathscr{H}}_O(t)$	$\tilde{\mathscr{H}}_P(t')$				
	δ_1	δ_2	δ_3	δ_4	δ_5
δ_1	0	$i\omega_1\Delta\omega\phi_x I_x\cos\theta$	0	$-i\omega_1\Delta\omega\phi_{\bar{x}}I_x\sin\theta$	0
δ_2	0	0	0	0	0
δ_3	0	$i\omega_1\Delta\omega\phi_x I_x\sin\theta$	0	$-i\omega_1\Delta\omega\phi_{\bar{x}}I_x\cos\theta$	0
δ_4	0	0	0	0	0
δ_5	0	$i\omega_1\Delta\omega\phi_x I_x\sin\theta$	0	$-i\omega_1\Delta\omega\phi_{\bar{x}}I_x\sin\theta$	0

and $\tilde{\mathcal{H}}_O$ being evaluated in the interval δ_3 would then be the negative of the above result.

In evaluating the entries of Table 5.6, we have used the angular momentum commutation rules (Chapter 2). Denoting by a_i the value of $\tilde{\mathcal{H}}_O(t)$ during time interval δ_i and by b_k the value of $\tilde{\mathcal{H}}_P(t)$ during time interval δ_k, as indicated in Table 5.5, we find

$$\frac{2it_c}{\omega_1 \Delta\omega} \bar{H}_{PO}^{(1)} = \sum_{j=1}^{n} \int_{\delta_j} dt \left[a_i, \sum_{k \leq j} \int_{\delta_k} dt' \, b_k \right]$$

$$+ \sum_{k=1}^{n} \int_{\delta_k} dt \left[b_k, \sum_{j \leq k} \int_{\delta_j} dt' \, a_j \right] \tag{5.63}$$

We see from Table 5.5 that only a_1, a_3, and a_5 are zero, as are b_2 and b_4. The nonvanishing first-order phase-error-offset cross terms, in detail then, are

$$\frac{2it_c}{\omega_1 \Delta\omega} \bar{H}_{PO}^{(1)} = \int_{\delta_2} dt \left[b_2, \int_{\delta_1} dt' \, a_1 \right] + \int_{\delta_3} dt \left[a_3, \int_{\delta_2} dt' \, b_2 \right]$$

$$+ \int_{\delta_4} dt \left[b_4, \left\{ \int_{\delta_3} dt' \, a_3 + \int_{\delta_1} dt' \, a_1 \right\} \right]$$

$$+ \int_{\delta_5} dt \left[a_5, \left\{ \int_{\delta_4} dt' \, b_4 + \int_{\delta_2} dt' \, b_2 \right\} \right]$$

We note that double integrals over identical time periods are zero, since a_i and b_i are nonzero in different periods. Therefore the double integrals of commutators are easy to evaluate, since limits on the inner integrals are constant (see Fig. 4.1).

For example, consider

$$\int_{\delta_2} dt \left[\tilde{\mathcal{H}}_P(t), \int_{\delta_1} dt' \, \tilde{\mathcal{H}}_O(t) \right] = \int_{\delta_2} dt \left[b_2, \int_{\delta_1} dt' \, a_1 \right]$$

$$= -i\omega_1 \Delta\omega\phi_x I_x \int_{\tau-\alpha}^{\tau+\alpha} dt \cos\theta(t) \int_0^{\tau-\alpha} dt'$$

$$= -i\omega_1 \Delta\omega\phi_x I_x (\tau - \alpha) \int_{\tau-\alpha}^{\tau+\alpha} dt \cos\theta(t)$$

If we assume that θ varies linearly from 0 to $\pi/2$ in period δ_2, we find $\theta = \pi(1 - \tau/\alpha + t/\alpha)/4$, or $dt = 4\alpha \, d\theta/\pi$, which yields

$$\int_{\tau-\alpha}^{\tau+\alpha} dt \cos\theta(t) = \frac{4\alpha}{\pi} \int_0^{\pi/2} d\theta \cos\theta = \frac{4\alpha}{\pi}$$

with the result

$$\frac{i}{2t_c} \int_{\delta_2} dt \left[\tilde{\mathcal{H}}_P(t), \int_{\delta_1} dt' \, \tilde{\mathcal{H}}_O(t') \right] = \frac{(\tau - \alpha)}{2\pi\tau} \omega_1 \Delta\omega\phi_x I_x$$

When evaluating the final result, remember that $2\alpha\omega_1 = \pi/2 = t_p\omega_1$, so we find

$$\bar{H}_{PO}^{(0)} = [(\phi_x + \phi_{\bar{x}})I_x\,\Delta\omega]/4 \qquad (5.64)$$

E. The Carr–Purcell–Meiboom–Gill Sequence with Ideal Delta-Function Pulses

The Carr–Purcell–Meiboom–Gill (CPMG) multiple-pulse experiment has already been treated classically in Chapter 1 and also discussed, in principle, in Section III of this chapter, in which it was found that $\bar{H}_O^{(0)}$ was zero to all orders for ideal delta-function pulses. Because of its historical importance and because of the pedagogical value of the discussion, we shall develop a much more extensive discussion of the experiment at this point. Initially a classical picture of the experiment, in terms of rotation of moments in a field, associated with $\pi/2)_x$ and $\pi)_y$ pulses is given. There are illustrated two possible quantum-mechanical methods of viewing the results of the experiment.

The CPMG sequence is utilized to measure transverse relaxation times, averaging to zero dc-field inhomogeneities at "spin echo" times between the 180° refocusing pulses. These field inhomogeneities are represented by an operator $\Delta\omega I_{zi}$ of the ith spin and have the effect of causing each spin in the ensemble to see a slightly different static field. Thus, the spins in the ensemble all precess with slightly different Larmor frequencies and lose phase coherence in the xy plane of the rotating frame. The sequence itself is simply one or more 180° pulses. In order to obtain an observable magnetization, a 90_x° preparation pulse is used to initially prepare the magnetization in the xy plane. For purposes of the ensuing calculations, we choose the cycle to be $[\tau, 180_y^\circ, 2\tau, 180_y^\circ, \tau]$, repeated n times. A spin echo, due to refocusing of spins in the xy plane, is observed at $4n\tau$, $n = 1, 2, 3, \ldots$, after the preparation pulse, which we choose to be 90_x°. An echo will, of course, occur at 2τ, after the first 180_y° pulse, etc., but its amplitude is sensitive to pulse-width errors, as we shall see. Thus we choose to consider the echoes between every pair, rather than between each 180_y° pulse. We assume all pulses to be ideal delta-function pulses. A classical picture of the behavior of an ensemble of nuclear spins under CPMG is shown in Fig. 5.15. After the preparation pulse (b), steps (c)–(f) are repeated and echoes observed in the window between the pairs of 180_y° pulses. If the first 180_y° pulse contains a pulse-width error ϵ, the odd echoes (first, third, . . .) will represent magnetization refocusing along y above and below the xy plane, at times $2\tau, 6\tau, \ldots$, as shown in Fig. 5.16. A subsequent error of ϵ in the second 180_y° pulse will commit the same mistake as the first but will have the effect of again placing the refocusing spins back in the xy plane, and so on for the even echoes, so the even echoes will have their highest possible values. An example of such a decay due to a

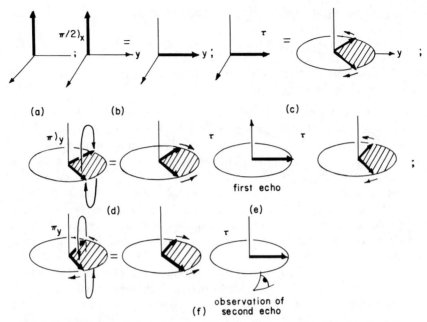

FIG. 5.15. The time evolution under the CPMG sequence, showing the refocusing effect.

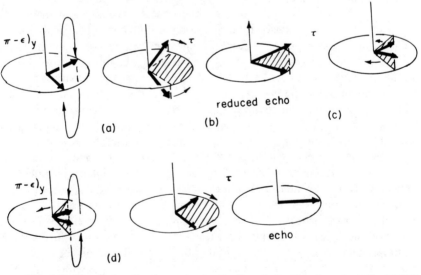

FIG. 5.16. The effect on the observed magnetization of missetting the pulse in the CPMG experiment.

CPMG sequence on water is shown in Fig. 5.17. The top scan of Fig. 5.17 shows the decay after the first 90°_x preparation pulse, as well as the succeeding echoes, so the first portion of the signal is not an echo but represents the on-resonance FID under the static field gradient. The bottom scan of Fig. 5.17 was triggered on the first 180° pulse, so only the response of the spin system associated with the 180° pulses is observed. It is appropriate to remark at this point that in the data shown, the mode of data accumulation was to trigger a transient recorder and then to let the recorder accumulate the data at times pre-set on the recorder and determined by a setting on the recorder, the instrumental "dwell time." Another possibility would have been to record data only at the peaks of the echoes, or "stroboscopically" at the echo times. Data taken in such a manner would only have exhibited the exponential decay associated with the limiting T_2 of the water sample, and not the dephasing and rephasing between 180° pulses. Such accumulation would have been possible by placing the recorder on "external time base" mode, with a trigger supplied every $4n\tau$ to signal data accumulation for a predetermined time during which the analog signal at the echo peak could be integrated for storage in memory as a single piece of digital information. Such a procedure allows T_2 to be obtained directly from the digitized data, without taking the trouble to pick out the echo amplitudes from a decay exhibiting dephasing and rephasing as well as echo amplitudes.

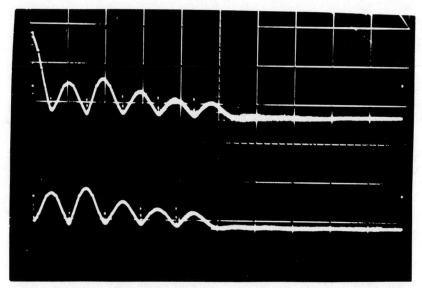

FIG. 5.17. Experimental observation of the missetting of a pulse's parameters in the CPMG experiment.

In the decay shown in Fig. 5.17, the width of the 180° pulse has been deliberately misadjusted to exhibit the effect upon the odd echoes. Note that the height of the first echo is not larger than that of the second, as it would be for perfect 180° pulses.

In order to calculate the effect of a dc-field gradient, with an internal Hamiltonian proportional to I_z, on the time development of the density matrix under this experiment, we may use at least two techniques. The first is to investigate the value of $U_{\text{int}}(Nt_c, 0)$. If the value of this operator is unity, then clearly by Eq. (5.23) the time development of the spin system is unaffected by that particular internal Hamiltonian of times $t = Nt_c$. A second approach is to realize that the Hamiltonians responsible for the time development of the system are time-independent in the observation frame, so that the time development of the density matrix may be directly calculated as a simple product of exponential operators. To help the student be more comfortable with such calculations, both approaches are illustrated in detail.

1. Calculation of $U_{\text{int}}(Nt_c, 0)$

The values of $\tilde{I}_z$ for this sequence have already been given by inspection in Fig. 5.4, the evaluations being made by using Eqs. (5.7) and (5.8). Equation (5.24) then gives by inspection that $U_{\text{int}}(t_c, 0)$ is unity, since the zeroth-order term directly integrates to zero, and all higher-order terms are zero by the fact that $\tilde{I}_z$ is $\pm I_z$ at all t and I_z commutes with itself. Therefore

$$U_{\text{int}}(4N\tau, 0) = 1 \tag{5.65}$$

and

$$\rho(4N\tau) = \rho(0) \tag{5.66}$$

which is to say that at times $t = 4N\tau$, the time development of the system is unaffected by I_z, as already discussed in Section III. An alternative statement is that dc-field inhomogeneities *do not* affect the damping constant measured under the CPMG sequence, neglecting, of course, relaxation effects as we have done. The reader will recognize that, since chemical shifts are also proportional to I_z, chemical-shift information will be lost if data are only taken at window times Nt_c.

It is easy to see that the time evolution of the system is similarly unaffected by I_z if one looks at the echo at $t = 2\tau$, after *a perfect π_y pulse*, as follows: by (5.18), we have, for times other than t_c, $t = t_c = 4\tau$,

$$\rho(t) = U_{\text{rf}}(t, 0)U_{\text{int}}(t, 0)\tilde{\rho}U_{\text{int}}^+(t, 0)U_{\text{rf}}^+(t, 0) \tag{5.67}$$

To calculate $U_{\text{rf}}(2\tau, 0)$, we utilize the Magnus expansion for U_{rf}:

$$U_{\text{rf}} = \exp\{-it_c(\bar{H}^{(0)} + \bar{H}^{(1)} + \cdots)\} \tag{5.68}$$

All terms higher than zero order vanish because they involve commutators of I_y with itself, or zero. Therefore, identifying $\tilde{\tilde{\rho}}$ as $\rho(0)$, we find

$$\rho(2\tau) = \exp\{-i\pi I_y\}U_{\text{int}}(2\tau, 0)\rho(0)U_{\text{int}}^{+}(2\tau, 0)\exp\{i\pi I_y\} \tag{5.69}$$

As before,

$$\bar{H}_{\text{int}}^{(0)}(2\tau) = \frac{1}{2}\left\{\int_0^\tau d\tau\, I_z + \int_\tau^{2\tau} d\tau(-I_z)\right\} = 0 \tag{5.70}$$

and

$$\bar{H}_{\text{int}}^{(i)} = 0, \qquad i > 1 \tag{5.71}$$

Therefore, $U_{\text{int}}(2\tau, 0) = 1$, and if $\tilde{\tilde{\rho}} \simeq I_y$,

$$\rho(2\tau) \sim \exp\{-i\pi I_y\}I_y\exp\{i\pi I_y\} = I_y \tag{5.72}$$

so the time development of the system is unaffected by the internal Hamiltonian $\mathscr{H}_{\text{int}} \sim I_z$. A redefinition of the cycle time as being between 3τ and 5τ yields, with the use of average Hamiltonian theory, the result that the odd echoes are similarly unaffected by I_z.

While average Hamiltonian theory is useful for many calculations, its limitations should not be misunderstood. It is important to understand that this device applies only at observation times t_c. For many purposes, we may wish to evaluate the behavior of the system at times other than these and indeed many experiments are noncyclic. Therefore a direct evaluation of the time development of the density matrix is desirable and is illustrated in the next discussion.

2. Direct calculation of the time development of ρ

In the discussion that follows, we recognize from Chapter 2 that ρ_{eq} is of the form $(1 - \beta H)/Z$. Here 1 is a unit matrix of the dimensionality of ρ; and in the high-temperature approximation, Z is a constant. Observables, which will be proportional to components of angular momentum, will be of the form $\text{Tr}\,\rho I_k$. Since $\text{Tr}\,1 \cdot I_k$ is zero, we may ignore the term 1. This leaves, for calculational purposes, $\rho_{\text{effective}} = -\beta H/Z$. We recognize therefore that ρ is proportional to I, and to ease notation in the ensuing calculations we shall write $\rho \sim I$ rather than $\rho = (1 - \beta H)/Z$ whenever appropriate. For example, with the system prepared by soaking in a static field, $\rho \sim I_z$. A 90_x° pulse will transform this situation into $\rho \sim I_y$, etc.

We wish to directly calculate the time development of the density matrix by utilizing the fact that with

$$i\frac{d\rho}{dt} = [\mathscr{H}, \rho]$$

and $\mathcal{H} \neq f(t)$,

$$\rho(t) = \exp\{-i\mathcal{H}t\}\rho(0)\exp\{i\mathcal{H}t\} \tag{5.73}$$

This result may be used for each segment of the cycle shown in Fig. 5.3. From $0 \leq t \leq \tau^*$, with $\tau^* = \lim_{\epsilon \to 0}(\tau - \epsilon)$, the system develops under the static, time-independent Hamiltonian $\mathcal{H}_{int} = \Delta\omega I_z$;

$$\rho(\tau^*) \sim \exp\{-i\Delta\omega I_z\tau^*\}I_y\exp\{i\Delta\omega I_z\tau^*\}$$
$$= -I_x\sin\Delta\omega\tau^* + I_y\cos\Delta\omega\tau^* \tag{5.74}$$

At $t = \tau$, there is the instantaneous transformation

$$\rho(\tau) \to \exp\{-i\pi I_y\}\rho(\tau^*)\exp\{i\pi I_y\} \sim I_x\sin\Delta\omega\tau + I_y\cos\Delta\omega\tau \quad (5.75)$$

For $\tau \leq t \leq 3\tau - \epsilon$ the system develops in time according to

$$\rho(3\tau - \epsilon) = \exp\{-i\Delta\omega I_z 2\tau\}\rho(\tau)\exp\{i\Delta\omega I_z 2\tau\}$$
$$\sim \sin\Delta\omega\tau(I_x\cos 2\Delta\omega\tau + I_y\sin 2\Delta\omega\tau)$$
$$+ \cos\Delta\omega\tau(-I_x\sin 2\Delta\omega\tau + I_y\cos 2\Delta\omega\tau) \tag{5.76}$$

At $t = 3\tau$, there is the instantaneous transformation

$$\rho(3\tau) \to \exp\{-i\pi I_y\}\rho(3\tau - \epsilon)\exp\{i\pi I_y\}$$
$$\sim -I_x(\sin\alpha\cos 2\alpha - \cos\alpha\sin 2\alpha)$$
$$+ I_y(\sin\alpha\sin 2\alpha + \cos\alpha\cos 2\alpha) \equiv aI_x + bI_y \tag{5.77}$$

where, for notational simplicity, we have a set $\alpha = \Delta\omega\tau$. For $3\tau \leq t \leq 4\tau$, the system develops in time according to

$$\rho(4\tau) = \exp(-i\alpha I_z)\rho(3\tau)\exp(i\alpha I_z)$$
$$\sim a(I_x\cos\alpha + I_y\sin\alpha) + b(-I_x\sin\alpha + I_y\cos\alpha) \tag{5.78}$$

By the use of standard trigonometric identities, we find

$$\rho(4\tau) = I_y \equiv \rho(0) \tag{5.79}$$

Thus, at window times $4N\tau$, the density operator is unaffected by static field gradients, and the magnitudes of spin echoes observed at these times will not be affected by these gradients. This result is strictly dependent upon the initial assumptions, i.e., that neither phase nor pulse-width errors are present and that time development during the periods of the pulses does not affect the experiment. The instructive calculation of the effect of pulse-width errors on the experiment is left as an exercise in Problems 5.8 and 5.9.

F. Pulse Sequences for Removal of Homonuclear Dipolar Interactions

In this section, we arrive at the destination to which we have been heading for the past four chapters with major excursions taken in order to be able to understand fully how, and why, the techniques in question work.

The pulse techniques described in this section allow removal of dipolar broadening from systems containing many nuclei of the same type, with relatively high gyromagnetic ratios (e.g., ^1H and ^{19}F). The use of NMR of such nuclei to extract chemical information in such systems, which are rigid (i.e., "solid"), is beset with an advantage and disadvantage, both of which are nonnegligible. The advantage has to do with the fact that NMR at room temperature is a relatively insensitive technique, since net populations of states leading to observable magnetizations are characterized by Zeeman splittings of the order of 0.01 K in temperature baths at 300 K. Therefore any feature of the system that will increase sensitivity is enormously desirable. One such feature is a high gyromagnetic ratio, since the magnetization, and therefore the observable signal, is proportional to the square of this number as discussed in Chapter 1. It is just this high gyromagnetic ratio, however, that leads to dipolar broadening that obscures the chemical-shift information; i.e., exactly that factor which makes those nuclei desirable as chemical probes in solids leads to their ability to obscure chemical information.

It is therefore desirable to counteract the features of nuclei that obscure chemical information while keeping those that enhance sensitivity. Such is the purpose of the experiments described in this section. The first such experiments that were proposed and executed actually had nothing to do with the use of rf pulses to average dipolar interactions. Rather, the technique used was that of "magic-angle rotation" in which the sample is spun about an axis making an angle of arc cos $3^{-1/2}$ to the magnetic field. The rotation frequency necessary for removal of dipolar interactions is, however, that which is in excess of the linewidth associated with such interactions. With dipolar linewidths in solids being commonly 20 kHz, therefore, achievement of sufficiently high rotational velocities to remove $\mathscr{H}_D$ was not possible without careful selection of systems to be studied. In addition, magic-angle rotation removes the anisotropy of the chemical shift; and in many cases, it is just this anisotropy that contains the most important chemical information. For achievement of high-resolution NMR in randomly oriented solids with strong homonuclear dipolar broadening, a coupling of magic-angle spinning and multiple-pulse homonuclear decoupling is necessary. It suffices to say at this point that it is the spatial portion of the dipolar Hamiltonian that is attacked by magic-angle spinning, and at present, we wish to continue the discussion of attacking the spin Hamiltonians with rf pulse sequences.

There are two types of pulse sequences that have been used to average homonuclear dipolar interactions. One, which focuses on the time development of $\mathscr{H}_D$ *between* pulses, has as its basic unit the two pulse sequence leading to the "solid echo," or "dipolar echo." This sequence is sketched in Fig. 5.18. In this approach, windows are left between pulses, and the magnetization of the abundant spins being decoupled is observed in specifically chosen windows. Another type of sequence that is conveniently used when it is desired

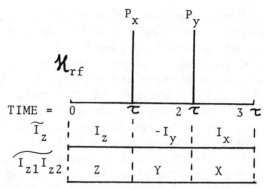

FIG. 5.18. The solid-echo sequence.

to decouple dipolar interactions between abundant spins, e.g., ¹H, while ob-
seving a rare spin, e.g., ¹³C, is the so-called windowless sequence [4]. The
windowless sequences accomplish homonuclear decoupling by arranging for
appropriate time development of $\mathscr{H}_D$ *during* the pulses. As the name implies,
however, such sequences do not allow observation of the spins being attacked
by the sequence. The process of decoupling using windowless sequences may
be understood by the same type of average Hamiltonian theory that we have
discussed for sequences with windows, and we pursue a description of the
latter for the purposes of this chapter.

Consider, therefore, the "dipolar echo" sequence of Fig. 5.18 [2, 5, and
6]. We immediately note that the dipolar-echo sequence is not cyclic, so if
one were to report this sequence with observation at $Nt_c = 3N\tau$, the condition of the vanishing of $\int_0^{t_c} \mathscr{H}_{rf}(t)\,dt$, which produces a cyclic sequence to
zero order, would be violated. An easy way to correct this violation is to
mirror the dipolar-echo sequence such that the radio frequency is phase-alternated. Such a four-pulse sequence [1] has been shown in Fig. 5.5. This sequence, *for purposes of properly adjusting $\bar{\mathscr{H}}_D(t)$ to average to zero over the
cycle time*, is a *symmetry equivalent* of the sequence shown in Fig. 5.18. The
concept of pulse-sequence symmetry has been extensively developed by
Mansfield [7], and the reader is referred to this work for more detailed treatment. Indicated in Fig. 5.5 are the identities of $I_z(t)$ in the frame of the rf $\tilde{I}_z$
and the identities of $I_{k1}I_{k2}(t)$, which for notational convenience are labeled
K. Since the scalar product $\mathbf{I}_1 \cdot \mathbf{I}_2$ is invariant to rotation, we immediately
see, as indicated in Fig. 5.18, that at cycle time $t_c = 3\tau$, the value of the zeroth-
order term for the homonuclear dipolar interaction

$$\bar{H}_D^{(0)} = \frac{1}{t_c} \int_0^{t_c} dt\, \mathscr{H}_D$$

vanishes.

For ideal delta-function pulses, we have shown that, under such four-pulse sequences, the value of $\bar{H}_D^{(0)}$, the zeroth-order average dipolar Hamiltonian, is zero. The zeroth-order value of the internal Hamiltonians associated with I_z, e.g., chemical shifts, dc-field inhomogeneities, and resonance offsets, is seen to be $(I_x - I_y + I_z)/3$ by inspection. As discussed in Section V.B, this value of $\bar{H}^{(0)}$ corresponds to an effective field of $\boldsymbol{\Omega} = (\mathbf{i} - \mathbf{j} + \mathbf{k})/3$, with absolute value $\omega = 3^{-1/2}$. The physical result that this implies is that the observed evolution under the WAHUHA sequence associated with the operator I_z will exhibit a frequency that is scaled by $3^{-1/2}$. This is to say that if a measurement of the chemical shift of the protons in toluene liquid were performed at 60 MHz utilizing the WAHUHA sequence, the splitting between the aromatic and aliphatic protons would be observed to be $273/\sqrt{3} = 158$ Hz. The use of this sequence *decreases* the resolution by a factor of 1.73. On the other hand, the dipolar broadening of protons in solid toluene leads to a single line spectrum roughly 30 kHz wide. The use of the line-narrowing sequence, therefore, could, in principle, buy a considerable amount of chemical information relative to the information available without multiple-pulse averaging.

The multiple-pulse averaging experiment has some associated quirks that must be faced before usable resolution may be attained. One is the fact that the experiment involves exposing the system to a sequence of pulses, collecting data at the sampling window, and repeating this process as many times as necessary to obtain the required spectral information. With the criterion $\|\mathcal{H}_{\text{int}}\| t_c \ll 1$ and t_c taken to be 3τ for the dipolar-echo sequence, a reasonable cycle time for the four-pulse experiment, 6τ, is calculated to be 20 μsec if dipolar interactions leading to linewidths of 30 kHz are to be narrowed. Suppose the inhomogeneously broadened linewidth due to residual chemical-shift anisotropy in a powder sample is 30 ppm, a value found for ^{1}H in solid H_2O. At a proton frequency of 60 MHz, this broadening would amount to 1.8 kHz. The resultant decay under the multiple-pulse experiment will be over in about $3T_2$, or roughly 500 μsec, using $2\delta f = (\pi T_2)^{-1}$ as a measure of T_2.

We note at this point that if single-phase detection is used, a resonance offset of 3 kHz will assure a representation of the signal in the frequency domain that does not contain artifacts due to reflections through the carrier frequency. This statement is not supposed to be obvious, and we have not discussed the sampling problem elsewhere in the text, so the uninitiated reader will simply want to note that a 3 kHz offset is necessary. A multiple-pulse experiment with a cycle time of 20 μsec will accumulate 25 data points before further accumulation leads to useless data. With a resonance óffset of 3 kHz, the maximum time between data points in the time domain is $(2 \times 3000)^{-1}$ sec $= 165$ μsec, so the 25 data points, each 20 μsec apart, will be more than sufficient for an accurate representation of the data in the frequency domain. With quadrature detection, the carrier may be in the center of the spectrum, and in the above example, the maximum frequency sampled would be 900 Hz. Note that the shielding anisotropy scales linearly with the magnetic field (see Problem 5.13).

A problem with the multiple-pulse technique in such an accumulation is that errors accumulate as well. If there exist pulse errors that are *not* averaged to zero by the sequence in question, these will rapidly accumulate and destroy the experiment, i.e., lead to broadening. A problem with the four-pulse experiment discussed here is that many cross terms between the dipolar Hamiltonian and pulse imperfections do not vanish under this sequence. In addition, when finite pulse widths are taken into account, *even* the zeroth-order average dipolar Hamiltonian is nonvanishing and (as the reader is given an opportunity to verify in Problem 5.11) is evaluated to be

$$\bar{H}_D^{(0)} = -\sum_{i<j} B_{ij} \frac{6}{\pi} \frac{t_\omega}{t_c} \left[I_{yi}(I_{xj} + I_{zj}) + (I_{xi} + I_{zi})I_{yj} \right] \tag{5.80}$$

Instrumental effects therefore become a serious consideration when attempting to remove homonuclear dipolar couplings, and, in fact, for maximum resolution under symmetry equivalents for four-pulse sequences, it is necessary to introduce deliberate pulse-width errors. A sequence that avoids many of these problems is shown in Fig. 5.19. Rhim *et al.* [8] used this sequence, with preparation pulse P_x, to enhance resolution of homonuclear dipolar-broadened lines in the solid state and demonstrated its utility by attaining a linewidth of 10 Hz for the ^{19}F spectrum of a spherical single crystal of CaF$_2$ with the external field along the $(1, 1, 1)$ direction of the crystal. The use of a spherical crystal allowed removal of broadening due to inhomogeneous demagnetization fields and alignment along $(1, 1, 1)$ placed nearest-neighbor fluorines at the magic angle to the external field with resultant

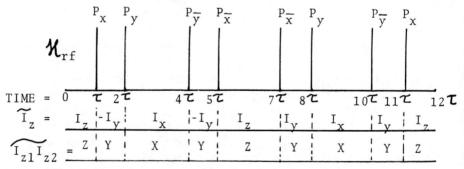

FIG. 5.19. The eight-pulse sequence first used by Rhim *et al.* [6a] to suppress dipole–dipole interactions. Note that this sequence is a specific combination of solid-echo sequences.

partial removal of dipolar interactions by virtue of geometry. As can be seen by inspection of Fig. 5.19, the zeroth-order dipolar Hamiltonian vanishes under this sequence, and the zeroth-order offset-plus-chemical-shift Hamiltonian becomes

$$\bar{H}^{(0)}_{O+cs} = \frac{1}{3} \sum_i (\Delta\omega + \omega_0 \sigma_{zzi})(I_{xi} + I_{zi}) \tag{5.81}$$

When finite pulse widths are taken into account, $\bar{H}^{(0)}_{O+cs}$ changes by a factor of $(1 + 2a)$, where

$$a = (3t_w/t_c)(4/\pi - 1) \tag{5.82}$$

In the limit of $t_w/t_c \ll 1$, therefore, chemical shifts are scaled by a factor of $\sqrt{2}/3$ under this sequence. In practice, the scaling factor is measured experimentally by summing time-domain scans under a given sequence on, e.g., ^{1}Hs in water that differ in offset frequency by, e.g., 1 kHz. The resultant sum is a beat pattern which is Fourier transformed to yield a series of peaks 1 kHz apart in actual frequency difference but less than 1 kHz apart in the digitized data. In these data, the number of hertz per channel is calculated from the dwell time and the number of points in the transform: hertz per point in Fourier transform equals (dwell time multipled by number of points in Fourier transform)$^{-1}$. For example, with an eight-pulse cycle time (which is the dwell time in the experiment) of 20 μsec and a 2048-point transform, there are 24.41 (scaled) Hz per point in the frequency domain of the eight-pulse experiment. If the peaks in the frequency spectrum under successive 1-kHz offsets are, on average, 20 channels apart in the digitized data, then the scaling factor is calculated to be $(20 \times 24.41)/1000 = 0.488$. This means that a 500-Hz spectral width under the multipulse experiment is $500 \div 0.488 = 1024$ real hertz. The calculation of average Hamiltonians under both four- and eight-pulse sequences becomes tedious when all possible energy and cross terms are taken into account to second order but have been published by Rhim et al. [9] and are listed in Table 5.7. The serious reader will wish to verify these results and should be able to do so by using the techniques and examples given previously in this chapter. As indicated in Table 5.7, not only does $\bar{H}^{(0)}_D$ vanish under the eight-pulse sequence discussed here, but so do the first-order dipolar term $\bar{H}^{(1)}_D$ as well as the first-order dipole-offset, dipole-phase-transient, and dipole-phase-error terms. Burum and Rhim [3] have subsequently developed 24-(the BR-24) and 52-pulse homonuclear decoupling sequences in which second-order dipolar terms are made to vanish. Since the 24- and 52-pulse sequences are essentially equally efficient in dipolar decoupling, only the BR-24 will be described here. If the symbol $1a$ is used to denote the four-pulse sequence $[\tau, P_x, \tau, P_y, \tau, P_{\bar{y}}, \tau, P_{\bar{x}}, \tau]$, with shorthand notation $(XY)(\bar{Y}\bar{X})$, and other four-pulse sequences are denoted as

TABLE 5.7

Average Hamiltonians (through Second Order) for the Four- and Eight-Pulse Cycles[a]

Average Hamiltonian	Four-pulse WAHUHA cycle	Eight-pulse MREV-8 cycle
$\bar{H}_0^{(0)b}$	$\frac{1}{3}\sum_i(\Delta\omega+\omega_0\sigma_{zzi})(1+a)(I_{xi}+I_{yi}+I_{zi})$	$\frac{1}{3}\sum_i(\Delta\omega+\omega_0\sigma_{zzi})(1+2a)(I_{xi}+I_{zi})$
$\bar{H}_0^{(1)}$	0	$\frac{1}{36}t_c\sum_i(\Delta\omega+\omega_0\sigma_{zzi})^2(I_{xi}-I_{zi})$
$\bar{H}_D^{(0)}$	$-\sum_{i<j}B_{ij}\dfrac{6}{\pi}\dfrac{t_w}{t_c}[I_{yi}(I_{xj}+I_{zj})+(I_{xi}+I_{zi})I_{yj}]$	0
$\bar{H}_D^{(1)}$	0	0^c
$\bar{H}_D^{(2)}$	$(t_c^2/648)[\mathcal{H}_D^{(x)}-\mathcal{H}_D^{(z)},[\mathcal{H}_D^{(x)},\mathcal{H}_D^{(y)}]]^d$	$(t_c^2/2592)[\mathcal{H}_D^{(x)}-\mathcal{H}_D^{(z)},[\mathcal{H}_D^{(x)},\mathcal{H}_D^{(y)}]]$
$\bar{H}_{DO}^{(1)}$	0	0
$\bar{H}_{DO}^{(2)e}$	f	$(t_c^2/864)[(\Delta\omega)[I_x,[\mathcal{H}_D^{(x)},\mathcal{H}_D^{(y)}]]+(t_c^2/432)(\Delta\omega)^2(\mathcal{H}_D^{(z)}-\mathcal{H}_D^{(x)})$
$\bar{H}_P^{(0)}$	$(1/t_c)[\phi_y-\phi_{-y}]I_x+(-\phi_x+\phi_{-x}-\phi_y+\phi_{-y})I_y+(\phi_x-\phi_{-x})I_z]$	$(2/t_c)[\phi_y-\phi_{-y}]I_x+(-\phi_x+\phi_{-x})I_y$
$\bar{H}_{PO}^{(1)}$	$\frac{1}{6}\sum_i(\Delta\omega+\omega_0\sigma_{zzi})[-(\phi_x+\phi_{-x})(I_{xi}-I_{yi})+(\phi_x+\phi_{-x}-\phi_y-\phi_{-y})I_{zi}]$	$\frac{1}{6}\sum_i(\Delta\omega+\omega_0\sigma_{zzi})[-(\phi_y+\phi_{-y})I_{xi}+2\phi_x I_{yi}]$
$\bar{H}_{PD}^{(1)}$	$\frac{1}{4}(-\phi_x-\phi_{-x}+\phi_y+\phi_{-y})\sum_{i<j}B_{ij}(I_{xi}I_{zj}+I_{zi}I_{xj})$	0
$\bar{H}_T^{(0)g}$	$(1/t_c)J_1(I_x+2I_y+I_z)$	$(2/t_c)J_1(I_x+I_z)$
$\bar{H}_{TO}^{(1)g}$	$\frac{1}{6}J_2\sum_i(\Delta\omega+\omega_0\sigma_{zzi})(I_{xi}-I_{yi})$	$\frac{1}{6}\sum_i(\Delta\omega+\omega_0\sigma_{zzi})[3J_1(I_{xi}-I_{zi})-J_2 I_{yi}]$
$\bar{H}_{TD}^{(1)}$	0	0
$\bar{H}_\delta^{(0)}$	$(1/t_c)[(-\delta_x+\delta_{-x})I_x+(\delta_y-\delta_{-y})I_z]$	$(2/t_c)(-\delta_x+\delta_{-x})I_x$

$\bar{H}_{\delta O}^{(1)}$	$-\frac{1}{6}\sum_i(\Delta\omega+\omega_0\sigma_{zzi})[(\delta_x+\delta_{-x})I_{zi}+(\delta_y+\delta_{-y})I_{yi}]$	$\frac{1}{6}\sum_i(\Delta\omega+\omega_0\sigma_{zzi})[-2\delta_{-x}I_{zi}+(\delta_y-\delta_{-y})I_{yi}]$
$\bar{H}_{\delta D}^{(1)}$	$\frac{1}{2}\sum_{i<j}B_{ij}(\delta_x+\delta_{-x})(I_{yi}I_{zj}+I_{zi}I_{yj})+(\delta_y+\delta_{-y})(I_{xi}I_{yj}+I_{yi}I_{xj})$	0
$\bar{H}_{\epsilon}^{(0)}$	0	0
$\bar{H}_{\epsilon O}^{(1)}$	$-\frac{1}{3}\sum_i\epsilon_i(\Delta\omega+\omega_0\sigma_{zzi})(I_{yi}+I_{zi})$	$-\frac{1}{3}\sum_i\epsilon_i(\Delta\omega+\omega_0\sigma_{zzi})I_{zi}$
$\bar{H}_{\epsilon D}^{(1)h}$	$\sum_{i<j}\epsilon_iB_{ij}[I_{yi}(I_{xj}+I_{zj})+(I_{xi}+I_{zi})I_{yj}]$	0
$\bar{H}_{d}^{(0)}$(nth cycle)[i]	$-\frac{\omega_s t_w}{t_c}\frac{2}{b}\exp(-6n/b)(2I_x+I_z)$	$\frac{\omega_s t_w}{t_c}\frac{12}{b^2}\exp(-12n/b)\sum_{i<j}B_{ij}[I_{yi}(I_{xj}+I_{zj})+(I_{xi}+I_{zi})I_{yj}]$
$\bar{H}_{dD}^{(1)}$(nth cycle)	$\omega_s t_w[1+(-1+3/b)\exp(-6n/b)]$ $\times\sum_{i<j}B_{ij}[I_{yi}(I_{xj}+I_{zj})+(I_{xi}+I_{zi})I_{yi}]$	$(3\omega_s t_w/b)\exp(-12n/b)\sum_{i<j}B_{ij}[I_{yi}(I_{xj}+I_{zj})+(I_{xi}+I_{zi})I_{yj}]$

[a] O, offset; D dipolar; P, phase transient; T, phase misadjustment; δ, pulse-width misadjustment; ϵ, rf inhomogeneity.

[b] $a=(3t_w/t_c)(4/\pi-1)$.

[c] For the eight-pulse cycle, even for pulses of finite width $\bar{H}_D^{(1)}=0$.

[d] $\mathscr{H}_D^{(k)}=I_1\cdot I_2-3I_{k1}I_{k2}$.

[e] This calculation assumes that all nuclei are chemically and geometrically equivalent.

[f] $\bar{H}_{DO}^{(2)}=(t_c^2/648)\{-3\Delta\omega[\mathscr{H}_D^{(y)},[\mathscr{H}_D^{(x)},I_x]]-3\Delta\omega[\mathscr{H}_D^{(z)},[\mathscr{H}_D^{(x)},I_x]]+(\Delta\omega)^2[I_y+I_z,[\mathscr{H}_D^{(y)},I_x]]+(\Delta\omega)^2[4I_x+3I_y+I_z,[\mathscr{H}_D^{(x)},I_y]]$
$+i(\Delta\omega)^2[\mathscr{H}_D^{(y)},I_x]-i(\Delta\omega)^2[\mathscr{H}_D^{(x)},I_y]+i(\Delta\omega)^2[\mathscr{H}_D^{(x)},I_z]\}$.

[g] $J_1=\int_0^{t_w}\omega_{rf}(t)(\sin\omega_1 t-\cos\omega_1 t)dt$; $J_2=\int_0^{t_w}\omega_{rf}(t)(\sin\omega_1 t+\cos\omega_1 t)dt$.

[h] It is assumed that ω_1 is constant over a scale of molecular dimensions.

[i] $\bar{H}_d^{(0)}$(nth cycle)$=\bar{H}_d^{(0)}[nt_c\to(n+1)t_c]$ and similarly for $\bar{H}_{dD}^{(1)}$. Exponential decay of $\omega_{rf}(t)$ with time constant b; $b\gg1$ is assumed.

$1b = (\bar{X}Y)(\bar{Y}X)$, $2a = (YX)(\bar{X}\bar{Y})$, $2b = (\bar{Y}X)(\bar{X}Y)$, then the BR-24 is denoted by $1a1b2a(\bar{Y}X)2a2b(\bar{X}Y)$.

Perhaps one of the greatest challenges (short of H_2 and F_2) to the ability of homonuclear decoupling sequences to reveal chemical-shift information in the solid state is that posed by the ^{1}Hs in linear, high-density solid polyethylene. The challenge results from the relative magnitudes of the shielding anisotropy $\sigma_{zz} - \bar{\sigma} = -2.1$ ppm relative to the 30-kHz dipolar broadening.

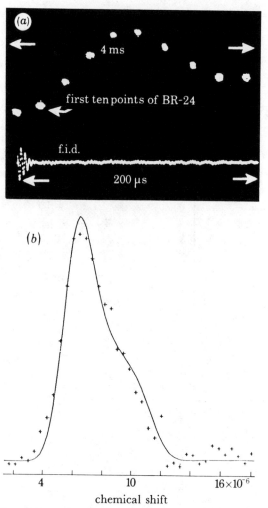

FIG. 5.20. The solid-state chemical-shift NMR spectrum of a sample of crystalline polyethylene, obtained by using dipole–dipole decoupling sequences: (a) top, first 10 points under BR-24, bottom, FID; (b) Fourier transform of decay under the eight-pulse cycle.

The results of homonuclear decoupling experiments on solid high-density linear polyethylene are shown in Fig. 5.20. The time response under the BR-24 compared to the response under a single-pulse excitation is shown in Fig. 5.20(a). The frequency spectrum under the previously discussed eight-pulse cycle, with an X prepulse and cycle time $12\tau = 21$ μsec, is shown in Fig. 5.20(b).

We note at this point that the choice of homonuclear decoupling sequence will depend upon the available rf power (i.e., the achievable cycle time) that sets the minimum dwell time and thus the spectral width (see Problem 5.14). Shorter cycle times reduce the effects of higher-order terms $\bar{H}_D^{(i)}$ and $\bar{H}_{D-X}^{(i)}$ on line width but are achieved, for given power, at the expense of increased Q, i.e., increased probe ringdown time, $\tau = Q/3f$ (see Chapter 1).

VI. The Effect of Motion upon Removal of Dipolar Interactions by Multiple-Pulse Techniques

In all of the discussion of the use of multiple-pulse techniques for removal of dipolar interactions, it has been, thus far, tacitly assumed that the dipolar interactions are time-independent. We know physically, however, that a spin system experiencing sufficiently rapid isotropic motion, e.g., molecules tumbling in a liquid, will of itself average the dipolar broadening to zero. In addition, systems exhibiting sufficiently rapid anisotropic motion will exhibit partial removal of the dipolar interaction. The effect of such motion upon the ability of the MREV-8 multiple-pulse sequence to remove dipolar broadening is considered in this discussion.

Until now, only the spin portion of the secular dipolar Hamiltonian has been considered in the discussion of multiple-pulse experiments. The spatial portion $(\gamma^2\hbar/r_{12}^3)(3\cos^2\theta_{12} - 1) = KY_{20}$ has been explicitly considered to be time-independent. When Y_{20} is allowed to be time-dependent, the calculations of the average dipolar Hamiltonian under given pulse sequences must be changed to accommodate this dependence. More specifically, $KY_{20}(t)$ must be incorporated into the integrals over time of $\tilde{\mathcal{H}}_D(t)$. A quick calculation of the zeroth-order average Hamiltonian in the Magnus expansion gives

$$\bar{H}_D^{(0)} = K\left\{\int_0^1 P_z(x)Y_{20}(xt_c)\,dz\,Z\right.$$

$$+ \int_0^1 P_y(x)Y_{20}(xt_c)\,dy\,Y$$

$$\left. + \int_0^1 P_x(x)Y_{20}(xt_c)\,dx\,X\right\} \tag{5.83}$$

where P_x, P_y, and P_z take on values of 1, 0, or -1 in the various "windows" of the sequence. It is obvious that, generally, $\bar{H}_D^{(0)}$ does not have a value of zero. If $Y_{20}(xt_c)$ is not dependent on x, the expression of (5.83) is zero. This is the case under which multiple-pulse NMR schemes were derived, and which was discussed previously. In the case that each integral in (5.83) vanishes identically, as when the dipolar vector is moving very fast and isotropically, $\bar{H}_D^{(0)}$ is zero. In the cases that are not at the extremes, one must reconsider the effect of molecular motion on the response to the multiple-pulse sequence.

We follow the approach of Vega and Vaughan [10] to the introduction of anisotropic motion into the time development. Anisotropy is introduced by separating the time-dependent dipole–dipole interaction into a static and fluctuating part:

$$\mathcal{H}_D(t) = \mathcal{H}_{stat} + \mathcal{H}_{fluc}(t) \tag{5.84}$$

The dipole–dipole Hamiltonian is time-dependent because the angles Ω are time dependent. If $P(\Omega, t)$ is the distribution function for Ω that arises from a Markovian process, the equation of motion is

$$\partial P(\Omega, t)/\partial t = \Gamma P(\Omega, t) \tag{5.85}$$

where Γ is the transition operator. The incorporation of the resulting time dependence into the Liouville–von Neumann equation evaluated in a frame rotating about the dc magnetic field and toggling with the rf sequence is

$$\frac{\partial}{\partial t} \rho^+ \cong -i[\bar{H}_{stat}^{(0)}, \rho^+] - \frac{g_0^2}{t_c} \int_0^{t_c} dt_2$$

$$\times \int_0^{t_c} dt_1 [\mathcal{H}_{fluc}^+(t_1), [\mathcal{H}_{fluc}^+(t_2), \rho^+]]\eta(t_2 - t_1) \tag{5.86}$$

where $\{g_n(\Omega)\}$ is the set of eigenfunctions of Γ and $\eta(t_2 - t_1)$ a transfer function [10]. The effect of the static part of the interaction may be treated as an average Hamiltonian. For this sequence, the static dipolar coupling is averaged to zero. Considering only the effect of the second term, one may calculate the time evolution of the stroboscopically observed magnetization:

$$\frac{d}{dt}\langle I_x \rangle = -\frac{1}{T_{1xz}} \langle I_x \rangle \tag{5.87a}$$

$$\frac{d}{dt}\langle I_y \rangle = -\frac{1}{T_{1y}} \langle I_y \rangle \tag{5.87b}$$

$$\frac{d}{dt}\langle I_z \rangle = -\frac{1}{T_{1xz}} \langle I_z \rangle \tag{5.87c}$$

where

$$\frac{1}{T_{1xz}} = \frac{2}{3}\Delta M_2 \left\{ \tau_c - \frac{3\tau_c^2}{t_c} \right.$$

$$\times \left[\frac{5 - 4\exp(-t_c/12\tau_c) + 4\exp(-t_c/4\tau_c) - 5\exp(-t_c/3\tau_c)}{1 + \exp(-t_c/6\tau_c) + \exp(-t_c/3\tau_c)} \right] \right\} \quad (5.87d)$$

$$\frac{1}{T_{1y}} = \frac{2}{3}\Delta M_2 \left\{ \tau_c - \frac{6\tau_c^2}{t_c} \right.$$

$$\times \left[\frac{1 + \exp(-t_c/12\tau_c) - \exp(-t_c/4\tau_c) - \exp(-t_c/3\tau_c)}{1 + \exp(-t_c/6\tau_c) + \exp(-t_c/3\tau_c)} \right] \right\} \quad (5.87e)$$

where an exponential correlation function is assumed and ΔM_2 is the difference between the rigid-lattice second moment and the sample's measured second moment and $t_c = 12\tau$.

The loss of resolution represented by Eqs. (5.87) is not the result of static higher-order terms in the Magnus expansion. They arise rather from a failure of the sequence to achieve removal of dipole–dipole coupling, because these couplings change independently of the rf excitation's imposed time dependence. Measurements of these spin–lattice contributions to broadening in multiple-pulse NMR can give information on the motion-modulating dipolar couplings, but care must be exercised in interpreting the results of such measurements since higher-order *static* terms may also contribute to the damping of the magnetization.

VII. Second Averaging

Thus far, average Hamiltonian theory has been used to show how a given Hamiltonian can be averaged to zero by attacking the interaction with a cyclic, periodic sequence of radio-frequency pulses. A generalization of this idea is made by utilizing *any* given average Hamiltonian to remove the effects of others. The basic requirements for such averaging are that the interaction being averaged be small compared to the interaction doing the averaging (e.g., a 5-G dipolar field can be averaged by a 60-Grf field) and that the integral of the averaged Hamiltonian in the interaction frame appropriate to the larger Hamiltonian go to zero over a cycle associated with the larger Hamiltonian:

$$\int_0^{t_c} \mathscr{H}_{int}(t)\, dt = 0 \quad (5.88)$$

As an example, consider the effect of the phase error, large compared to a dc-field inhomogeneity on the phase-tuning flip-flop cycle discussed in Section IV.A. We previously calculated that the average Hamiltonian associated with a phase error in this cycle is given by

$$\bar{H}_{\mathrm{P}}^{(0)} = (1/t_{\mathrm{c}})(\phi_{\bar{x}} - \phi_x)(I_y + I_z) \tag{5.89}$$

with associated Ω vector (see Section V.B).

$$\Omega_{\mathrm{P}} = (1/t_{\mathrm{c}})(\phi_{\bar{x}} - \phi_x)(\mathbf{j} + \mathbf{k})$$

In the limit of infinitesimal pulse widths, i.e., $t_{\mathrm{w}}/t_{\mathrm{c}} \ll 1$, the average Hamiltonian associated with a dc-field inhomogeneity leading to an offset $\Delta\omega$ is

$$\bar{H}_{\mathrm{O}}^{(0)} = (\Delta\omega/2)(I_z - I_y) \tag{5.90}$$

with associated Ω vector

$$\Omega_0 = (\Delta\omega/2)(-\mathbf{j} + \mathbf{k})$$

We inquire about the effect of deliberately introducing a phase error in the $\bar{x}$ pulse that will cause an interaction large compared to the offset interaction of a dc inhomogeneity. This condition is that

$$|\phi_{\bar{x}}/t_{\mathrm{c}}| \gg |\Delta\omega| \tag{5.91}$$

We first note that Ω_{p} and Ω_0 are perpendicular. This fact leads to a physically intuitive manner of thinking about the second-averaging phenomenon as follows: if $|\Omega_{\mathrm{P}}| \gg |\Omega_0|$, the observed motion of the system will be the projection on the axis of observation of precession of the moment about the vector $\Omega_{\mathrm{P}} + \Omega_0$. The contribution of Ω_0 to the observed frequency of rotation is obtained from the value of

$$\omega = |\Omega| = \sqrt{\Omega_{\mathrm{P}}^2 + \Omega_0^2} \tag{5.92}$$

If $\Omega_{\mathrm{P}}/\Omega_0 = 10$, for example, Ω_0 will perturb the observed frequency by a factor of $\frac{1}{200}$. The formulation of this effect utilizing average Hamiltonian theory is as follows. Under the flip-flop sequence the zeroth-order average Hamiltonian for an offset $\Delta\omega$ in the limit of $t_{\mathrm{w}}/t_{\mathrm{c}} \ll 1$ is given by Eq. (5.90). The zeroth-order average Hamiltonian for a phase error along $\bar{x}$ is

$$\bar{H}_{\mathrm{P}}^{(0)} = \phi_{\bar{x}}(I_y + I_z)/t_{\mathrm{c}} \tag{5.93}$$

When $\phi_{\bar{x}}/t_{\mathrm{c}}$ is deliberately set much larger than $\Delta\omega$ for the sequence, the motion of the system will be dominated by $\bar{H}_{\mathrm{P}}^{(0)}$, so it is natural to transform into the interaction frame of $\bar{H}_{\mathrm{P}}^{(0)}$ and inquire about the motion of the system associated with $\bar{H}_{\mathrm{O}}^{(0)}$ in this frame. It is important to keep in mind that the frame of $\bar{H}_{\mathrm{P}}^{(0)}$ is not the frame of observation. Hence the observed signal will contain information about $\bar{H}_{\mathrm{P}}^{(0)}$.

There are times $t' = Nt_c$ at which the observation frame and the inter-action frame coincide. These are defined by the times for which

$$U_P = T \exp\left\{-i \int_0^{t_c} \bar{H}_P^{(0)}(t') \, dt'\right\} \tag{5.94}$$

is unity. At these times, we may treat the evolution under all other Hamil-tonians in the average-Hamiltonian sense. But this is a second-average Hamiltonian:

$$\bar{\bar{H}}_O^{(0)} = \frac{1}{t_c'} \int_0^{t_c} \tilde{\bar{H}}_O^{(0)}(t') \, dt' \tag{5.95}$$

where

$$\tilde{\bar{H}}_O^{(0)}(t) = U_P^{-1}(t)\bar{H}_O^{(0)}U_P(t) \tag{5.96}$$

Because of the manner in which data are taken, the rf observation frame is *not* always coincident with the interaction frame of $\bar{H}_P^{(0)}$. It turns out that the oscillation in the rf frame induced by $\bar{H}_P^{(0)}$ allows us a way to identify these points. They are the points separated by a cycle of the oscillation.

Since $\bar{H}_P^{(0)}$ is time-independent, this evolution operator is

$$U_P = \exp(-i\bar{H}_P^{(0)}t) = \exp[-i\phi_{\bar{x}}(I_y + I_z)t/t_c] \tag{5.97}$$

The resultant motion is simply an oscillation about $\mathbf{\Omega}_P$ with frequency

$$\omega = |\mathbf{\Omega}_P| \equiv \sqrt{2}\,\phi_{\bar{x}}/t_c \quad \text{rad sec}^{-1} = \sqrt{2}\,\phi_{\bar{x}}/2\pi t_c \quad \text{Hz} \tag{5.98}$$

Since the motion of the spin system under $\bar{H}_P^{(0)}$ is periodic, with period

$$t_c' = \sqrt{2}\,\pi t_c/\phi_{\bar{x}}$$

a good bet is to inquire upon the effect of $\tilde{\bar{H}}_O^{(0)}$ at times Nt_c'. The result is the "double" average Hamiltonian for the offset operator

$$\bar{\bar{H}}_O^{(0)} = \frac{1}{t_c'} \int_0^{t_c} \tilde{\bar{H}}_O^{(0)} \, dt \tag{5.99}$$

where

$$\begin{aligned}\tilde{\bar{H}}_O^{(0)} = U_P^{-1}\bar{H}_O^{(0)}U_P &= \exp\{i\phi_{\bar{x}}(I_y + I_z)t/t_c\} \\ &\times [(I_z - I_y)\Delta\omega/2] \exp\{-i\phi_{\bar{x}}(I_y + I_z)t/t_c\}\end{aligned} \tag{5.100}$$

Since I_y and I_z do not commute, the exponential operators are not a simple product of exponentials. The standard trick, as seen previously in the text, is to transform to a coordinate system in which the exponential operator contains only one component of angular momentum, perform the required transformation, and then invert to the original coordinate system. A bit of reflection reveals that a rotation about $\mathbf{i}$ by $\pi/4$ in either direction will accomplish this end; i.e., rotating the vector $\mathbf{k} + \mathbf{j}$ by $\pi/4$ about $\mathbf{i}$ converts it into either $\mathbf{k}$ or $\mathbf{j}$. A $\pi/4$ rotation about x that will place $\mathbf{k} + \mathbf{j}$ along the $\mathbf{k}$ direction is performed by the operator $\exp(-iI_x\pi/4)$; i.e.,

$$\exp(-iI_x\pi/4)(I_z + I_y)\exp(iI_x\pi/4)$$
$$= I_z\cos\pi/4 - I_y\sin\pi/4 + I_y\cos\pi/4 + I_z\sin\pi/4 = \sqrt{2}I_z' \quad (5.101)$$

Similarly, this rotation carries $(I_z - I_y)$ into $-\sqrt{2}I_y'$. In the primed coordinate system, therefore

$$\tilde{\bar{H}}_O^{(0)} = \exp\{i\phi_{\bar{x}}I_z'\sqrt{2}t/t_c'\}(-\Delta\omega I_y'/\sqrt{2})\exp\{-i\phi_{\bar{x}}I_z'\sqrt{2}t/t_c'\}$$
$$= -\frac{\Delta\omega}{\sqrt{2}}\left[I_y'\cos(\phi_{\bar{x}}\sqrt{2}t/t_c') + I_x'\sin(\phi_{\bar{x}}\sqrt{2}t/t_c')\right] \quad (5.102)$$

Inverting back into the unprimed coordinate system by a rotation corresponding to the operator $\exp\{iI_x\pi/4\}$ yields

$$\tilde{\bar{H}}_O^{(0)} = \frac{\Delta\omega}{\sqrt{2}}\left[\frac{I_z - I_y}{\sqrt{2}}\cos(\phi_{\bar{x}}2t/t_c) - I_x\sin(\phi_{\bar{x}}2t/t_c)\right] \quad (5.103)$$

Since the integrals of both the sine and cosine terms vanish at times $t = t_c' = 2\pi t_c/\phi_{\bar{x}}$, (they vanish at times $t_c'/2$ as well) the result is

$$\bar{H}_O^{(0)} = \frac{2}{t_c'}\int_0^{t_c'/2}\tilde{\bar{H}}_O^{(0)}\,dt = 0 \quad (5.104)$$

Therefore the spin system *observed at window times* $t_c' = \pi t_c/\sqrt{2}\phi_{\bar{x}}$ will be unaffected by offset. What does this mean physically? If a phase error in the $\bar{x}$ channel is deliberately introduced into the x, $\bar{x}$ flip-flop sequence and the signal is detected along y in the window after the $\bar{x}$ pulse, then according to the view in Section V.B, the result will be a projection along y of the moment, starting along z and precessing about $\Omega_p \propto \mathbf{k} + \mathbf{j}$. The precession frequency will be $\omega = |\Omega_p| = \sqrt{2}\phi_{\bar{x}}/t_c$. The observed signal will therefore be proportional to $[1 + \cos(\sqrt{2}\phi_{\bar{x}}t/t_c)]$. When this oscillating signal is observed at times $N\pi t_c/\sqrt{2}\phi_{\bar{x}}$, e.g., at the upper and lower extremes of the oscillations, no decay due to a dc-field inhomogeneity is observed, *provided that the offset associated with the inhomogeneity is small compared to the ratio of the phase error to the flip-flop cycle time*. A similar consideration of the signal observed after the x pulse shows it to be proportional to $[1 + \sin(\sqrt{2}\phi_{\bar{x}}t/t_c)]$, so the total observed pattern will be superposition of a sine and a cosine with all signals ≥ 0 and extrema exhibiting no T_2^* decay. The effect of a dc-field inhomogeneity has been "second averaged" by a phase error!

In summary, second averaging a given Hamiltonian utilizing another requires

(a) that the Hamiltonian to be averaged should correspond to an Ω that is perpendicular (and cannot be parallel) to the Ω of the averaging Hamiltonian,

(b) that the averaging Hamiltonian must be made large compared to the interaction that is to be averaged, and

(c) that the averaging condition is only valid at times corresponding to cycle times associated with precession of the moment about the Ω vector of the averaging Hamiltonian.

VIII. Combined Rotation and Multiple-Pulse Spectroscopy: Addition of Magic-Angle Spinning

Thus far in this chapter we have seen how rotations of spin angular momenta can be used to average internal (and in the case of resonance offset, externally supplied) interactions at stroboscopically detected window times t_c. We have mentioned that rotations in real space can also be used to average selected internal interactions. In Chapter 3 we discussed the effect of sample spinning upon the shielding Hamiltonian as a special example. We saw that sample spinning at an angle β to $\mathbf{B}_0$ scales shielding powder patterns by $(3\cos^2\beta - 1)/2$. When β is the "magic angle," the remaining signal is the isotropic value of the shielding. To avoid the problem of side-bands, the spinning frequency is large compared to $\delta = \sigma_{zz} - \bar{\sigma}$. Most samples of chemical interest contain many chemically inequivalent protons (e.g., the three chemically distinct protons in 4,4'-dimethylbenzophenone) or ^{19}Fs. In this case, removing dipolar broadening is insufficient for unique identification of shielding components because of the problem of overlapping tensors. We have discussed (Section VI) the effect of motion upon the homonuclear decoupling experiment. We saw that motion could destroy the effect of the radio-frequency decoupling. Consider the case in which homo-nuclear dipolar decoupling is combined with motion externally supplied by sample spinning at an angle β to $\mathbf{B}_0$ and a frequency $\omega_r > \delta\omega_0 = (\sigma_{zz} - \bar{\sigma})\omega_0$. The critical question with regard to the combined experiment, in which both the spin and space portions of the space–spin internal Hamiltonian are rotated, is "what is the period of sample rotation $2\pi/\omega_r$ compared to the multiple-pulse cycle time t_c?" or "under what conditions can the pulse decoupling be considered to act on a static sample?" Easily achievable

sample-spinning frequencies are about 3 kHz, although 10 kHz is possible
if the sample is the rotor itself. The period of sample rotation is therefore
330 μsec. With homogenous dipolar broadening of 30 kHz, we have seen
that $3\tau = 10$ μsec would appear to satisfy the averaging criterion (5.26) at
least for a rough guess. To remove zero- and first-order terms of dipolar
coupling, an eight-pulse cycle with a cycle time $12\tau = 40$ μsec would seem
to be implied. However, it is the authors' personal experience that a value
of 25 μsec for t_c of the eight-pulse sequence is necessary under such conditions in order to minimize the effects of second- and higher-order terms.
For removal of even-order terms to $\bar{H}_D^{(2)}$ and *all* odd-order terms, a 48-pulse
cycle time of $72\tau = 150$ μsec would be used. Which of these is "short" compared to a sample rotation period of 330 μsec, such that sample rotation
does not destroy the effect of pulse dipolar decoupling? It would appear that
a 24-pulse cycle time of $36\tau = 90$ μsec is sufficiently short such that under
homonuclear decoupling and magic-angle spinning, the three sets of chemically shifted protons in 4,4'-dimethylbenzophenone may be distinguished,
as shown in Fig. 21 [11]. The proton frequency for this experiment was
56 MHz. Higher frequencies would of course lead to better resolution and
have been used by the group at Jena [12]. This cycle time included the fact
that the sample *plus* the sample coil have been tilted to an angle of 54.7°
to $\mathbf{B}_0$, (see Section X for experimental considerations leading to this necessity),
which means 90° pulse widths had to be increased, relative to a condition
with $\mathbf{B}_1$ perpendicular to $\mathbf{B}_0$ by a factor of $[\cos(90 - 54.7)]^{-1} = 1.23$.

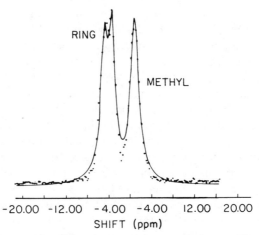

RING

METHYL

-20.00   -12.00   - 4.00   - 4.00   12.00   20.00
SHIFT (ppm)

FIG. 5.21. The proton NMR spectrum of solid polycrystalline 4, 4' dimethylbenzothenone
under combined rotation and multiple-pulse spectroscopy; BR-24 used for homonuclear dipolar
decoupling.

We note that if shielding anisotropies are scaled by $(3 \cos^2 \theta - 1)/2$ under magic-angle spinning (MAS), and if isotropic values of shielding are sufficiently different in a given sample, off-magic-angle spinning (OMAS) can be used to recover both anisotropic and isotropic shielding components. This has been accomplished for the acid proton in 2,4-dimethylbenzoic acid [13] and, as we shall see in Chapter 6 for rare nuclei such as ^{13}C, ^{113}Cd, and ^{31}P in selected solids. Other methods of recovering shielding anisotropies while spinning slower than the anisotropy are discussed in Chapter 6.

IX. Tuning the Spectrometer for Multiple-Pulse Experiments

One of the aspects of multiple-pulse NMR that has made its use something to be avoided by some and a delight to others is the fact that results are dependent upon spectrometer tuning. Intelligent spectrometer tuning depends, in turn, upon an understanding of how Hamiltonians affect time development. Where multiple-pulse sequences are used in such a manner that many sequences are needed for a single experiment, e.g., 1024 eight-pulse sequences used to remove homonuclear dipolar coupling, small tuning errors may rapidly accumulate. The effects of phase transients ("glitch"), which, as discussed in Section IV of this chapter, are unavoidable, must be contended with and the effects of these transients upon tuning must be understood. Fortunately at this point, these effects are easily described; and with an understanding of the second-averaging effect of one Hamiltonian upon another, errors associated with phase, pulse width, and glitch may be systematically minimized.

A spectrometer such as the one diagrammed in Fig. 5.1 will have four channels, x, y, $\bar{x}$, $\bar{y}$, successively differing in phase by 90°. To attain spectrometer tuning appropriate for multiple-pulse experiments, three adjustments must be made. First, phase transients must be reduced to a minimum and symmetrized (*vida infra*). Second, the phases between channels must be adjusted such that, e.g., x and $\bar{x}$ differ by 180°. Third, pulse widths must be adjusted to provide the appropriate rotation (e.g., $\pi/4$, $\pi/2$, π, etc.) of moments in the rotating frame. Because of drift of factors such as dc field and transmitter power, tuning is an iterative process and is continued until consistent results are obtained in all checks of pulse width, phase, and glitch. The final tuning is attained by small adjustments made while observing the response of the spin ensemble to the actual sequence in question, as will be illustrated by one possible tuning procedure for the eight-pulse sequence. A preliminary adjustment of tuning $n\pi/2$ phase shifts between channels for the spectrometer illustrated in Fig. 5.1 is simply accomplished by adjusting the lengths of transmission line in each channel such that the difference in length between

x and y is $\lambda/4$ for the frequency in question, the difference between x and $\bar{x}$ is $\lambda/2$, and the difference between y and $\bar{y}$ is $\lambda/2$. For the spectrometer shown in Fig. 5.1, this adjustment is made at a wavelength corresponding to a frequency of 30 MHz, and the spectrometer is broadband by mixing 30 MHz with some other frequency with subsequent filtering. For example, to operate at 56 MHz, 30 MHz could be mixed with 86 MHz, producing signals at 116 and 56 MHz. The 116-MHz signal is easily filtered with a series tuned L–C filter at 56 MHz before the combined signal reaches the transmitter. (Note that mixing with 26 MHz would also produce 56 MHz, but 4 MHz is harder to filter than 116 MHz.). For fine adjustments of phase differences relative to x, which we take as the reference channel, one may use variable-length constant-impedance transmission lines or some constant-impedance trimming device, such as the Merrimac PS-30 unit. A rough guide to phase differences between channels may be attained by producing the pulse sequence x, τ, $\bar{x}$, τ, y, τ, $\bar{y}$, τ,[†] and phase detecting the envelope of the pulses thus produced. The circuit needed for this experiment is available from components normally used in the spectrometer. The receiver in Fig. 5.1 is replaced by an aerial with the end near (not in!) the inductor in the probe. The resulting signal is observed by leading the output of the dc amplifier to an oscilloscope rather than a transient recorder. The oscilloscope must be properly triggered externally by the pulse one wishes to observe.

If one applies an x pulse and detects the response along y, the resulting signal is the dc envelope of the transient (glitch) in the y direction. One possible example of such a transient has been shown in the lower trace of Fig. 5.10. Note that in Fig. 5.10 the phase transient is symmetric with respect to the time at the center of the pulse [cf. Eq. (5.37)]. The symmetry of the transient will be seen to be an important criterion (with a symmetric transient being best) for removing the effects of phase transients upon the MREV-8 sequence to zero order, as will be discussed shortly. Such symmetry is attained by carefully matching the input impedance of the probe to the output impedance of the transmitter (usually 50 ohms) by using a vector impedance meter and an adjustable π circuit between the transmitter and probe for fine tuning. This transient can be used to adjust the phase difference between the x and $\bar{x}$ channels, since when these two channels are 180° out of phase, the transients will be mirror images of each other through a plane perpendicular to the plane of the image, and containing the zero line. Next, the phases of y and $\bar{y}$ may be set relative to x by detecting the response to minimize the signal from the x channel while adjusting the phases of the y and $\bar{y}$ channels to maximize the dc envelopes of the y and $\bar{y}$ pulses.

[†] It is assumed that a pulse programmer capable of allowing any desirable sequence to be produced is available.

With this rough adjustment of glitch and phase, pulse widths for all channels are adjusted by utilizing the $[\tau, \pi/2), \tau]_n$ sequence, illustrated in Fig. 5.22(a), on a sample containing only one chemical species of the nucleus of interest, e.g., $H_2O(l)$ for ^1H or C_6F_6 for ^{19}F. The flip-flop sequence, shown in Fig. 5.22(b), is then used to minimize the glitch and adjust the phase of $\bar{x}$ relative to x. By utilizing knowledge of the average Hamiltonian associated with phase errors and with transients listed in Problem 5.3 at the end of the chapter, when neither glitch nor phase error is present, the on-resonance response to the flip-flop sequence would be expected to be positive and constant after the x window, and zero after the $\bar{x}$ window. In practice, a decay of responses after both x and $\bar{x}$ pulses is seen due to dc drift of the receiver system. If a phase error in the $\bar{x}$ channel is present, however, the signal observed after both x and $\bar{x}$ windows will be associated with projection of the oscillation of the moment about $\Omega = \phi_{\bar{x}}(\mathbf{k} + \mathbf{j})/t_c$ upon the y axis. This projection will remain positive for the signals after both x and $\bar{x}$ windows. Since the initial value of the observed magnetization is a maximum after the first $\bar{x}$ window and is zero after the first x window, however, the signals observed after those two windows will cross, as shown in the bottom scan of Fig. 5.22(b).

The effect of a phase transient, on the other hand, as seen in the table accompanying Problem 5.3, is to cause a precession of the moment about $\Omega = J_1(\mathbf{k} - \mathbf{j})/t_c$ after the $\bar{x}$ pulse. Here,

$$J_1 = \int_0^{t_w} \omega_T(t)(\sin \omega_1 t - \cos \omega_1 t) \, dt \tag{5.105}$$

With the moment along z after the $\bar{x}$ pulse, the expected response is an oscillation with frequency $\omega = |\Omega| = \sqrt{2} J_1/t_c$ varying between zero and negative values. On the other hand, the signal after an x pulse is affected by the average Hamiltonian for transients of

$$\bar{H}_T^{(0)} = J_1(I_z + I_y)/t_c \tag{5.106}$$

With the initial value of the magnetization after the x pulse lying along y, the signal associated with a transient error would start at a maximum and oscillate between this maximum and zero, again with frequency $\omega = \sqrt{2} J_1/t_c$. The signal expected from observation after both windows is shown in the center scan of Fig. 5.22(b). Clearly the signals associated with phase errors and with glitch are distinguishable via observation of whether signals after x and $\bar{x}$ channels cross or track with the same dc offsets. Fine adjustments of phase and π-circuit tuning between transmitter and probe may be made at this point in an attempt to obtain signals similar to that observed in the top scan of Fig. 5.22(b). A further check of tuning of transients may be made by noting that both the offset and the glitch are proportional to $(I_z - I_y)$. The

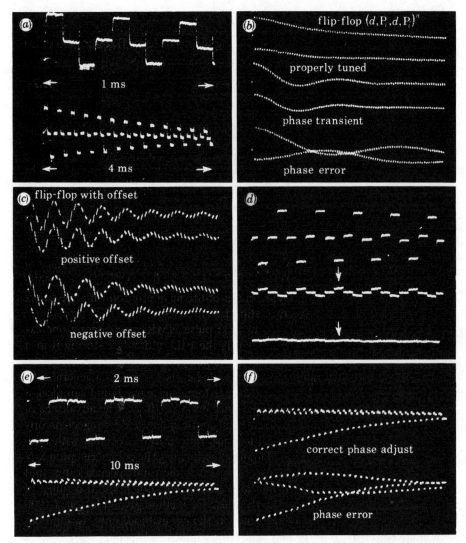

FIG. 5.22. Time development of the magnetization of protons in a water sample under various sequences to tune a multiple-pulse spectrometer for pulse widths and phases: (a) $[90_k^\circ, \tau]x^n$; top, $n = 10$; bottom, $n = 39$; (b) the flip-flop cycle $[90_k^\circ, \tau, 90_{-k}^\circ, \tau]$; (c) flip-flop with offset; (d) tuning phase detector for orthogonality between x and y channels; (e) $[90_k^\circ, \tau, 90_{k \pm 90^\circ}^\circ \tau]$ to adjust relative phases of $\pm x$ to $\pm y$; (f) phases properly and improperly adjusted. (From Gerstein [17].)

sign of the offset may be changed, however, by making $\Delta\omega$ positive or negative. Therefore, if the response to the flip-flop cycle is observed, first with a
positive offset and then with negative offset, the number of beats counted in
a given time interval will be the same if the effect of the glitch is zero. With
nonzero glitch, the number of beats for a given offset above and below resonance will not be identical, as shown in Fig. 5.22(c). The difference in offsets
for $\Delta\omega = \pm 1$ kHz, illustrated in Fig. 5.22(c), has been found to be indicative
of sufficiently small transient imperfection to perform useful homonuclear
dipolar averaging experiments.

As a further remark regarding minimizing the cumulative effect of transients in removal of homonuclear-dipolar broadening via the eight-pulse
sequence, note that the integral J_1 defined by Eq. (5.105) is zero if $\omega_T(t)$ is
symmetric about the center of the pulse, since with $\theta = \omega_1 t_w$,

$$J_1 = \int_0^{t_w} dt\, \omega_T(t)[\sin\omega_1 t - \cos\omega_1 t] = \frac{1}{\omega_1}\int_0^{\pi/2} d\theta\, \omega_T(\theta)[\sin\theta - \cos\theta]$$

$$= \frac{1}{\omega_1}\int_0^{\pi/2} d\theta\, \omega_T(\theta)\sin\theta - \int_0^{\pi/2} d\theta\, \omega_T(\theta)\cos\theta \equiv I_1 - I_2 \qquad (5.107)$$

If $I_1 = I_2$, J_1 vanishes, and the transient term in both the flip-flop and
eight-pulse (Table 5.7) cycle vanishes. We illustrate I_1 and I_2 for a choice
of $\omega_T(\theta)$ symmetric about the center of the pulse in Fig. 5.23. We see that
the products of $\omega_T(\theta)$ and $\sin\theta$, $\cos\theta$ can be adjusted to have nearly equal
areas over the pulse width, so the difference in the two areas I_1 and I_2 cancels.
If the transients are determined only by the responses of probe and transmitter (and not by gates in individual channels), symmetrization of transients
in any channel of the transmitter should be sufficient for symmetrization in all
channels. Similarly, removal of the effect of glitch for the x, $\bar{x}$ flip-flop
sequence should remove the effect of glitch on the eight-pulse sequence described earlier, since J_1 is the same for both.

At this point, a fine adjustment of the phases of y relative to x may be
made by using the sequence $[\tau, \pi/2)_x, 2\tau, \pi/2)_y, \tau]^n$. First, in using the sequence
$[\tau, \pi/2)_x]^n$, the reference phase is adjusted [Fig. 5.22(d)] until a zero response
is obtained. Then the x, y phase-comparison sequence is used to adjust the
phase of the y channel relative to x. A bit of thought about the response to
be expected from this sequence, if perfectly tuned, will reveal that, on resonance, with detection along x, the signal for a cycle of six pulses should
consist of three zero responses, two positive responses, and a zero response.
Such a set of responses to two such tuning cycles (twelve pulses) is illustrated
for proper phase adjustment of y relative to x in Fig. 5.22(e) and for improper
adjustment in Fig. 5.22(f). Replacing y by $\bar{y}$ in this sequence will test the phase
of $\bar{y}$ relative to x. The phasing of y relative to $\bar{y}$ can be checked by using the

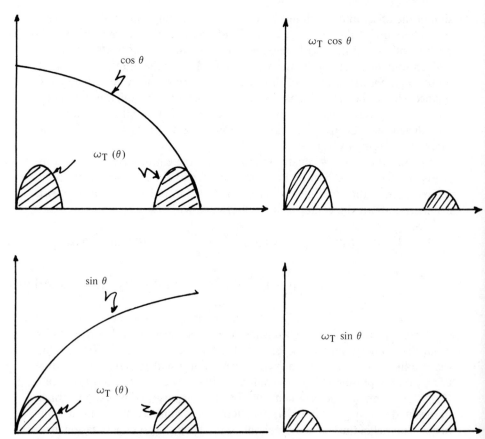

FIG. 5.23. The phase transients for a nonideal rf pulse, showing the error, ω_T.

flip-flop sequence. The phase of $\bar{x}$ relative to y and $\bar{y}$ can then be checked to ensure internal consistency. Finally, the response of a *liquid* sample to the eight-pulse sequence is used to check overall tuning. At this point, the widths of all pulses may be simultaneously adjusted slightly to minimize the oscillations in the signal observed on resonance, and small adjustments of the π circuit at the output of the transmitter and/or the probe can be further used for the same purpose. Oscillations in the response to the eight-pulse sequence of a sample observed on resonance imply some average Hamiltonian with associated Ω, about which the moment is precessing. Minimization of the frequency of these oscillations means that the errors leading to Ω have been minimized. With care, tuning may reduce these errors such that the eight-pulse cycle, observed on resonance, appears to be no more than 100 Hz off resonance. Other tune-up cycles have been discussed in the literature [14].

X. Experimental Considerations

This text is meant as a practical guide to the theory and the practice of pulse NMR in solids. In line with that intent, we have attempted throughout to include details of experimental parameters necessary for successful performance of experiments for which the theory has been developed. An important example was the consideration of cycle time versus sample rotation speed in the section on combined multiple-pulse decoupling and sample spinning. Two parameters previously alluded to, but not specifically discussed, are those affecting minimum cycle times in multiple-pulse experiments. These are transmitter plus probe ringdown and receiver recovery. As a limitation, we only wish to deal with the case in which probe ringdown, and not transmitter ringdown, is the limitation in recovery. This condition may be achieved (1) by the use of an expensive broadband transmitter with a minimum power of 400 W in pulse mode or (2) via a high-power (1–2-kW), tuned (and therefore inexpensive) transmitter with Q appropriately modified to achieve a relatively flat response. An easy method of lowering the transmitter Q is to place a 50-ohm load in parallel with the output, the load consisting of four 2-W, 200-ohm carbon resistors in parallel to ground. The time constant of the transmitter alone may be monitored by utilizing a high-power 50-ohm load in place of the tuned circuitry of the probe. The pulses, properly attenuated, may be viewed at the load, with a high-frequency oscilloscope triggered on an x pulse using the tuning sequence $[10 \ \mu sec, P_x]^{50}$.

If ringdown of a power pulse is limited by the time constant of the probe, then with a probe having quality factor Q, a peak voltage V_0 will ring to $1/e$ of its value in a time $\tau = Q/3f$, with f in hertz; $\ln(V/V_0) = -t/\tau$. A 500-W pulse into a 50-ohm load will correspond to an rms voltage of $500 = V_{rms}^2/50$ and $V_{rms} = 150$ V corresponds to a peak voltage of $150 \times \sqrt{2} = 223$ V. The number of time constants n necessary for this voltage to ring to 1 μV (a reasonable lower limit for reception of 20–30-μV NMR signals) is calculated as

$$-\ln(10^{-6}/223) = n\tau/\tau = n = 19$$

The value of n generally quoted for safety is $n = 21$. A probe with a Q of 30 operating at 60 MHz will therefore have a ringdown time of $21Q/3f = 3.5 \ \mu sec$. This value, with a given pulse width, automatically sets the cycle time in stroboscopically observed multiple-pulse experiments, because the observation window (cf. Fig. 5.20) is 2τ, defined to be the time between the trailing edge of the last pulse of one sequence and the leading edge of the first pulse of the next. With a minimum time of 1 μsec for integrating the voltage from the dipolar echo in the 2τ windows, the value of τ is set by $2\tau = t_w + 21\tau + 1$.

With a pulse width of 1.5 μsec and $Q = 30$, we find $\tau = 3 \ \mu sec$, indicating an eight-pulse cycle time of 36 μsec. An implicit assumption here is that the receiver has a recovery time after an extended chain of saturating pulses that is short compared to probe ringdown. A scheme that protects the receiver and limits receiver saturation is shown in Fig. 5.24. The boxes with a slash through them are crossed diodes. The Lowe–Tarr scheme of using quarter-wavelength cables shorted to ground (which act as very high impedances to the frequency in question) between the transmitter and receiver reduces the 250-V transmitter pulse (which would destroy a transistor receiver) to a value below 2 V at the input of the first stage of the broadband receiver. Tuning elements at the Larmor frequency of the observed nucleus, consisting of simply constructed L–C series circuits, are helpful in reducing saturation through the receiver chain. In addition, the second and third stages of the receiver chain are limiting amplifiers, which have the characteristic of cutting the gain at an output voltage of 75 mV, thus preventing saturation. With such a receiver chain, at 15-db gain per stage, the receiver response indicated in Fig. 5.25 was obtained in our laboratory. The bottom scan of Fig. 5.25 shows the pulses detected by an antenna near the probe inductor. The top scan shows the receiver response, which is divided into four regions, a–d; region a, encompassing two pulses from the transmitter, shows the 1-V pulses washing completely through the receiver. Region b, after the pulse turn-off, and during probe ringdown, shows the dead space of the receiver during which the power pulse has rung down sufficiently so as not to wash through the receiver, but the receiver is still dead. Region c shows the receiver beginning to recover, amplifying the tail of the power pulse. Region d shows the receiver noise, unaffected by any remnant of the power pulse. It is a region that can safely be used for signal accumulation.

As a final note, we mentioned in the section on magic-angle spinning the necessity of rotating the coil with the sample. The considerations leading to this requirement have to do with the homogeneity of the rf field, the signal-to-noise ratio (S/N) in multiple-pulse experiments with stroboscopic detection, and the available rf power. For efficient decoupling, as noted earlier, a $\pi/2$ pulse must not deviate appreciably from 90° or pulse-width errors

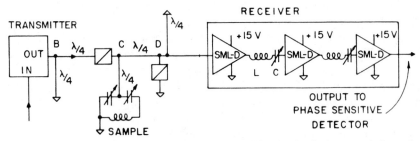

FIG. 5.24. A circuit placed before the video amplifier (receiver) to protect it from the overload during a pulse.

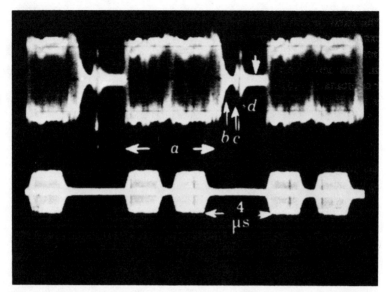

FIG. 5.25. The detected response to a string of rf pulses. An antenna is placed near the coil and the voltage that is received is detected in a manner similar to that in which the response of the sample is detected.

can accumulate to destroy the experiment. For maximum signal-to-noise ratio, all spins in the sample must be exposed to the same intensity of rf field. One means of accomplishing this is to have the length of the inductor be large compared to its diameter. Typical dimensions of inductors used with 500-W transmitters are 2 cm in length by 0.65 cm in diameter. In such a coil, the rf homogeneity over the central 5 mm can be adjusted to be uniform to 5%, and a sample in a 5-mm-diameter rotor will reasonably "fill" the center of the coil (the number describing the ratio of filled to unfilled coil is termed the filling factor). Of course, S/N depends critically upon the filling factor. It is possible to have 5-mm rotors in a configuration for which the rotation axis may be adjusted with respect to $\mathbf{B}_0$ without changing the coil orientation [15]. In this case, however, the coil diameter becomes large relative to 5 mm.

PROBLEMS

5.1 In multiple-pulse type experiments, why is it convenient to use an rf sequence that is cyclic? What happens if it is not cyclic?

5.2 Why is it convenient to use an rf sequence that is periodic? How would you have to modify expressions like (5.21) to accommodate this change?

5.3 Verify the entries in the accompanying table:

	$\pi/2_x - \pi/2)_x - \pi/2)_x - \pi/2)_x$	Flip-flop cycle (window after $\bar{x}$ pulse)
$\bar{H}_0^{(0)a}$	0	$(1/2)\Delta\omega(1 + 2a)(I_z - I_y)$
$\bar{H}_0^{(1)}$	$-(1/16)t_c(\Delta\omega)^2 I_x$	0
$\bar{H}_P^{(0)}$	0	$(1/t_c)(\phi_{\bar{x}} - \phi_x)(I_z + I_y)$
$\bar{H}_{PO}^{(1)}$	0	$[(\phi_x + \phi_{\bar{x}})/4]I_x$
$\bar{H}_T^{(0)b}$	0	$(1/t_c)J_1(I_z - I_y)$
$\bar{H}_{TO}^{(1)}$	$-(1/2)\Delta\omega J_1 I_x$	$-(1/4)\Delta\omega J_2 I_x$
$\bar{H}_\delta^{(0)}$	$-(4/t_c)\delta_x I_x$	$(1/t_c)(-\delta_x + \delta_{\bar{x}})I_x$
$\bar{H}_{\delta O}^{(1)}$	$(1/2)\Delta\omega\delta_x I_z$	$-(1/4)\Delta\omega(\delta_x + \delta_{\bar{x}})I_z$
$\bar{H}_\epsilon^{(0)}$	$-(4/t_c)\sum_i \epsilon_i I_{xi}$	0
$\bar{H}_{\epsilon O}^{(1)}$	$(1/2)\Delta\omega\sum_i \epsilon_i I_{zi}$	$-(1/2)\Delta\omega\sum_i \epsilon_i I_{zi}$

^{a} $a = (t_w/t_c)(4/\pi - 1)$.
^{b} $J_1 = \int_0^{t_w} \omega_T(t)(\sin\omega_1 t - \cos\omega_1 t)\,dt$;
$\quad J_2 = \int_0^{t_w} \omega_T(t)(\sin\omega_1 t + \cos\omega_1 t)\,dt$.

5.4 What is the frequency scaling factor for the flip-flop phase-tuning sequence given in Fig. 5.12, and how does your measured result compare with the result given in the text?

5.5 Why is the cycle time of the pulse-width tuning cycle 8τ (rather than 4τ, as for the flip-flop cycle)?

5.6 In a certain pulse experiment investigating NMR of protons at 56 MHz, a probe with $Q = 30$ is used. The power of the pulse corresponds to a peak voltage of 500 V. The signal from the sample is 20 μV. How long does one have to wait after a power pulse in order to be sure that at least 90% of the observed signal at the receiver is due to the sample, if the only pertinent voltages are sample voltage and probe ringdown voltage?

5.7 The NMR linewidth of some species of protons in vitrain coals is found to be 30 kHz. A multiple-pulse sequence is devised that averages the interaction responsible for this linewidth to zero in the times chosen for observation of the spin system. What will be an upper limit on the cycle time used for this pulse sequence if narrowing is to be achieved?

5.8 Using direct evaluation of the time development of the density matrix, show that the Carr–Purcell–Meiboom–Gill sequence with a pulse-width error in the 180° pulses will not give the correct echo magnitude at time 2τ.

5.9 Using average Hamiltonian theory, show that the CPMG sequence with pulse–width errors will give the correct echo magnitude at cycle times $t_c = 4\tau$.

5.10 The total time scan for the CPMG results shown in Fig. 5.17 is 20 msec. What is T_2 for the doped water sample on which these results were obtained?

5.11 Show that the zeroth-order average dipolar Hamiltonian under the WAHUHA four-pulse sequence is given by

$$\bar{H}_D^{(0)} = \frac{1}{2} \sum_{i>j} B_{ij} \frac{6}{\pi} \frac{t_w}{t_c} \left[I_{yi}(I_{xj} + I_{zj}) + (I_{xi} + I_{zi})I_{yj} \right]$$

when finite pulse widths are taken into account.

5.12 Calculate $\bar{H}_P^{(0)}$ for the response observed after the $\bar{x}$ pulse in the flip-flop cycle. Sketch the analog of Fig. 5.13 for this Hamiltonian to determine the effective field for this Hamiltonian under the flip-flop cycle. Verify that Fig. 5.22(c) illustrates the responses to be expected after x and $\bar{x}$ pulses if there is a phase error in one of the pulses.

5.13 Suppose that a ^1H shielding anisotropy is 20 ppm and a cycle time of 30 μsec is used in an eight-pulse homonuclear narrowing experiment. What is the maximum value of Zeeman field under which such an experiment could be used, without violating Nyquist's sampling theorem, or allowing spectral artifacts in the frequency domain for (a) single-phase detection and (b) quadrature detection.

5.14 A 24-pulse sequence, with cycle time $36\tau = 108$ μsec, is used for homonuclear decoupling. In a measurement of the scaling factor, it is found that summing responses under the multipulse experiment that are 1 kHz apart in offset leads to a 4096-point transformed frequency spectrum with peaks 170 channels apart.

   (a) What is the experimental scaling factor in this experiment?

   (b) How does this value compare with that calculated for the BR-24 pulse sequence described in Section V?

5.15 Create a pulse sequence under which homonuclear dipolar plus chemical-shift Hamiltonians are averaged to zero in zeroth order (see C. R. Dybowski and R. G. Pembleton [16]). Under such a sequence residual lifetimes under homonuclear decouplings may be determined.

5.16 Show that the signal-to-noise ratio from data collected in a four-pulse homonuclear decoupling experiment will be twice that for data collected in an eight-pulse experiment for equivalent number of scans.

5.17 The eight-pulse experiments described in this chapter are preceded by a preparation pulse along x. Show that in such an experiment, with phase detection along y, data may be accumulated every 6τ in the windows between the four-pulse sequences from which the eight-pulse sequence is formed.

5.18 (a) Consider the following sequence:

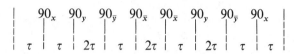

Calculate $\bar{H}_{CS}^{(0)}$ for this sequence to show $\bar{H}_{CS}^{(0)} = (\omega_0\sigma/3)(I_y + I_z)$

(b) Suppose one uses a $-y$ pulse as preparation to this sequence. Calculate the behavior expected for a system subject to the average Hamiltonian of part (a).

(c) Consider the usual eight-pulse sequence:

$$
\begin{array}{ccccccccc}
| & 90_x & 90_y & 90_{\bar{y}} & 90_{\bar{x}} & 90_{\bar{x}} & 90_y & 90_{\bar{y}} & 90_x & | \\
| & \tau & \tau & 2\tau & \tau & 2\tau & \tau & 2\tau & \tau & \tau |
\end{array}
$$

which has the average Hamiltonian $\bar{H}_{CS}^{(0)} = (\omega_0\sigma/3)(I_x + I_z)$. Suppose that one uses a $-x$ pulse as preparation; what is the expected behavior?

(d) Show that the preparation pulse plus the first two pulses of the sequence of part (c) have the same effect as a $-y$ pulse. What property could this suggest about the relationship of the detected signals in the windows of the eight-pulse sequence?

REFERENCES

1. J. S. Waugh, L. M. Huber, and U. Haeberlen, *Phys. Rev. Lett.* **20**, 180 (1968).
2. I. J. Lowe, *Bull. Amer. Phys. Soc.* **2**, 344 (1957).
3. M. Matti Maricq, *Phys. Rev. B* **25**, 6662 (1982).
4. D. P. Burum, M. Linder, and R. R. Ernst, *J. Magn. Reson* **44**, 173 (1981).
5. J. G. Powles and P. Mansfield, *Phys. Lett.* **2**, 58 (1962).
6. D. Burum and W.-K. Rhim, *J. Chem. Phys.* **71**, 944 (1979).
7. P. Mansfield, *Philos. Trans. R. Soc. London* **A299**, 479 (1981).
8. W.-K. Rhim, D. D. Elleman, and R. W. Vaughan, *J. Chem. Phys.* **58**, 1772 (1973).
9. W.-K. Rhim, D. D. Elleman, L. Schreiber, and R. W. Vaughn, *J. Chem. Phys.* **60**, 4595 (1974).
10. A. J. Vega and R. W. Vaughn, *J. Chem. Phys.* **64**, 1958 (1978).
11. L. M. Ryan, R. E. Taylor, A. J. Paff, and B. C. Gerstein, *J. Chem. Phys.* **72**, 508 (1980).
12. G. Scheler, V. Haubenreisser, and H. Rosenberger, *J. Magn. Reson.* **44**, 134 (1981).
13. R. E. Taylor, R. G. Pembleton, L. M. Ryan, and B. C. Gerstein, *J. Chem. Phys.* **71**, 4541 (1979).
14. D. P. Burum, M. Linder, and R. R. Ernst, *J. Magn. Reson.* **43**, 463 (1981).
15. C. A. Fyfe, H. Mossbruger, and C. S. Yannoni, *J. Magn. Reson.* **36**, 61 (1979).
16. C. R. Dybowski and R. G. Pembleton, *J. Chem. Phys.* **70**, 1962 (1979).
17. B. C. Gerstein, *Philos. Trans. R. Soc. London* **A299**, 521 (1981).

CHAPTER 6

HETERONUCLEAR PULSE
EXPERIMENTS

I. Introduction

In Chapter 5 the principal focus was elimination of the effects of homonuclear dipole–dipole interactions by abrupt changes in the state of the system produced by the use of sequences of strong rf pulses. This removal of homogeneous dipolar broadening allowed the observation of weaker interaction such as the shielding tensor. In this chapter, we discuss spin systems in which two different spin species are present, e.g., ^{1}H and ^{13}C. In addition to the homonuclear dipole–dipole coupling we have discussed, there is for such systems the possibility of dipolar and scalar couplings between the two different kinds of spins (see Fig. 6.1). As in homonuclear dipolar coupled spin systems, this heteronuclear dipole–dipole coupling may obscure weaker interactions that are useful in the characterization and analysis of the chemical structure. Such will very often be the case in investigation of the NMR spectra of spins in low abundance ("rare" spins) whose dominant interaction is via dipolar coupling to an abundant spin. An example of importance is the coupling of a ^{13}C nucleus to a neighboring proton, whose dipolar interaction, as we saw in Chapter 3 results in a broad ^{13}C resonant absorption. We describe decoupling experiments that permit the suppression of the effect of dipolar coupling of this heteronuclear type on the rare-spin spectrum.

There is a special problem with rare spins that has been addressed because of the dipolar coupling to abundant spins. Because spins like ^{13}C in natural abundance are few (1.1% for ^{13}C), the response in a single experiment is small relative to that for abundant spins like ^{1}H. Coaddition of responses increases the overall signal-to-noise ratio permitting observation of the resonance of such spins. The rate of coaddition is determined by the rate of repolarization of the rare nuclei. This rate may be tens of minutes or even hours for rare nuclei in the solid state that are repolarized by spin–lattice relaxation. The coupling of abundant to rare spins produces a relatively rapid mechanism for the repolarization of the rare spins by a process known as cross-polarization, described in this chapter.

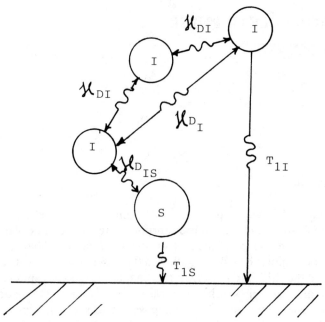

FIG. 6.1 The various interactions in a system of two spin-$\frac{1}{2}$ species that have an effect on the solid-state NMR spectrum.

The decoupling of spins and the enhancement of signals by cross-polarization would have restricted use in solid-state NMR, because most samples are powders, with spectra broadened by shielding anisotropies. As discussed in Chapter 5, the combination of strong spin decoupling and magic-angle spinning (MAS) allows the attainment of liquidlike spectra. We discuss some of the limitations of resolution and sources of residual broadening in this chapter. Magic-angle spinning does suppress information that is useful—the anisotropic parts of the chemical-shift interaction. Regaining this information is an important endeavor in characterizing the chemical systems as completely as possible, as is the correlation of properties like the dipolar coupling with the chemical-shift coupling, the mechanisms for which we describe at the end of the chapter.

The Hamiltonians that affect a heteronuclear spin-$\frac{1}{2}$ system are similar to those we have seen for a homonuclear spin-$\frac{1}{2}$ system. We take as an example a pair of spins-$\frac{1}{2}$ (an abundant spin I and a rare spin S) such as a proton and a ^{13}C nucleus in a typical organic substance. The Hamiltonian for this case consists of several terms:

$$\mathcal{H} = \mathcal{H}_{Z_I} + \mathcal{H}_{cs_I} + \mathcal{H}_{Z_S} + \mathcal{H}_{cs_S} + \mathcal{H}_{D_I} + \mathcal{H}_{D_{IS}} + \mathcal{H}_{rf} \qquad (6.1)$$

Here Z stands for Zeeman, cs is chemical shift, D is dipolar, and rf has its usual meaning. We assume that the rare spin is not strongly dipolar coupled to other rare spins and that no scalar couplings are present. For a ^{13}C–^{1}H pair, the *heteronuclear* interaction typically has a magnitude of 20 kHz and dominates the rare-spin spectrum. The likelihood of having a pair of ^{13}C nuclei close enough to produce a comparable broadening due to *homonuclear* ^{13}C–^{13}C coupling is negligible because of the low natural abundance. In extensively enriched chemical systems, the homonuclear coupling may not be neglected in calculating the NMR spectrum of the "rare" spin; in effect, the enrichment has produced an abundant-spin system. Such enrichment is useful in two respects: (1) It increases the signal-to-noise ratio, and (2) by adding pairs of rare nuclei at high enrichments, one may accentuate the effect of the homonuclear interaction, which, as we saw in Chapter 3, contains information on the geometric relationship of the two nuclei.

II. Pulse Decoupling

When heteronuclear dipolar coupling is dominant in broadening the rare-spin resonance, it is necessary to devise a mechanism for suppressing this interaction if observation of smaller interactions is desired. For example, consider a heteronuclear pair of spins-$\frac{1}{2}$ subject to the Hamiltonian of Eq. (6.1), a more explicit form of which is

$$\mathscr{H} = \omega_{0_I}\sigma_I I_z + \omega_{0_S}\sigma_S S_z + \mathscr{H}_{D_I} - \omega_D I_z S_z + \mathscr{H}_{rf} \qquad (6.2)$$

Here only the secular part of heteronuclear dipolar coupling has been retained. In addition, we view the spins from the frame in which the Zeeman interactions of both spins have been removed—the doubly rotating frame. We see that only three terms in the Hamiltonian affect the rare (S) spin: the chemical shift, the heteronuclear dipolar interaction and the applied rf fields. The use of appropriate rf excitations to average the effect of the dipolar Hamiltonian would produce a situation in which the evolution of the S magnetization would be governed primarily by the chemical-shift interaction, allowing observation of this interaction.

Drawing on the results of the experiments carried out on homonuclear coupled systems, we note that a sequence of rf pulses organized into a *cycle* may produce the requisite averaging of the dipolar Hamiltonian. However, the secular part of the heteronuclear dipolar Hamiltonian is different in form from the homonuclear dipolar Hamiltonian. In fact, from the point of view of the S spin, it has the same form as the chemical shift, being proportional to S_z. This analogy suggests that the Carr–Purcell (CP) sequence of Fig. 6.2 might effect such a heteronuclear decoupling. In Fig. 6.2, $P_{\theta x}$ indicates

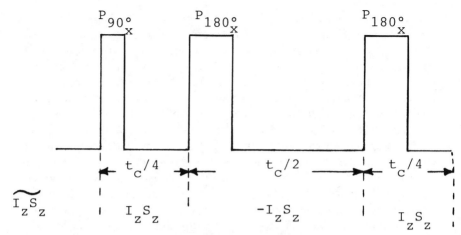

FIG. 6.2 A pulse cycle for decoupling two spins-½ subject to the heteronuclear dipole–dipole interaction. This cycle also refocuses the chemical-shift interaction.

a pulse of nutation angle θ about the x axis. This sequence is carried out repetitively with selective sampling at times Nt_c. Provided that t_c is sufficiently short (*vide infra*), one may apply average Hamiltonian theory (Chapter 5) to calculate the time evolution of the S magnetization. The operator portion of the secular heteronuclear dipole–dipole Hamiltonian is switched, in the course of the cycle, through the values given at the bottom of Fig. 6.2. The zero-order Hamiltonian is given by the time average, as discussed in Chapter 5.

$$\bar{\mathcal{H}}^{(0)}_{\mathrm{D}IS} = \frac{1}{t_c} \int_0^{t_c} \tilde{\mathcal{H}}_{\mathrm{D}IS}(t)\, dt$$

$$= \frac{1}{t_c} \int_0^{t_c/4} \tilde{\mathcal{H}}_{\mathrm{D}IS}(t)\, dt + \frac{1}{t_c} \int_{t_c/4}^{3t_c/4} \tilde{\mathcal{H}}_{\mathrm{D}IS}(t)\, dt + \frac{1}{t_c} \int_{t_c/4}^{t_c} \tilde{\mathcal{H}}_{\mathrm{D}IS}(t)\, dt$$

$$= -\frac{\omega_{\mathrm{D}} I_Z S_Z}{4} + \frac{\omega_{\mathrm{D}} I_Z S_Z}{2} - \frac{\omega_{\mathrm{D}} I_Z S_Z}{4}$$

$$= 0 \tag{6.3}$$

It is clear that application of a Carr–Purcell cycle to the observed spins decouples the I and S spins. The Carr–Purcell cycle also affects the other Hamiltonians. To describe the time evolution of the S magnetization, it is necessary to calculate the zeroth-order average Hamiltonians for all the terms in the expansion of Eq. (6.2). Because of the difference in resonance frequency, application of a Carr–Purcell sequence at the S-spin resonance frequency has a negligible effect on parts of the Hamiltonian that depend only on the I-spin components; however, such a sequence does affect the chemical-shift

Hamiltonian of the S spin. The average Hamiltonian for the chemical shift is similar to the calculation of the average Hamiltonian of the heteronuclear dipolar Hamiltonian:

$$\bar{\mathscr{H}}^{(0)}_{\text{css}} = \frac{1}{t_c} \int_0^{t_c} \mathscr{H}_{\text{css}}(t)\, dt$$

$$= \frac{1}{4} \sigma_S \omega_{0S} S_z - \frac{1}{2} \sigma_S \omega_{0S} S_z + \frac{1}{4} \sigma_S \omega_{0S} S_z$$

$$= 0 \tag{6.4}$$

The fact that this average Hamiltonian is zero implies that, under this sequence, the magnetization does not evolve under the chemical-shift interaction. If the intention is to achieve a spectrum that depends only on the chemical shift of the S spin, the Carr–Purcell sequence of Fig. 6.2 is not appropriate.

The average Hamiltonian theory also requires the calculation of the higher-order terms, as we have seen in Chapter 5. For this Hamiltonian, the calculation is trivial. The operator of the transformed dipole–dipole Hamiltonian is always $I_z S_z$. The commutator for the first-order correction (and for all higher-order terms)

$$[\tilde{\mathscr{H}}_{\text{D}_{IS}}(\tau), \tilde{\mathscr{H}}_{\text{D}_{IS}}(\phi)]$$

is zero because $I_z S_z$ commutes with itself. Thus, this and all higher-order dipole–dipole terms vanish. Similarly, the higher-order cross terms due to the chemical shift also vanish. Indeed, the cross terms between the chemical shift and the heteronuclear dipolar Hamiltonian also vanish because S_z commutes with itself.

The Carr–Purcell sequence on the observed spin exactly suppresses both heteronuclear dipolar and chemical-shift interactions for that spin. Figure 6.2 shows that I–S decoupling is achieved by having the heteronuclear Hamiltonian during one period of time cause evolution that is opposite to that occurring at other times in the cycle. It is clear that the excitation sequence that produces this reversal also reverses the effect of the S-spin chemical shift as well. What is needed if the shielding of the rare spin is to be observed is a sequence that will produce appropriate reversal of the heteronuclear Hamiltonian's effect while simultaneously leaving the S spin's chemical shift inviolate. What is needed is an rf excitation sequence that affects the I spins, since their state also determines the extent of averaging of the heteronuclear dipolar interaction, but irradiation of the I spins does not affect the S-spin chemical-shift interaction. The sequence of Fig. 6.3 is capable of producing decoupling in a manner analogous to the Carr–Purcell sequence of Fig. 6.2 in the sense that it causes the heteronuclear dipolar Hamiltonian to progress through the same states as does the Carr–Purcell sequence, but it leaves the

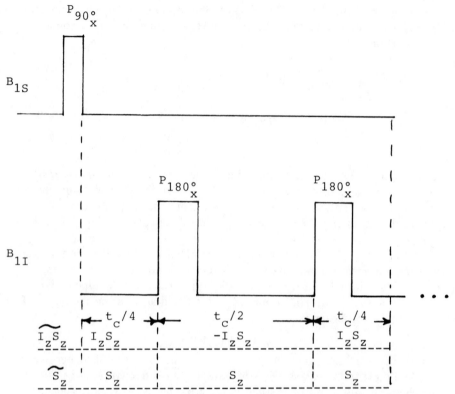

FIG. 6.3 A pulse cycle for decoupling two spins-$\frac{1}{2}$ subject to the heteronuclear dipole–dipole interaction. This cycle does not suppress the chemical-shift interactions of the S spins.

S-spin chemical-shift interaction unaffected. The sequence of Fig. 6.3 is applied repetitively, with selective detection at times Nt_c as with the Carr–Purcell sequence. A calculation of the zero-order Hamiltonians that affect the S spin under the sequence gives the average Hamiltonian

$$\bar{\mathscr{H}}^{(0)} = \int_0^{t_c} \{\tilde{\mathscr{H}}_{DIS}(t) + \tilde{\mathscr{H}}_{css}(t)\} \, dt$$

$$= \frac{1}{4}(-\omega_D I_z S_z + \sigma_S \omega_{0S} S_z)$$

$$+ \frac{1}{2}(\omega_D I_z S_z + \sigma_S \omega_{0S} S_z)$$

$$+ \frac{1}{4}(-\omega_D I_z S_z + \sigma_S \omega_{0S} S_z)$$

$$= \sigma_S \omega_{0S} S_z = \mathscr{H}_{css} \qquad (6.5)$$

Hence the S magnetization evolves at times Nt_c as if it were subject to only a chemical-shift interaction.

Sequences of π pulses are not the only ones that will decouple S from I spins. Consider a sequence of $\pi/2$ pulses such as shown in Fig. 6.4. It should be clear from the preceding example that application of the decoupling sequence to the I spin does not affect the S-spin chemical-shift interaction and the average Hamiltonian will contain the unscathed chemical shift. The contribution due to the heteronuclear dipolar coupling, to zeroth order, can be calculated from the transformed Hamiltonians indicated in Fig. 6.4:

$$\bar{\mathscr{H}}^{(0)}_{DIS} = \frac{1}{t_c} \int_0^{t_c} \tilde{\mathscr{H}}_{DIS}(t)\, dt = 0 \qquad (6.6)$$

Thus, a sequence of $\pi/2$ pulses will also produce average evolution of the S magnetization governed only by the chemical-shift interaction. Indeed, a sequence of $2n$ pulses of nutation angle, π/n, applied to the I spins will produce decoupling to allow observation of the S magnetization and its evolution under the influence of the chemical-shift interaction (but see Problem 6.4).

In the preceding discussion we have assumed that the pulse decoupling sequences are "set" in such a manner that decoupling will be successful. It should be noted that if the spacing between pulses is sufficiently long, in fact no decoupling of the I and S spins occurs, in much the same way that long cycle times for the homonuclear-decoupling sequences cause the cycle to be

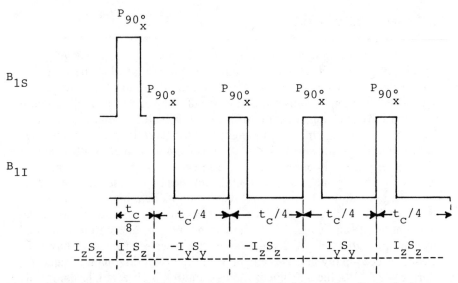

FIG. 6.4 A sequence of $\pi/2$ pulses used to suppress the heteronuclear dipole–dipole coupling between two spins-$\frac{1}{2}$, which retains chemical-shift information.

ineffective. The question is: At what rate must the I spins be excited for the decoupling to be effective? Consider one of the I spins that is coupled strongly to the S spin (Fig. 6.1) but that is also strongly coupled to its surrounding I-spin neighbors. After the preparation pulses, the S spin begins to evolve under the influence of both its own chemical-shift interaction and the heteronuclear dipole–dipole coupling to this I spin. During this period, the I spin is also subject to interactions with its I-spin neighbors. If during this period before the inversion pulse spin information about the state of the I spin is transferred from the particular I spin to other I spins, the result is that the evolution of the system after the inverting pulse cannot completely compensate for the evolution due to the heteronuclear dipolar interaction between this I spin and the S spin. The net effect is to cause decay of the S-spin magnetization, an effect that one is specifically trying to avoid. To ensure that this effect does not occur appreciably, the pulse spacing must be shorter than the times over which mutual I–I spin flips occur. Here T_{2I} (or $M_{2II}^{1/2}$) is a measure of this time, so that one has the requirement on the pulse spacing that

$$t_c/2 < T_{2I} \qquad (6.7)$$

For typical protons in an organic molecule, T_{2I} is about 20 μsec. This implies that to produce effective decoupling of these protons from ^{13}C in this molecule for observation of the ^{13}C chemical-shift interaction, one must have sequences of π pulses such that $t_c < 40$ μsec. This condition is fairly stringent, but can be met with modern NMR spectrometers.

III. Continuous-Wave Decoupling

In the preceding section we calculated the effect of sequences of pulses on a simple heteronuclear-coupled spin-$\frac{1}{2}$ system. We saw that the appropriate string of pulses with stroboscopic detection permits the detection of weaker interactions such as the chemical shift. The effect of the pulses is to produce a cyclic change of state, such that any evolution due to the action of the heteronuclear dipolar coupling in one (or several) time intervals is exactly cancelled by time evolution under this Hamiltonian at another time. A similar experiment is one in which the interfering nucleus is continuously excited [1]. Under such an excitation the nucleus of interest will also appear to undergo evolution that does not depend on its heteronuclear interaction with the interfering nucleus. The sequence for this continuous-wave decoupling is shown in Fig. 6.5. First, a transient S magnetization is created by the application of a $\pi/2$ pulse, as in the other sequences we have discussed. A decoupling field is then applied at the I-spin resonance frequency, during which time the S magnetization's evolution is observed. Keeping only the secular terms (Chapter 3) of the heteronuclear dipolar interaction, the

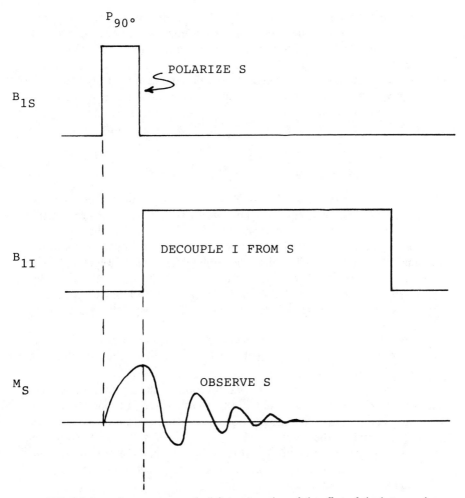

FIG. 6.5 A continuous-wave method for suppression of the effect of the heteronuclear dipole–dipole coupling on the spectrum of the S spins.

total Hamiltonian, expressed in the doubly rotating frame specified by the operators $\exp(-i\omega_{0I}tI_z)$ and $\exp(-i\omega_{0S}tS_z)$ based on the Zeeman Hamiltonians for both spins, is given by Eq. (6.2). We ignore the homonuclear I–I coupling term because it is not a part of the present problem. The last term in Eq. (6.2) is the Hamiltonian for interaction with the decoupling field. For decoupling to be effective, this term must be sufficiently large to be a major factor in determining the eigenstate of the system. As discussed in Section IV of Chapter 4, when a nonsecular time-independent term dominates the time evolution in the frame of observation, it is instructive to view the motion of the system from a frame in which this term is secular. A further transformation to the frame of this dominant Hamiltonian will remove its effect on the time evolution. If in this frame the other Hamiltonians such as the dipolar interaction are strongly time-dependent, they will have practically no observable effect on the transient magnetization. Of course, one always has to be aware of what frame of reference is used to detect the magnetization. In this case, the S magnetization is observed in the last frame, as illustrated in Fig. 6.6. In this frame the Hamiltonian has the form

$$\tilde{\tilde{\mathcal{H}}} = \omega_{0_s}\sigma_S S_z + \omega_{0_I}\sigma_I[I^- \exp(i\omega_1 t) + I^+ \exp(-i\omega_1 t)]$$
$$- \omega_D[I^- \exp(i\omega_1 t) + I^+ \exp(-i\omega_1 t)]S_z \qquad (6.8)$$

All terms except the chemical shift of the spin S oscillate at frequency ω_1, which depends on the strength of the rf decoupling field. To the extent that ω_1 is large compared to the spectral width for observation of the S magnetization, the effects of these terms will not be seen. With a 100-W proton transmitter and a 10-mm coil, typical decoupling field amplitudes applied to protons can be around 10 G corresponding to a frequency of 42.7 kHz. The spectral width of the chemical-shift spectrum of carbon in a magnet of 1.5 T for ^{13}C is less than 6 kHz. Therefore, one can easily accomplish cw decoupling under such conditions.

It would have been difficult to have seen this result without viewing the system of the I and S spins from this peculiar "doubly rotating–tilted rotating" frame of reference. The property of the experimental design that permitted the system to be viewed from this frame is the fact that we assumed that the term of largest magnitude in the doubly rotating frame is the rf excitation of the I spins. This assumption means that there are lower bounds on the magnitude of the decoupling field that will permit this approximation. In particular, the amplitude of the rf field in frequency units must be larger than both the magnitude of the heteronuclear and the homonuclear dipole–dipole couplings of the I spins.

$$\omega_1 > \omega_{D_I} \quad \text{and} \quad \omega_{D_{Is}} \qquad (6.9)$$

One may estimate the minimum required field for decoupling from this inequality for a ^{13}C–^{1}H pair, with a homonuclear coupling (the larger of the two) for the protons of 30 kHz. Thus, ω_1 must be greater than 30 kHz, which gives a minimum rotating field amplitude of about 7 G, a field which is certainly achievable in modern NMR spectrometers. A useful relation between the rf field amplitude in an inductor such as is used in a probe circuit that has an internal volume V (in cubic centimeters) and a quality factor Q and the power P (in watts) is

$$B_1 \approx 3(PQ/f_0 V)^{1/2} \qquad (6.10)$$

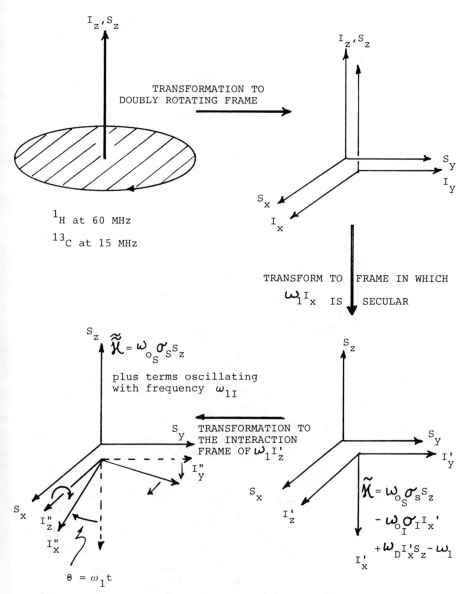

FIG. 6.6 The transformation involved in going to the frame in which the S spins are observed while the decoupling of the I spins is occuring.

For a typical arrangement, a 10-G decoupling field can be achieved with a nominal power of 100 W. Problem 6.6 gives an example of this calculation for a particular coil geometry.

We saw in Section I that the requirement on pulse decoupling is couched in terms of the cycle time relative to T_{2I}. In this section we note that, for continuous-wave decoupling, the requirement is couched in terms of the rotating-frame amplitude of the decoupling field relative to the amplitude of the dipole–dipole interaction. Since these two processes produce the same result, one might expect that these criteria represent similar situations. Indeed they do. The important criterion from the point of view of decoupling is that an I spin's state be switched rapidly compared to the time it takes for it to interact substantially with other I spins, via its homonuclear dipole–dipole interaction. Whether this switching is by the jerky motion of short rf pulses or by the continuous movement of the spin by a continuous-wave excitation, it must occur quickly enough so that no substantial change has occurred by the effect of other interactions, which is what Eqs. (6.7) and (6.9) both imply.

It is important to note that an analogous calculation is applied to the scalar spin–spin decoupling sequences used in high-resolution liquid-state NMR spectroscopy of ^{13}C and other nuclei coupled to abundant spins such as protons. The difference from the present case is that the magnitude of the scalar-coupling interaction is given by the spin–spin coupling constant J rather than the dipolar interaction. Since J is generally (but not always) of the order of a few to a few hundred hertz, the requirements on the amplitude of the decoupling field are much less stringent. It is typical, under such circumstances, to use much lower power than is used for dipole–dipole decoupling. This has an added practical advantage: It is less likely that a mistake will produce smoke signals. Another difference between the scalar decoupling and the dipolar decoupling sequences is the type of system that is irradiated. In scalar decoupling experiments, one frequently wishes to decouple a number of spins in different environments that are coupled. The main concern in these experiments is how to distribute the available power over the spectrum to excite all of the transitions in the I-spin system, e.g., how to irradiate both phenyl and aliphatic carbons in the decoupled mode. There are schemes to produce this decoupling, such as broadband noise decoupling, which are used in these cases. In a strongly dipolar-coupled system, on the other hand, the I spins are so strongly coupled that excitation of one of the spins is quickly transferred to many others. That is, a strongly coupled dipolar system such as the protons of a typical organic solid is *homogeneously* broadened, whereas the protons in an organic molecule in the liquid state comprise an *inhomogeneously* broadened spectrum (in this case the spectrum is inhomogeneously broadened by the differences in chemical shift), each isochromat of which must be excited independently of the others.

There is one feature of decoupling that will be important in future discussions of enhancement. The excitation of the I spins can be accomplished in two ways: (1) by irradiation of the I-spin system with no preparation and (2) by irradiation of the I-spin system after a preparation that leaves the I spin in a state in which it can be *spin locked* by the decoupling field. Both procedures produce heteronuclear decoupling, but the first method causes rapid destruction of the I magnetization. For reasons that will become obvious in the next section, this manner of decoupling is not desirable because one requires an I magnetization in multiple-contact cross-polarization experiments. By applying a $\pi/2$ pulse to the I spins, followed by a decoupling field with an rf carrier 90° out of phase with the preparation pulse, one achieves spin locking in a manner analogous to the measurement of $T_{1\rho}$ discussed in Chapter 1. Typically such sequences (Fig. 6.7) are used for performing decoupling, although the preparation is not essential to the decoupling process itself.

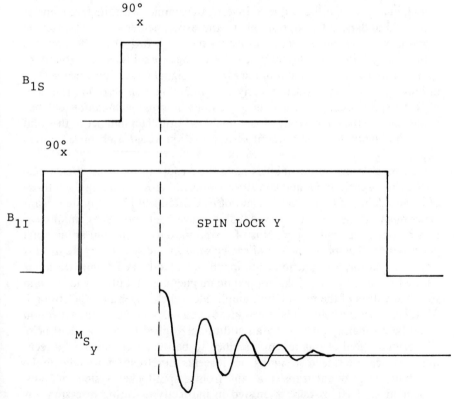

FIG. 6.7 Spin-lock decoupling that preserves the magnetization of the I-spin system.

IV. Signal Enhancement by Cross-Polarization

The study of chemical shielding of rare spins such as ^{13}C in solids using heteronuclear decoupling experiments has an inherent feature—the rare spins are few. This is both a blessing and a curse. The fact that the spins are rare means that they are not subject to significant homonuclear dipole–dipole interactions. Practically, this fact means that homonuclear dipolar decoupling sequences like those discussed in Chapter 5 are unnecessary and that the problems of adjusting the spectrometer system inherent in them need not be as much of a concern to the experimenter. However, this benefit is offset by the loss of signal-to-noise ratio suffered due to the limited number of resonant nuclei. Enrichment of the sample in nuclei can help. For example, doubling or even increasing the number of ^{13}C nuclei in a sample tenfold increases the signal without appreciably introducing strong $^{13}C-^{13}C$ homonuclear dipolar interactions. There are samples that cannot be enriched without destroying the characteristics of the sample that NMR is to probe, and it would, in any case, be desirable to investigate samples with the rare spins at natural abundance. The option left to the experimenter is to increase the signal from these rare nuclei by other means. As it turns out, the presence of the strong dipole–dipole coupling between the rare spins and the abundant spins permits an enhancement of the rare-spin signal under appropriate conditions. This process, known properly as *polarization transfer*, or in the jargon of NMR spectroscopy under pseudonyms such as *cross-polarization* (cp), has made possible the direct study of rare nuclei in natural abundance in the solid state, thus increasing the amount of information about a chemical system [2].

Cross-polarization techniques have their roots in the experiments performed in the early 1960s and late 1950s by several groups, among them those of Erwin Hahn, of C. P. Slichter, and of A. G. Redfield [2]. In quite elegant experiments, these workers showed that it is possible to produce communication between the consitiuent spins of a spin-locked system and an external perturbation. The phenomenon of *rotary saturation* discussed by Redfield is a resonant absorption phenomenon in which a spin-locked magnetization is affected by modulation of the applied dc magnetic field, but only for certain specified values of the modulation amplitude, which depend on the strength of the spin-locking field. Thus, it was known that a spin-locked magnetization could be resonantly affected by an additional applied time-dependent field. The primary goal of this research seemed to be the development of a technique to measure rf field strengths, but it seems, in retrospect, to indicate the possibility of resonant transfer to and from a spin-locked system of spins. Hartmann and Hahn were interested in indirectly detecting rare spins, as were Lurie and Slichter, and the resonant transfer of magnetization from one spin system to another was the method of choice. Interestingly, the buildup of rare-spin magnetization interfered with the saturation of the abundant-spin magnetization that was to be detected, so both of these groups built into the detection schemes ways to periodically destroy the rare-spin magnetization. A variant of such indirect detection schemes has been demonstrated by Mansfield, but it has not been exploited [3]. In the current manner of investigating rare spins [2], the experimenter detects the rare spin directly.

A. Spin Temperature

Before discussing the resonant transfer of magnetization, it is necessary to review the concept of spin temperature that is used to model this transfer. As seen in Chapter 2, the equilibrium spin density matrix for spin systems in the high-temperature limit is given by

$$\rho = 1 - \hbar \mathscr{H}/kT \tag{6.11}$$

The magnetization, energy, and entropy are given by appropriate traces:

$$M_k = \gamma \hbar \operatorname{Tr} \rho I_k \qquad (k = x, y, z) \tag{6.12a}$$

$$E = \operatorname{Tr} \rho \mathscr{H} \tag{6.12b}$$

$$\mathscr{S} = -k \operatorname{Tr} \rho \ln \rho \tag{6.12c}$$

For spin systems in the high-temperature limit subject to the Zeeman interaction with $\mathbf{B}_0$ along z [see Eqs. (2.72) and (2.73)],

$$M_0 = CB_0/T \tag{6.13a}$$

$$E = CB_0^2/T \tag{6.13b}$$

$$\mathscr{S} = (CB_0^2/T^2) + \text{const} \tag{6.13c}$$

Consider the thermodynamics and kinetics of polarization transfer between N_I spins, with magnetogyric ratio γ_I, at spin temperature T_I, and N_S spins characterized by γ_S and T_S (Fig. 6.8). For purposes of discussing polarization transfer in terms of spin thermodynamics, the I and S spin systems form reservoirs that can be in contact with each other and with the lattice, which is effectively an infinite reservoir. The rates of spin–lattice relaxation under some condition x are given in terms of time constants T_{1x_I} and T_{1xs} (e.g., under spin-lock conditions on spin I, T_{1x_I} would be $T_{1\rho_I}$). The rate of polarization transfer between the I and S reservoirs is characterized by a time constant T_{IS}, which under conditions of cross-polarization we call T_{cp}.

There are three steps in a polarization transfer experiment: (I) reduction of the spin temperature of the I reservoir, (II) transfer of polarization between

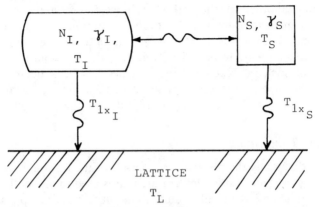

FIG. 6.8 The Zeeman reservoirs and the lattice. Arrows indicate transfer routes among the various reservoirs by cross-polarization or spin–lattice relaxation.

the two reservoirs, and (III) observation of the S spin. There are many methods of cooling the I reservoir, of which we discuss two: (1) spin-locking (SL), in which the Zeeman order of the I spin is increased, and (2) adiabatic demagnetization in the rotating frame (ADRF), in which the order in the I-spin system is transferred to dipolar order. The rates of destruction of this order by spin–lattice processes under these two conditions are given by $(1/T_{1\rho I})^{-1}$ and $(1/T_{1D})^{-1}$, respectively. Here T_{1D} is the spin–lattice relaxation time of the dipolar order. The choice of which method to use in cooling the I system will depend on the rates of loss to the lattice relative to the rate of cross-polarization T_{cp}^{-1}.

B. Spin-Lock Polarization Transfer

Consider the spin-lock experiment to cool I, with polarization transfer by simultaneous irradiation of I and S. The rf irradiation of both I and S spins in this experiment is shown in Fig. 6.9, with regions I–III, described in Section IV.A, indicated. Region IV, the $90_{\bar{x}}^{\circ}$ "flip-back" pulse on the I spin, has been added to place the remaining coherent transverse I magnetization back along $\mathbf{B}_0$. This addition to the original sequence decreases the time between signal averages for a single-contact experiment, which is limited by T_{1I}.

Immediately after the 90_{x}° pulse is applied to the I-spin system (region I in Fig. 6.9) and at the point when the spin lock of the I system along y begins, the I spins have a spin temperature T_0 in $\mathbf{B}_{1I}$, the I spin-lock field. This spin temperature is very low compared to that of the I spins T_L at equilibrium in the field $\mathbf{B}_0$. This result is clear from the fact that the 90_{x}°

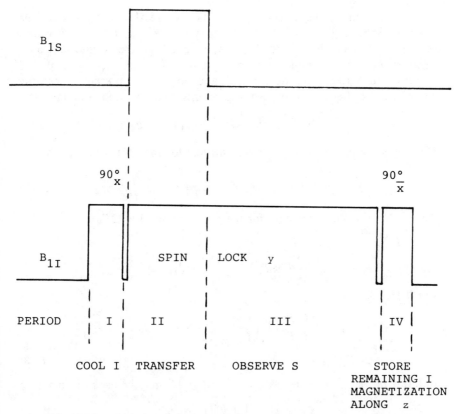

FIG. 6.9 The spin-lock polarization transfer experiment, showing the three domains. Note that the *I* magnetization is restored along the dc magnetic field at the end of this sequence by a $90^\circ_{\bar{x}}$ pulse.

pulse is sufficiently rapid that the change in the *I*-spin reservoir can be considered to be adiabatic. In the limit that this change is adiabatic and reversible, the entropy of the *I*-spin system, given by (6.13c), is conserved. Therefore, from (6.13c), applied to the initial and final states (at T_L and T_0, respectively), we find

$$B_0/T_L = B_1/T_0 \qquad (6.14)$$

This relationship implies that, at the beginning of the spin lock, T_0 for the *I*-spin reservoir is very low relative to the lattice temperature T_L. For $B_1 = 60$ G and $B_0 = 10^4$ G, we calculate that $T_0 = 1.8$ K for a lattice temperature of 300 K. Consider the region of polarization transfer (region II). Under conditions of spin-lock fields B_{1I} and B_{1S} on the *I* and *S* systems, polarization transfer will occur in a manner yet to be defined. If the rates of the

I and S systems' being warmed by the lattice are slow compared to the cooling of the S-spin reservoir by the I-spin reservoir T_{cp}^{-1}, conservation of energy between the I- and S-spin systems allows the equilibrium temperature $T(1)$ of the coupled I–S spin system to be calculated. After period I, the I and S reservoirs are isolated, and no work is present. Therefore, energy transfer between I and S involves only heating I and cooling of S

$$q_I = \int_{T_0}^{T(1)} c_B^I \, dt = \int_{T_L}^{T(1)} c_B^S \, dt = -q_S \tag{6.15}$$

Here c_B^j is the heat capacity at constant field for the jth reservoir, given by

$$c_B^j = \frac{\partial E^j}{\partial T} = \left(\frac{\partial}{\partial T}\right)_B \text{Tr} \, \rho \mathcal{H}_j = \frac{\partial}{\partial T}\left(-\frac{C_j B^2}{T}\right) = \frac{C_j B^2}{T^2} \tag{6.16}$$

where C_j is the Curie constant for the jth reservoir [Eq. (2.73)].
   Therefore

$$C_I B_{1I}^2 \int_{T_0}^{T(1)} \frac{dT}{T^2} = C_S B_{1S}^2 \int_{T_L}^{T(1)} \frac{dT}{T^2} \tag{6.17}$$

which yields

$$T(1) = T_0\left(1 + \frac{C_S B_{1S}^2}{C_I B_{1I}^2}\right)\bigg/\left(\frac{C_S B_{1S}^2}{C_I B_{1I}^2}\frac{T_0}{T_L} + 1\right) \simeq T_0\left(1 + \frac{C_S B_{1S}^2}{C_I B_{1I}^2}\right)$$

$$\equiv T_0(1 + \delta) \tag{6.18}$$

In terms of the lattice temperature T_L, Eq. (6.14) may be used to obtain

$$T(1) = (1 + \delta)T_L B_{1I}/B_0 \tag{6.19}$$

   The preceding discussion has dealt with the thermodynamics, but not the kinetics, of polarization transfer. Unless some specific action is taken to break the I–S interaction, e.g., decoupling, in the presence of B_{1I} and B_{1S} there will *always* be polarization transfer to some extent. It may be shown that the polarization-transfer rate is a maximum when the precession frequencies of I and S, in their respective rotating frames, are equal:

$$\omega_I^{\text{eff}} = \gamma_I B_{1I} = \gamma_S B_{1S} = \omega_S^{\text{eff}} \tag{6.20}$$

Equation (6.20) is known as the *Hartmann–Hahn condition*. Under this condition of matched rotating-frame spin-lock frequencies, the value of δ in Eq. (6.18) is found to be

$$\delta = \frac{C_S B_{1S}^2}{C_I B_{1I}^2} = \frac{\gamma_I}{\gamma_S} \frac{N_S}{N_I} \frac{S(S + 1)}{I(I + 1)} \tag{6.21}$$

If S is a rare spin, by definition $N_S/N_I \ll 1$ and δ is a number generally small compared to unity. The enhancement of S magnetization associated with the cooling of the S-spin reservoir to $T(1)$ is calculated from Curie's law. At the temperature $T(1)$ [Eq. (6.19)] the magnetization of the S-spin reservoir in a field B_{1I} is, under the Hartman–Hahn condition,

$$M_S = \frac{C_S B_{1I}}{T(1)} = \frac{C_S B_0}{T_L(1 + \delta)} \frac{\gamma_I}{\gamma_S} = M_L \frac{\gamma_I}{\gamma_S} \qquad (6.22)$$

This magnetization is larger than the magnetization obtained by allowing the rare-spin system to come to equilibrium in the applied dc field $\mathbf{B_0}$ by roughly the ratio of the gyromagnetic ratios. In the case of ^{13}C and ^1H, as in a typical organic molecule, the maximum gain in rare-spin signal intensity is four. The magnetization is a spin-locked transverse magnetization. Turning off the spin-locking field results in a free-induction decay during period III that can be recorded. By maintaining the spin lock on the I spins, this free-induction decay will not be influenced by the heteronuclear dipolar interaction, for the reasons discussed in the previous sections. Hence, Fourier analysis of this response will give the spectrum of the rare-spin system, subject only to the chemical shift and other smaller Hamiltonians.

An enhancement by a factor of γ_I/γ_S is certainly an achievement; however, it is possible to accumulate spectra several times because the cross-polarization process does not eliminate all of the abundant-spin magnetization. Thus, the spectroscopist can use the multiple-contact cross-polarization sequence shown in Fig. 6.10, which comes very close to being a Lurie–Slichter experiment. The sequence consists of repeated coupling of the two-spin systems by Hartmann–Hahn matching, followed by a period of recording, without allowing the abundant spins to repolarize. These S-magnetization transients are co-added as one would usually do in typical

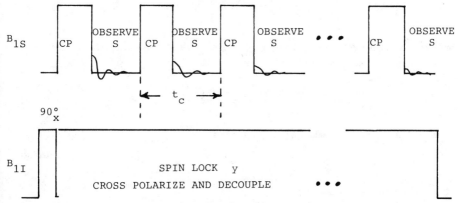

FIG. 6.10 The multiple-contact cross-polarization experiment. Detection of the S-spin free-induction decay occurs in the intervals between the polarizing pulses.

data collection. The product of this process is a signal that is the co-addition of N cross-polarized free-induction decays. The question is how much improvement in signal-to-noise ratio can be obtained by a multiple-contact cross-polarization over a given time, when compared with collection of free-induction decays under the usual set of circumstances in which one allows complete repolarization by spin–lattice relaxation processes between the free-induction decays. For this multiple-contact version of the cross-polarization sequence one may extend the spin-thermodynamic arguments to predict that the overall co-added signal is

$$M_S(N) = M_L \frac{\gamma_I}{\gamma_S} \sum_{n=1}^{N} e^{-n\delta} \tag{6.23}$$

This quantity can be made arbitrarily large by taking N arbitrarily large; however, that requires a longer time Nt_c and risks the destruction of the probe circuitry.

The important consideration in using such enhancement sequences is the *gain* in signal-to-noise ratio in a given amount of time using the enhancement sequences. The signal-to-noise ratio is defined as

$$\phi_a = E_a/E_{noise} = V_a^2/V_{noise}^2 \tag{6.24}$$

where V_a is the total signal voltage and V_{noise} is the associated accumulated noise voltage. The voltage observed in a single-pulse experiment is

$$V = M_S L \tag{6.25}$$

where L is the inductance of the coil

Hence, for multiple-contact cross-polarization,

$$V_{cp} = LM_L \left\{ \frac{\gamma_I}{\gamma_S} \right\} \sum_{n=1}^{N} e^{-n\delta} \tag{6.26}$$

The accumulated noise voltage is the number of scans multiplied by the noise voltage for a single scan. Therefore,

$$\phi_{cp} = \left\{ \frac{LM_L}{V_{noise}} \right\}^2 \frac{1}{N} \left\{ \frac{\gamma_I}{\gamma_S} \right\}^2 \left\{ \sum_{n-1}^{N} e^{-n\delta} \right\}^2 \tag{6.27}$$

In a single-pulse experiment, the voltage is

$$V_{FID} = LM_L \tag{6.28}$$

and so

$$\phi_{FID} = \left\{ \frac{LM_L}{V_{noise}} \right\}^2 \tag{6.29}$$

Hence, the effective signal-to-noise gain over a single free-induction-decay experiment is

$$\frac{\phi_{cp}}{\phi_{FID}} = \frac{1}{N} \left\{ \sum_{n=1}^{N} e^{-\delta n} \right\}^2 \left\{ \frac{\gamma_I}{\gamma_S} \right\}^2 \tag{6.30}$$

We may calculate the optimum value for the number of repetitions per cycle by differentiation of (6.30) with respect to N. Pines $et\ al.$ [2] quote this optimum value as $N\delta \approx 1.3$, which gives

$$\phi_{cp}^{opt} = \frac{0.41}{\delta} \left\{ \frac{\gamma_I}{\gamma_S} \right\}^2 \left\{ \frac{LM}{V_{noise}} \right\}^2 \tag{6.31}$$

A more appropriate measure of resolution enhancement is the time required to achieve a given signal-to-noise ratio with cross-polarization compared to co-addition of free-induction decays. This time depends on how often one may repeat the experiment. In the case of free-induction decays, one must wait a time long enough to repolarize the S system before repeating the experiment. As discussed in Chapter 1, that time is $\sim 5T_{1S}$, where T_{1S} is the S spin–lattice relaxation time. On the other hand, the cross-polarization experiment can be repeated at a rate that is limited by the repolarization time of the I spins T_{1I}. In a specified time T, one can repeat the FID experiment $T/5T_{1S}$ times and the cross-polarization experiment $T/5T_{1I}$ times, assuming that T_{1S} and T_{1I} are much longer than T_{2S} or t_c (Fig. 6.10). The improvement in signal-to-noise ratio for each of these cases is then

$$\phi_{cp} = \frac{T}{T_{1I}} \phi_{cp}^{opt} = \frac{T}{T_{1I}} \frac{0.41}{\delta} \left\{ \frac{\gamma_I}{\gamma_S} \right\}^2 \left\{ \frac{LM_L}{V_{noise}} \right\}^2 \tag{6.32a}$$

$$\phi_{FID} = \frac{T}{T_{1S}} \phi_{FID} = \frac{T}{T_{1S}} \left\{ \frac{LM_L}{V_{noise}} \right\}^2 \tag{6.32b}$$

The ratio of these two quantities is a measure of the sensitivity improvement per unit time when using the cross-polarization experiment instead of the standard free-induction-decay technique.

$$I_E = \frac{\phi_{cp}}{\phi_{FID}} = \frac{T_{1S}}{T_{1I}} \frac{0.41}{\delta} \left\{ \frac{\gamma_I}{\gamma_S} \right\}^2 \tag{6.33}$$

Consider a practical example in which the I spins are protons and the S spins are ^{13}C. The ratio of protons to carbons-13 in an organic solid would be 100 if for each carbon there were one proton. In fact, there usually are more. Assume that there are 200 protons for every ^{13}C. For purposes of illustration, assume that $T_{1S} \approx 5T_{1I}$, not an unrealistic approximation. Here

δ is defined by (6.18) but $S = I$ for this case and $\delta = N_S/N_I$. Therefore the improvement factor is

$$I_E = 0.41 \left\{\frac{T_{1S}}{T_{1I}}\right\} \left\{\frac{N_I}{N_S}\right\} \left\{\frac{\gamma_I}{\gamma_S}\right\}^2 \qquad (6.34)$$

For protons and carbons $\gamma_I/\gamma_S \approx 4$, so

$$I_E = 0.41\,(5)\,(200)\,(4)^2 \approx 6.5 \times 10^3$$

which is a substantial improvement in signal-to-noise ratio for the rare-spin resonance. Even if $T_{1I} = T_{1S}$, the gain would still be about 10^3. This sensitivity improvement relative to the usual means of increasing signal-to-noise ratio by co-addition of FIDs constitutes one of the important advantages of the cross-polarization technique for signal enhancement of rare spins.

C. ADRF Transfer

Cooling of the I reservoir by ADRF (adiabatic demagnetization in the rotating frame) transfers order to the dipolar reservoir for the I-spin system. Polarization transfer between I and S is then effected by Hartmann–Hahn matching of the dipolar frequency of I, ω_{DI}, with ω_{1S}. The process begins by the usual spin lock of the I spins after a $90°$ preparation pulse. The spin temperature of the I system is calculated in accordance with Eq. (6.14), in which the dipolar interaction between I spins is explicitly included:

$$\mathscr{H}_I = -\omega_{1I} I_x + \mathscr{H}_D \qquad (6.35)$$

At the beginning of the spin lock, the spin temperature of the I reservoir [by converting fields into frequencies in Eq. (6.14)] is given by

$$\frac{1}{T} = \frac{1}{T_L} \left[\frac{\omega_0^2}{\omega_{1I}^2 + \omega_D^2} \right]^{1/2} \qquad (6.36)$$

Unlike the spin-lock case, however, the amplitude of the locking field is decreased, but at a rate such that the spins have sufficient time to readjust to the total instantaneous field in the rotating frame. Under these conditions, there exists a spin temperature, and the net entropy change is zero. We divide the I-spin system into a Zeeman reservoir, associated with the Zeeman states of the ensemble, and a dipolar reservoir, associated with the dipolar states. The isentropic ADRF converts "Zeeman order" into "dipolar order." Stirring the rare-spin system with a cross-polarization field in a subsequent step transfers order from the dipolar reservoir of the I spins to the rare-spin Zeeman system, from which a free-induction decay may be accumulated. The transfer from the dipolar system to the Zeeman system occurs for only specified values of the cross-polarization field; i.e., the transfer is optimal for a Hartmann–Hahn condition given by

$$\gamma B_{1S} = \omega_D \qquad (6.37)$$

The transfer rates for this process is given by (*vide infra*)

$$(T_{cp})^{-1}_{\text{ADRF}} = \frac{B^2_{1S}}{B^2_{1S} + \Delta B^2_S} M^{IS}_2 J_{\text{ADRF}}(\omega_{cs}) \qquad (6.38)$$

After the enhancement process, one may observe a free-induction decay of the rare spins by suddenly turning off the cross-polarization field. To eliminate the effects of the heteronuclear dipolar interaction it is necessary to provide decoupling during this period. The application of strong decoupling fields such as described earlier will produce decoupling; however, they may also produce unwanted side effects. The dipolar order of the abundant spins that was so carefully created in the first step may, for example, be destroyed by the action of the decoupling field, as shown in Fig. 6.11(a). Thus, the potential for multiple-contact or repolarization of the *I*-spin system from a partially relaxed state may not be possible if the dipolar order has been destroyed. One way to perform the decoupling without destroying dipolar order is by performing a "magic sandwich" decoupling experiment as shown in Fig. 6.11(b). The decoupling spin-lock is preceded by a $\pi/2$ pulse that is specified by a density operator that commutes with the succeeding rf irradiation, producing no destruction of the dipolar order. At the end of the decoupling, the $\pi/2$ pulse returns the system to a state, the density operator of which commutes with the applied dc field, and the process may be repeated, or if one wishes, the *I*-spin system may be repolarized to allow relaxation processes to rebuild the *I*-spin magnetization.

D. Phenomenological Treatment of Transfer Kinetics

The cross-polarization experiments we have been sketching change the distribution of energy in the spin system. An important consideration is the detailed manner in which this evolution occurs. A phenomenological treatment of this problem is as follows. Consider the system as divided into a rare-spin part and an abundant-spin part. The energy of the system is given by the sum of the energies of the two parts:

$$\langle \mathcal{H} \rangle = \langle \mathcal{H}_I \rangle + \langle \mathcal{H}_S \rangle \qquad (6.39a)$$

During the polarization transfer part of the cp experiment, we consider the *I* and *S* spin reservoirs to be isolated, and energy is conserved:

$$d\langle \mathcal{H} \rangle/dt \qquad (6.39b)$$

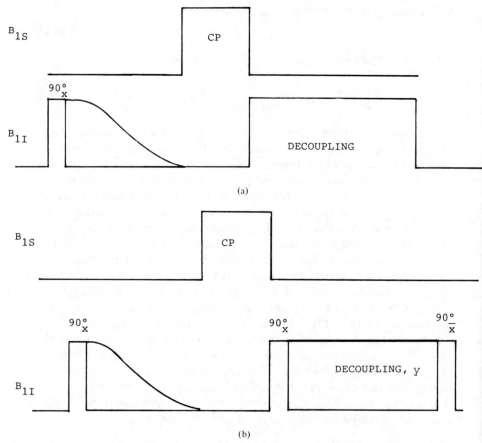

FIG. 6.11 Cross-polarization via adiabatic demagnetization in the rotating frame: (a) with decoupling that destroys any residual dipolar order and (b) with decoupling using the "magic sandwich" that preserves dipolar order during decoupling.

For each of the individual components of the spin energy, the general relaxation equation, assuming first-order kinetics, is

$$d\langle \mathcal{H}_I \rangle/dt = -(1/T_{IS})\langle \mathcal{H}_S \rangle - (1/T_{II})\langle \mathcal{H}_I \rangle \qquad (6.40a)$$

$$d\langle \mathcal{H}_S \rangle/dt = -(1/T_{SS})\langle \mathcal{H}_S \rangle - (1/T_{SI})\langle \mathcal{H}_I \rangle \qquad (6.40b)$$

By comparison of (6.39b) and (6.40), the following conditions must hold:

$$1/T_{II} = -1/T_{SI} \qquad (6.41a)$$

$$1/T_{SS} = -1/T_{IS} \qquad (6.41b)$$

It must also be true that, at long times, the two subsystems are in equilibrium and that the energies $\langle \mathcal{H}_I \rangle$ and $\langle \mathcal{H}_S \rangle$ have their equilibrium values calculated by spin thermodynamics $\langle \mathcal{H}_I \rangle^{eq}$ and $\langle \mathcal{H}_S \rangle^{eq}$. This requirement gives a relationship between the relaxation times:

$$\frac{1}{T_{II}} = (\langle \mathcal{H}_S \rangle^{eq}/\langle \mathcal{H}_I \rangle^{eq})(1/T_{SS}) = \delta/T_{SS} \qquad (6.42)$$

These conditions yield

$$d\langle \mathcal{H}_I \rangle/dt = -(1/T_{cp})\{\delta\langle \mathcal{H}_I \rangle - \langle \mathcal{H}_S \rangle\} \qquad (6.43a)$$

$$d\langle \mathcal{H}_S \rangle/dt = -(1/T_{cp})\{\langle \mathcal{H}_S \rangle - \delta\langle \mathcal{H}_I \rangle\} \qquad (6.43b)$$

and the cross-polarization rate T_{cp}^{-1} is identified with the phenomenological constant T_{SS}, and δ is defined from Eq. (6.18).

The systems are never completely isolated, of course. Equations (6.43) represents the manner in which energy is transferred within the *isolated* systems. To that we must add a term that represents the transfer of energy into the lattice.

The abundant I-spins usually relax most effectively with the lattice, and we assume that spin–lattice relaxation of the S-spin reservoir may be neglected. The time dependence of the I-spin reservoir then includes a term to account for spin–lattice relaxation,

$$d\langle \mathcal{H}_I \rangle/dt = -(1/T_{cp})\{\delta\langle \mathcal{H}_I \rangle - \langle \mathcal{H}_S \rangle\} - (1/T_{1xI})\langle \mathcal{H}_I \rangle \qquad (6.43a')$$

where T_{1xI} is the time constant for the spin–lattice relaxation of the I-spin bath. This relaxation time will be $T_{1\rho}$ if the spin-lock transfer is used or it will be T_{1D}, the spin–lattice relaxation time of the dipolar bath, if the ADRF transfer is used. Assuming no rare-spin magnetization initially, Eqs. (6.43a') and (6.43b) may be solved to give the distribution of energy as a function of the cross-polarization time:

$$\langle \mathcal{H}_I(t) \rangle = \frac{\langle \mathcal{H}_I(0) \rangle}{1 + \delta + T_{cp}/T_{1x}} \left\{ \frac{2 + \delta + T_{cp}}{T_{1x}} e^{-\lambda^+ t} - e^{-\lambda^- t} \right\} \qquad (6.44a)$$

$$\langle \mathcal{H}_S(t) \rangle = \frac{\langle \mathcal{H}_I(0) \rangle}{1 + \delta + T_{cp}/T_{1x}} \left\{ e^{-\lambda^+ t} - e^{-\lambda^- t} \right\} \qquad (6.44b)$$

where

$$\lambda^{\pm} = \frac{1}{2}\left[\frac{1+\delta}{T_{cp}} + \frac{1}{T_{1x}} \right]\left\{ 1 \pm \sqrt{1 - \frac{4T_{1x}T_{cp}}{[(1+\delta)T_{1x} + T_{cp}]^2}} \right\}$$

Figure 6.12 shows the predicted time dependence of the energies of the subsystems according to Eqs. (6.44). In Fig. 6.12(a) are shown spectra for

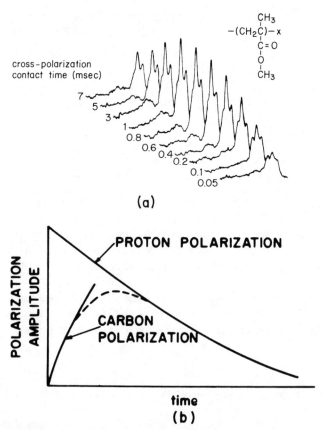

(a)

(b)

FIG. 6.12 (a) The dependence of signal intensity upon cross-polarization (contact) time. (b) A plot of the magnetization of (a) that shows the cross-polarization build-up and the decay due to $T_{1\rho}$ processes. Calculated curves are shown as solid line. (From Shaefer *et al.* [4]. Reprinted with permission from *Macromolecules* Vol. 4, p. 1341. Copyright 1977 American Chemical Society.)

various cross-polarization times. It is clear that the S Zeeman energy, which is directly proportional to the magnetization, builds up and decays, as shown in Fig. 6.12(b).

One of the problems of using cross-polarization NMR techniques to study rare spins is quantitation. Because the cross-polarization rates T_{cp}^{-1} depend on the nature of the interactions, they may differ for different nuclei in the same molecule. The intensities found for a given time of cross-polarization will not necessarily correspond to the relative amounts of various nuclei. Quantitative measurements must, therefore, include some knowledge of the rate of cross-polarization of the various spins.

In the absence of a knowledge of these rates, one can always check the *consistency* of intensities under, e.g., a spin-lock cp experiment by varying the contact times. A useful sequence with regard to this problem is shown in Fig. 6.13. In Fig. 6.13. periods I and II are the same as for the previously described parts of the spin-lock cp experiment. However, in period III, the I and S spins are placed parallel to z, thus preserving them from transverse relaxation. During the contact time t_1, the I spins will have lost magnetization according to $\exp(-t_1/T_{1\rho I})$. If there is a question about whether the S spins are equally polarized during this period, the period t_2, which satisfies the inequalities $t_2 \simeq T_{1I} \ll T_{1S}$, may be used: (a) to store the S magnetization accumulated during the cp period t_1 and (b) to repolarize the I-spin reservoir. During period V, the I and S magnetizations are placed along the y axes of their respective rotating frames, and the polarization transfer is again repeated for a time t_1. The storage and repolarization of I may be continued as long as the total time before observation is short compared to $5T_{1S}$. The number of contacts between I and S is increased until relative intensities in the S spectrum remain constant.

In the development of (6.44), in which simple first-order kinetics were assumed to apply to all relaxation, the model was purely phenomenological. The actual calculation of T_{cp} is quite involved. We quote the solution given by Mehring [5]. The system can be divided into two subsystems and the terms in the Hamiltonian may be classified as either Hamiltonians of the I-spin system, of the S-spin system, or as perturbation Hamiltonians that couple the two subsystems. The calculation of the time evolution of the system involves a perturbation treatment, the results of which are specific forms of the cross-relaxation times. The results are

$$(T_{cp}^{-1})_{SL} = \frac{B_{1S}^2 B_{1I}^2 M_2^{IS} J_x(\omega_{eS} - \omega_{eI})}{2(B_{1S}^2 + \Delta B_S^2)(B_{1I}^2 + \Delta B_S^2)} \tag{6.45a}$$

$$(T_{cp}^{-1})_{ADRF} = \frac{B_{1S}^2 M_2^{IS} J_z(\omega_{eS})}{(B_{1S}^2 + \Delta B_S^2)} \tag{6.45b}$$

where ω_{eK} is the frequency associated with the effective field on K in the rotating frame of K, and ΔB_S is the offset field on the S spin. Here $J_i(\omega)$ is the cosine transform of a correlation function (see Section XIV, Chapter 2):

$$J_i(\omega) = \int_0^\infty \cos \omega \tau C_i(\tau)\, dt \tag{6.46}$$

and $C_i(\tau)$ is a correlation function for a spin operator of the I-spin system. The forms of the correlation functions are found empirically. Demco *et al.* [6], and MacArthur *et al.* [7] have used the following forms in studies of the transfer in CaF_2:

$$C_z = 1/[1 + (\tau/\tau_c)^2] \tag{6.47a}$$

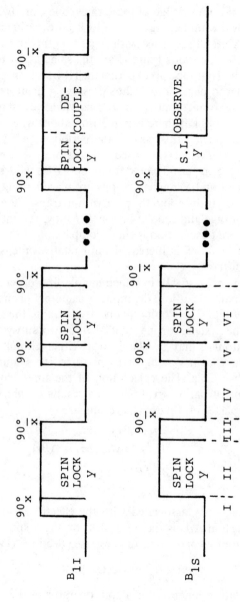

FIG. 6.13 A sequence for storage and repolarization of magnetization.

and

$$C_x(\tau) = \exp(-\tau^2/\tau_c^2) \qquad (6.47b)$$

from which J_x and J_z could be calculated. The calculation satisfactorily reproduced the dependence of the experimental T_{cp}^{-1} of CaF_2 on the irradiation strength, the mismatch of the coupling condition, and the irradiation frequency compared to the resonance frequency of the S spin.

V. Magic-Angle Spinning

The techniques of dipolar decoupling, both homonuclear and heteronuclear, and cross-polarization would be only interesting cases of spin dynamics of solids if magic-angle spinning (or some other similar technique) did not affect the NMR spectrum in the manner it does to increase the resolution. The reason is obvious: Most solid materials are powders. Powders containing only a single type of spin exhibit broad lines under dipolar decoupling sequences because of the dispersion of chemical shift (see Chapter 3). The typical broadening is an appreciable fraction of the entire range of isotropic chemical shifts of a spin. Thus, complex powders exhibit broad, complex spectra due to the overlap of many resonances. Magic-angle spinning applies equally to the improvement of resolution of abundant- and rare-spin spectra. Our discussion of the mechanism of resolution here concentrates on the effects on rare-spin spectra. In Chapter 5, we discuss the experimental results for abundant-spin spectra.

Figure 6.14(a) is an example of the dipolar-decoupled static ^{113}Cd spectrum of a decanuclear Cd salt. It is obvious that there is something complex about the sample; however, from the static spectrum, one would have to guess the relative numbers and types (as determined by the different chemical shifts) of cadmium present. With the combined use of high-power decoupling and magic-angle spinning, the spectrum of Fig. 6.14(b) results. From this spectrum, one can infer the kinds of cadmium species in this sample and the relative amounts of each in the sample. This effect of magic-angle spinning allows identification of the isotropic chemical shift of each spin in the sample. The result (and a very useful one) is a pseudoliquidlike spectrum.

Oddly enough, magic-angle spinning was originally developed by Lowe and by Andrew as a means of suppressing dipole–dipole couplings. For this decoupling, MAS proved to be a theoretical success, but a practical success only in cases in which the magnitude of the dipole–dipole coupling is less than the attainable spinning speeds (typically 4–5 kHz). Many of the chemically important spin systems experience homogeneous dipolar broadening

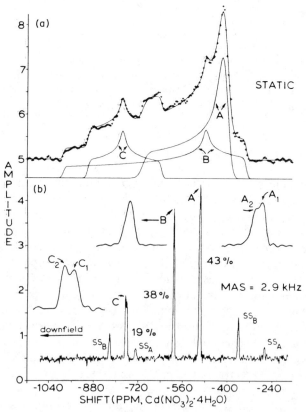

FIG. 6.14 Decoupled ^{113}Cd NMR spectra of $Cd_{10}(SCH_2CH_2OH)_{16}^{4+}$. (a) The static spectrum shows the several overlapping powder patterns of cadmium in different chemical environments. (b) The magic-angle-spinning spectrum gives the isotropic chemical shifts of the various cadmium species. (From Murphy *et al.* [8]. Reprinted with permission from *J. Am. Chem. Soc.*, Vol. 103, p. 4400. Copyright 1981 American Chemical Society.)

that is much larger than this upper limit and cannot be effectively suppressed by this technique. We have seen earlier, however, that homogeneous dipolar broadening can be removed by rotations in spin space using radiofrequency decoupling. Magic-angle spinning would have been relegated to the toybox of interesting NMR techniques had it not been for the realization (pointed out by Andrew [9] very early in these studies) that MAS affects other interactions of the spins, most notably the chemical shift. The combination of high-power decoupling and magic-angle spinning first demonstrated for rare spins by Schaefer *et al.* [4] and for abundant spins by Gerstein *et al.* [10] brought the advantages of this technique into sharp focus and ensured that it could be commonly used for analytical studies.

To focus on how the magic-angle spinning produces narrowing of the chemical-shift dispersion, consider a single spin-$\frac{1}{2}$ nucleus like ^{13}C or ^{113}Cd, subject to an anisotropic chemical-shift interaction. We assume that no strong dipole–dipole couplings affect this nucleus, either by virtue of strong decoupling or by the absence of such interactions. In the rotating frame, the Hamiltonian affecting the time evolution of the system is just the secular part of the chemical shift:

$$\mathcal{H}_{cs} = \omega_{0I}\sigma I_z \tag{6.48}$$

The chemical-shift-tensor element appropriate to this interaction is given by

$$\sigma = \sigma_{iso} + (\delta/2)(3\cos^2\theta - 1) - (\delta\eta/4)\sin^2\theta(e^{i2\phi} + e^{-i2\phi}) \tag{6.49}$$

where we have written

$$\sigma_{iso} = \tfrac{1}{3}(\sigma_{xx} + \sigma_{yy} + \sigma_{zz})$$

$$\delta = \tfrac{2}{3}\sigma_{zz} - \tfrac{1}{3}(\sigma_{xx} + \sigma_{yy})$$

$$\eta = 3(\sigma_{yy} - \sigma_{xx})/(2\sigma_{zz} - \sigma_{yy} - \sigma_{xx})$$

One sees explicitly how the chemical shift of this single spin depends on the orientation of the chemical-shift principal axes (θ, ϕ) relative to the magnetic field. A sample of spins subject to this Hamiltonian would have a resonance at the frequency defined by $\omega_0\sigma$, where σ is given by Eq. (6.49). When the sample is rotated, this Hamiltonian becomes time-dependent, and the spectrum of such a sample is different from that of the static sample. A convenient way to describe these effects is to rewrite Eq. (6.49) in terms of the spherical harmonics, for which the transformation under rotation is well known.

$$\sigma = \sigma_{iso}(4\pi)^{1/2}Y_{0,0} + \delta\left(\frac{4\pi}{5}\right)^{1/2}Y_{2,0}(\theta,\phi) + \frac{\delta\eta}{2}\left(\frac{8\pi}{15}\right)^{1/2}$$
$$\times (Y_{2,2}(\theta,\phi) + Y_{2,-2}(\theta,\phi))$$

$$= \sum_{l=0}^{1}\sum_{m=-2l}^{2l}\sigma_{2l,m}Y_{2l,m}(\theta,\phi) \tag{6.50}$$

A powder of spins, each having a different chemical shift due to a dispersion of orientations (θ, ϕ) of the principal axes in the powder, has a broad resonance line shape that is a superposition of the resonances of all of the spectra of the spins for the various orientations. In order to calculate the effects of rotation on the spectra of such samples, we need to know the instantaneous form of the chemical-shift Hamiltonian during rotation about an axis inclined at an angle β to $\mathbf{B}_0$ for each initial orientation. We know the general form of $\mathcal{H}_{cs}$ in a coordinate system with z along $\mathbf{B}_0$ by Eq. (3.53).

The *value* of any interaction, a scalar quantity, must be independent of the choice of coordinate system, a choice made at the convenience of the scientist. We know how to describe rotation about any axis $\mathbf{k}$ by the rotation matrix $\mathbf{R}_k$ (see Appendix 3), and specifically we know how to describe rotation about a z axis by $\mathbf{R}_z$. We therefore choose our coordinate system such that the z' axis is along the axis of physical rotation for convenience in viewing the form of $\mathcal{H}_{cs}$ with rotation. The transformation which places the coordinate system with z along $\mathbf{B}_0$ into a configuration with z' along the axis of physical rotation with z' inclined at an angle β to z, rotating about z with angle $\gamma = \omega_r t$ (ω_r is the rotation frequency) is given by the matrix $\mathbf{R}(0, \beta, \omega_r t)$. The Euler angles α, β, γ, which specify such a rotation by α about the initial z axis, by β about the y axis, and by γ about the resulting z' axis, are 0, β, and $\omega_r t$ in this case. A rotation by β about y of the original coordinate system places z along z', the axis of rotation. A rotation by $\gamma = \omega_r t$ about z' then specifies the instantaneous form of $\mathcal{H}_{cs}$ in the system rotating at ω_r about z';

$$\mathcal{H}'_{cs} = R^{-1}(0, \beta, \omega_r t)\mathcal{H}_{cs}R(0, \beta, \omega_r t) \tag{6.51a}$$

or, in terms of the spherical-tensor description of σ,

$$\mathcal{H}'_{cs} = \omega_0 \sum_{l=0}^{1} \sum_{m=-2}^{2l} \sigma_{2l,m}R^{-1}(0, \beta, \omega_r t)Y_{2l,m}(\theta, \phi)R(0, \beta, \omega_r t)I_z \tag{6.51b}$$

The matrix multiplications implied in (6.51b) are tedious and algebraically messy to perform in detail. Fortunately, the work has already been done in a general form in terms of Wigner rotation matrices $D_{m,m'}^{2l}$, the component of which may be obtained from a table;

$$\mathcal{H}'_{cs} = \omega_0 \sum_{l=0}^{1} \sum_{m,m'=-2l}^{2l} \sigma_{2l,m}D_{mm'}^2(0, \beta, \omega_r t)Y_{2l,m}(\theta, \phi)I_z \equiv \omega_0\sigma(t)I_z \tag{6.51c}$$

Equation (6.51c) is the desired form of $\mathcal{H}_{cs}$ with physical rotation.

The observables in an NMR experiment are the expectation values of transverse components of angular momentum: $R = x, y$

$$\langle I_R(t)\rangle = \mathrm{Tr}\,\rho(t)I_R \tag{6.52}$$

$\rho(t)$ is determined by the time-dependent Schrödinger equation (2.60), the solutions of which (cf. Chapter 2) are

$$\rho(t) = U^{-1}(t)\rho(0)U(t)$$

where the Liouville operator $U(t)$ is determined by $\mathcal{H}$ through equation (6.48), with solution given by (4.64);

$$U = T\exp\left(-i\int_0^t \mathcal{H}(t')\,dt'\right)$$

$\mathcal{H}(t')$ is given by (6.51c), just developed, so

$$\langle I_R(t) \rangle = \text{Tr } \rho(t)I_R = \text{Tr } U^{-1}(t)\rho(0)U(t)I_R = \text{Tr } \rho(0)U(t)I_R U^{-1}(t)$$

where the last equality was obtained by cyclic permutation. We realize that $T \exp[-i\int_0^t \omega_0\sigma(t')I_z\,dt'] \equiv e^{-i\theta(t)}I_z$, so the result of (6.52b) is rotation of I_R about z by $\theta(t)$, which immediately leads to

$$\begin{aligned}
\langle I_x(t) \rangle &= \text{Tr } \rho(0)\exp[i\theta(t)I_z]I_x\exp[-i\theta(t)I_z] \\
&= \text{Tr } \rho(0)[I_x\cos\theta(t) + I_y\sin\theta(t)] \\
&= \langle I_x(0) \rangle \cos\theta(t) - \langle I_y(0) \rangle \sin\theta(t)
\end{aligned} \qquad (6.53a)$$

and similarly

$$\langle I_y(t) \rangle = \langle I_y(0) \rangle \cos\theta(t) + \langle I_x(0) \rangle \sin\theta(t) \qquad (6.53b)$$

Let us assume an initial state with the spin along the x axis of the rotating frame $\rho(0) \sim I_x$, so the observed response of the magnetization in the time domain is

$$\langle I_x(t) \rangle = \langle I_x(0) \rangle \cos\theta(t) \qquad (6.53a')$$

$$\langle I_y(t) \rangle = \langle I_x(0) \rangle \sin\theta(t) \qquad (6.53b')$$

The frequency spectrum is given by the Fourier transform

$$\begin{aligned}
I(\omega) &= \int_0^\infty [\langle I_x(t) \rangle + i\langle I_y(t) \rangle]e^{-i\omega t}\,dt \\
&= \langle I_x(0) \rangle \int_0^\infty e^{i\theta(t)}e^{-i\omega t}\,dt
\end{aligned} \qquad (6.54)$$

From the definition of $D_{mm'}^{2l}(0, \beta, \omega_r t)$

$$\theta(t) = \omega_0 \sum_{l=0}^{+2l} \sum_{m,m'=-2l} \sigma_{2l,m}Y_{2l,m'}(\theta,\phi)d_{mm'}^{2l}(\beta)\int_0^t \exp(-im'\omega_r t')\,dt' \qquad (6.55)$$

Thus, $\theta(t)$ is an oscillatory function of time. The relative importance of each of the terms depends on the amplitude $\omega_0\sigma_{2l,m}Y_{2l,m'}(\theta,\phi)d_{mm'}^{2l}(\beta)/\omega_r m'$. These amplitudes can be made arbitrarily small by making ω_r arbitrarily large. In addition, there is also a nonoscillating term for $m' = 0$:

$$\theta(t) = \sigma_{\text{static}}\omega_0 t \sum_{l=0}^{1} \sum_{m=-2l}^{+2l} \sigma_{2l,m}Y_{2l0}(\theta,\phi)d_{m0}^{2l}(\beta) \qquad (6.56)$$

the amplitude of which does not depend on the value of ω_r. Hence, if ω_r is large enough that contributions from the oscillatory terms may be neglected, the observed Fourier transform is given by Eq. (6.57)

$$\begin{aligned}
F(\omega) &\simeq \langle I_x(0) \rangle \int_0^\infty e^{i\omega_0\sigma_{\text{static}}t}e^{-i\omega t}\,dt \\
&= \langle I_x(0) \rangle \delta(\omega - \omega_0\sigma_{\text{static}})
\end{aligned} \qquad (6.57)$$

Under these conditions the spectrum appears to be a single, sharp resonance at $\omega = \sigma_{static}\omega_0$. Evaluation of σ_{static} using Eq. (6.57) gives

$$\sigma_{static} = \sigma_{iso} + \frac{(3\cos^2\beta - 1)}{2}\frac{\delta}{2}[(3\cos^2\theta - 1) + \eta\sin^2\theta\cos 2\phi] \quad (6.58)$$

In particular, if $3\cos^2\beta_m - 1 = 0$, the result collapses to

$$\sigma_{static} = \sigma_{iso} \quad (6.59)$$

and for *each* initial orientation (θ, ϕ) the resonance position is the same, namely, the isotropic resonance position $\sigma_{iso}\omega_0$. This angle is the magic angle and we see why it is so special: For a powder sample spun around an axis inclined at β_m relative to $\mathbf{B}_0$, all nuclei seem to resonate at only the isotropic resonance position. Such a spectrum is extremely useful for analyzing powder samples with nuclei that have a number of different isotropic chemical shifts, a case frequently encountered when one wishes to identify by ^{13}C NMR spectroscopy an organic powder.

Figure 6.14(b) shows the ^{113}Cd spectrum of the salt under magic-angle spinning conditions. From this spectrum one may easily discern the various chemically distinct types of cadmium present, corresponding to ^{113}Cd in the various coordinations shown. It would clearly be quite difficult to obtain this information from the spectrum under conditions of no spinning [Fig. 6.14(a)].

Although the overlap of resonances in a complex spectrum makes analysis difficult, it is important to realize that the principal values of the chemical-shift tensor provide detailed and valuable information about the chemical state of the system. Thus, a great amount of effort has been put into designing experiments that can retrieve this information for each resonance in a complex spectrum. There are three ways to obtain this information and each has its advantages: (1) off-magic-axis spinning [10]; (2) Herzfeld–Berger [11] analysis of sideband intensities under conditions of slow spinning, and (3) separation of the isotropic and anisotropic components of the line shape [12].

Off-magic-axis spinning $(\beta \neq \beta_m)$, even at high speeds, causes the exact resonance position to depend on the initial position (θ, ϕ) [see Eq. (6.58)]. For a single-crystal sample, this process would result in an apparent shift of the resonance line away from the isotropic shift. For a powder in which all initial orientations (θ, ϕ) are equally likely, this off-magic-axis spinning effects a broadening that depends on θ. This broadening of the resonance may be estimated by calculating the difference between the minimum and maximum frequencies possible for a crystallite:

$$\omega_{max} = \omega_{iso} + \omega_0\delta(3\cos^2\beta - 1)/2 \quad (6.60a)$$

$$\omega_{min} = \omega_{iso} - \omega_0\delta(1 + \eta)[(3\cos^2\beta - 1)/2] \qquad (6.60b)$$

$$\Delta\omega = \omega_{max} - \omega_{min} = \omega_0\delta(\tfrac{3}{2} + \eta)(3\cos^2\beta - 1) \qquad (6.60c)$$

A calculation of the line shape under rapid spinning about an axis inclined at β to the magnetic field is analogous to the calculation of the line shape of the spectrum of a nonspinning sample, except that the values of $\omega - \sigma_{iso}\omega_0$ at the edges and at the singularities are scaled by the factor $\tfrac{1}{2}(3\cos^2\beta - 1)$ compared to the static case. By spinning the sample not at the magic angle, but slightly off, it may be possible to scale the patterns of overlapping lines such that the edges and the singularities are visible for all lines. From these values and the value of β, it is possible to calculate the values of the principal tensor elements for each of the resonances in a spectrum.

The spectrum of a powder sample subject to spinning about an axis at the magic angle relative to the applied dc magnetic field at a rate that is not fast compared to the width of the line in frequency units is quite different in appearance from either the spectrum of a static powder sample or of a powder rotated at a much faster rate. It should be obvious that the structure of the spectrum under such conditions contains information about the chemical-shift principal values (see Fig. 6.15). Such "slow" spinning provides an opportunity to regain information about these principal elements from experiments on powders. The resolution is sufficiently high that it should be possible to extract this information for each resonance in a complex spectrum, the limitation being that one has to associate the various lines in a spectrum to a resonance. Once this association is made, the tensor principal values must be extracted from the available information, i.e., the intensities of the various lines associated with each resonance. The problem of making the association of the lines is not an easy task; however, we shall defer the discussion of this part of the problem for the moment. We concentrate on the recovery of the principal values from the available information for a single resonance as discussed by Herzfeld and Berger [11].

As we saw in Eq. (6.55), the function $\theta(t)$ is an oscillatory function of time with a variety of frequencies that are displaced from σ_{iso}, ω_0 by integral multiples of the rotation frequency. A powder sample contains nuclei with all initial orientations. Thus, to find the spectrum of a powder, Eq. (6.54) must be integrated over all initial orientations.

Equation (6.55) shows that the function $\theta(t)$ is a sum of sines and cosines of the various angles. The exponential of a sine or cosine can be expanded by the use of Bessel functions:

$$e^{iR\sin\alpha} = \sum_{k=-\infty}^{+\infty} e^{ik\alpha}J_k(R) \qquad (6.61)$$

where $J_k(R)$ is a Bessel function of the first kind. The result of applying this

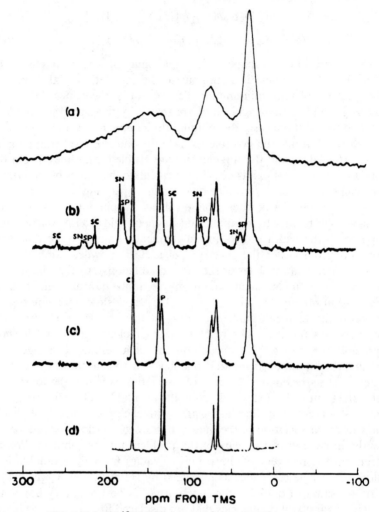

FIG. 6.15 MAS ^{13}C NMR spectra of a segmented co-polymers,

$$\left[(O\!-\!\overset{\overset{\displaystyle O}{\|}}{C}\!-\!O\!-\!\overset{\overset{\displaystyle O}{\|}}{C}\!-\!OCH_2CH_2CH_2CH_2\!-\!O)_{24}\{O\!-\!\overset{\overset{\displaystyle O}{\|}}{C}\!-\!O\!-\!\overset{\overset{\displaystyle O}{\|}}{C}\!-\!O[(CH_2)_4\!-\!O\!-\!]_{12}\}_1 \right]_n$$

(a) A static spectrum: (b) MAS spectrum obtained under the condition that the spinning rate is not fast compared to the chemical-shift anisotropy of the various lines: (c) spectrum with the sidebands artificially removed; (d) liquid-state spectrum of the same polymer. (After Jelinski [13]. Reprinted with permission from *Macromolecules*, Vol. 14, p. 1341. Copyright 1981 American Chemical Society.)

to Eq. (6.55) gives for the spectrum

$$F(\omega) = \int_{-\infty}^{+\infty} \exp(-i\omega t) \sum_{N=-\infty}^{+\infty} \exp(i\omega_0 t\sigma) I_N \exp(iN\omega_r t)\, dt \qquad (6.62)$$

It should be obvious from Eq. (6.62) that the resulting spectrum will consist of a series of spikes that have intensities I_N. These intensities depend on integration of sums of products of Bessel functions over the initial random distribution of orientations. Herzfeld and Berger show clearly that intensities of the sidebands relative to the center band (which is at the isotropic value) depend on the principal values of the chemical-shift tensor. They define two parameters μ and ρ as

$$\mu = (\omega_0/\omega_r)(\sigma_{zz} - \sigma_{xx}) \qquad (6.63a)$$

$$\rho = (\sigma_{xx} + \sigma_{zz} - 2\sigma_{yy})/(\sigma_{zz} - \sigma_{xx}) \qquad (6.63b)$$

for which they generate curves for the relative intensities of the Nth sidebands compared to the intensity of the isotropic peak $I_{\pm N}/I_0$. An example of such curves is shown in Fig. 6.16. By determining the values of the ratios for the various sidebands of a line, one may overdetermine the values of μ and ρ, from which the principal values of the chemical-shift tensor may be determined.

The Herzfeld–Berger technique, or the moment technique of Maricq and Waugh [12], which we have not discussed, can be applied in the cases in which one may identify the sidebands associated with a single resonance. For many cases, the isotropic lines lie in such close proximity to each other that a clear separation of the lines associated with one resonance and those associated with a second resonance cannot be readily made. Under those conditions, it is desirable to identify the lines that separately belong to the individual resonances. To perform such a separation, one must use two-dimensional Fourier-transform spectroscopy.

In 2-D NMR spectroscopy, spectra are taken as a function of some time variable that is incremented. If a technique is devised in which the system evolves under the influence of one Hamiltonian during the first interval and under a different Hamiltonian in the second interval, then a double integration will lead to a mapping that correlates the Hamiltonian active during the first time period with the Hamiltonian during the second. Obviously, to separate the isotropic and the anisotropic parts of the chemical shift interaction, one needs to develop a sequence for which the evolution in one time period is the isotropic chemical-shift Hamiltonian and for the second period, the system evolves under the full anisotropic chemical-shift Hamiltonian. The sequence used by Aue et al. [14] to accomplish this separation is shown in Fig. 6.17. As with many of these sequences, the preparation may

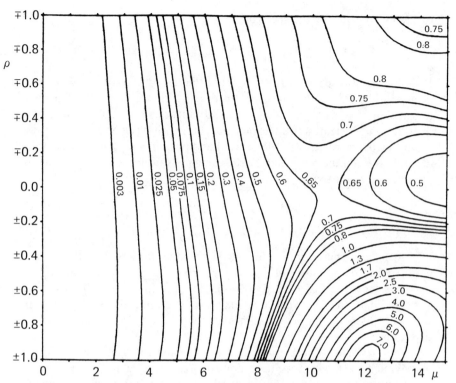

FIG. 6.16 Plots of the relative intensities of I_n/I_0 for $n = \pm 3$ as a function of ρ and μ in the Herzfeld–Berger scheme for evaluating chemical shift elements in the slow-spinning regime. (From Herzfeld and Berger [11].)

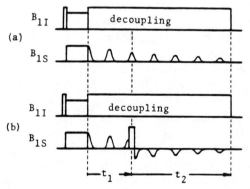

FIG. 6.17 The scheme of Aue et al. [14] to separate the isotropic and anisotropic parts of the chemical shift in a two-dimensional experiment. (After [14].)

consist of a variety of processes that create transverse magnetization. We assume a $\pi/2$ pulse is used, although cross-polarization may also be used to generate rare-spin magnetization. It is based on the fact that the rotation of the sample produces *rotational echoes*, at which point the evolution is independent of the anisotropic parts of the chemical shift. Thus, sampling the second half of this echo will produce a spectrum that is dependent on the full chemical-shift Hamiltonian. However, the exact amplitudes in such a spectrum will depend on the amplitudes at the echo maximum. These amplitudes are modulated by the *isotropic* chemical shift only. The amplitude modulation, however, also produces a mixing of dispersion and absorption modes that is undesirable. To circumvent the problem, an inversion is obtained by insertion of a 180° pulse at the height of the rotational echo. The combination of these two spectra suppresses this problem and produces spectra like those of Fig. 6.18, in which the isotropic and anisotropic chemical shifts are separated in two dimensions. After achieving this resolution, one may use the Herzfeld–Berger technique to extract the information on the principal values of the chemical-shift tensors of the separated resonances.

Another technique for producing anisotropic and isotropic shielding via a 2-D spread of the information involves magic angle "flipping" [15]. In this method, the sample is spun at an angle perpendicular to the static field for a fixed time, then tilted to the magic angle in a time short compared to the appropriate T_1 for another period, as indicated in the top portion of Fig. 6.19. Also indicated are the pulse sequences on the S and I spins used

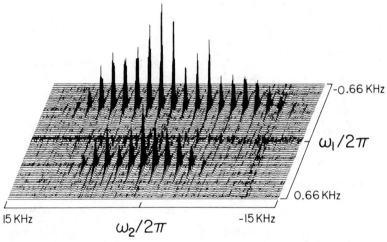

FIG. 6.18 The 2-D ^{31}P NMR spectrum of a physical mixture of barium diethylphosphate and brushite, showing the isotropic chemical shifts in one dimension and the anisotropic chemical shift spectrum in the second dimension. (After Aue *et al.* [14].)

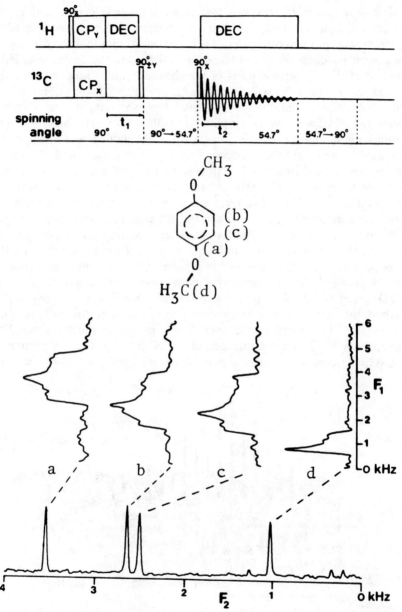

FIG. 6.19 The "magic-angle flipping" experiment [15] to extract isotropic and anisotropic chemical shifts via 2-D separation.

to cross-polarize, store, and read out the S-spin magnetization. At the bottom of Fig. 6.19 are shown the isotropic and anisotropic shieldings of para-dimethoxybenzene obtained from such data acquisition with a 2-D Fourier transformation. The technique requires a spinning system that can be accurately and rapidly flipped during the experiment [16].

Magic-angle spinning, together with techniques like off-magic-angle spinning, Herzfeld–Berger analysis and, if necessary, two-dimensional separation of the spectra, offers the promise of being able, in a reasonable number of experiments, to get all of the information about the chemical shift of a complex molecule, even though only a powder is available. Of course, detailed information on the orientation of such chemical-shift tensor axes can only be obtained generally through the study of single crystals.

VI. Transient Dipole–Dipole Oscillations

Transfer of magnetization between the two subsystems of a heteronuclear spin system occurs by modulation of the dipole–dipole interactions among the spins. Cross-polarization processes are designed to allow mutual spin flips between unlike spins to occur at appreciable rates. Despite the fact that each rare spin is, in principle, coupled to every abundant spin in the system, its dominant coupling is to one or a few spins whose proximity makes the coupling stronger than the couplings to relatively more distant spins. In the extreme limit, all other spins are very far away so that the system consists of isolated small groups of spins, each of which is subject only to dipolar interactions among its constituents. For such a small group, we know that the behavior is not exponential at all, but rather oscillatory in nature. Thus, it should not be surprising that, under appropriate conditions of isolation, the transfer begins at early times with the observation of transient oscillation of the magnetization. The resulting oscillation is due to the fact that the correlation function that enters the calculation of the cross-polarization time has an oscillatory character from this strong dipole–dipole coupling. The observation of these oscillations depends on how strongly isolated the groups are from each other. It is often the case in the systems of interest that the groups are coupled to each other mainly through the abundant-spin homonuclear dipole–dipole couplings. Thus, the couplings of the I and S spins within the group can be observed only if the time of transfer among the I spins is long. In certain cases, this isolation occurs naturally. For example, Ernst and co-workers [17] first observed such oscillations in ferrocene (see Fig. 6.20) in which the homonuclear spin diffusion among the protons is sufficiently ineffective for specific orientations of the single crystal that it permits observation of this effect.

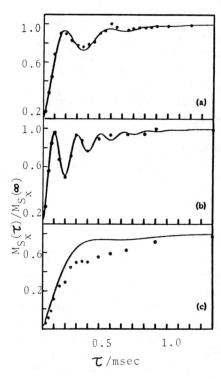

FIG. 6.20 The oscillatory buildup of ^{13}C magnetization in a single crystal of ferrocene, for three different orientations of the crystal relative to the magnetic field direction: (a) an orientation that results in a modest dipole–dipole coupling (b) an orientation that results in a strong dipole–dipole coupling, and (c) an orientation for which the dipole–dipole coupling is rather weak. (After Müller *et al.* [17].)

This observation of transient oscillations suggested that it might be worthwhile to isolate dipolar coupled groups by the application of appropriate rf excitation with the goal of enhancing these oscillations. Like the dipole–dipole oscillations for isolated groups such as the protons on isolated waters of hydration, the transient dipolar oscillations have important information about the structure and geometry of the spin environment. There are a number of ways to produce this homonuclear decoupling, but all of them have in common the goal of suppressing the homonuclear dipole–dipole interactions among the abundant spins, which results in the effective isolation of the spins into groups of a rare spin and one or a few abundant spins: (1) orientation of a single crystal, such as was done by Müller *et al.* [17]; (2) decoupling of the abundant spins by an off-resonance spin-locking sequence [18]; or (3) decoupling by use of a multiple-pulse decoupling sequence such as was discussed extensively in Chapter 5 [19]. As an example, we discuss the results of the experiments of Stoll *et al.* [19], using the last technique. The sequence of excitations used is shown in Fig. 6.21.

The experiment consists of three time periods: (1) preparation, (2) dipole–dipole evolution, and (3) detection. During the preparation period, a spin-

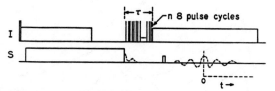

FIG. 6.21 The sequence used by Stoll et al. [19] to observe transient dipolar oscillations due to suppression of the abundant-spin dipole–dipole interaction that interferes with the observation of the heteronuclear dipole–dipole coupling. (After Stoll et al. [19].)

lock cross-polarization transfer produces an enhanced magnetization of the rare spins. This is not a necessary process; the preparation could have been as simple as a $\pi/2$ pulse applied to the S spins. After completion of this preparation, the spin-lock on the I spin is shut off, initiating a free-induction decay. After that decay, the spin lock on the S spins is shut off, initiating an S-spin free-induction decay. During this period, the I spins are excited by n eight-pulse sequences that suppress the effects of homonuclear $I–I$ dipolar couplings. At the end of this period, strong decoupling of the I spins from the S spins is initiated while the S magnetization is detected. In order to eliminate the effects of S-spin chemical shifts during the delay, a spin echo produced by the application of a π pulse to the S spin resonance gives an FID detected from the echo maximum forward in time. Fourier transformation of the resulting transient produces a chemical-shift spectrum, as shown in Fig. 6.22 for the ^{13}C resonance in a powder of solid benzene. The amplitude at any point in the spectrum has mapped into it the dipolar evolution of the spins of that chemical shift during period 2. The resulting spectra for various lengths τ of the evolution period are shown on the right side of Fig. 6.22, and the calculated spectra from the known orientation of the chemical-shift tensor relative to the dipole–dipole vector and the known dipole–dipole coupling are shown on the left side of the figure. The resulting spectra are given mathematically by

$$\langle S_x(t,\tau)\rangle = \sum_j \prod_{i=1}^{6} \cos(\Omega_{ji}\tau)\cos\sigma_j\omega_0 t \tag{6.64}$$

$$\langle S_y(t,\tau)\rangle = \sum_j \prod_{i=1}^{6} \cos(\Omega_{ji}\tau)\sin\sigma_j\omega_0 t \tag{6.65}$$

$$\Omega_{ji} = \frac{\alpha\gamma_i\gamma_j}{3\sqrt{2}r_{ij}^3}(1-3\cos^2\theta_{ij}) \tag{6.66}$$

where the sum over j is over the S spins and the product over i is taken over all six I spins of the molecule in which the S spin resides. It is possible to display the data in a different way, by taking a second transformation at

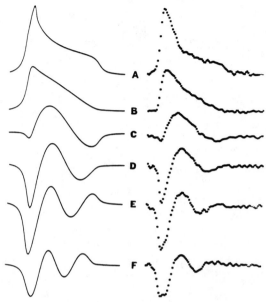

FIG. 6.22 Transient oscillations in the powder ^{13}C NMR spectrum of benzene. Several different values of the dipolar evolution time were used. Different parts of the powder spectrum oscillate at different rates because they are subject to different dipole–dipole couplings. The calculated behavior of the benzene powder spectrum is shown at the left. (After Stoll *et al.* [19].)

constant chemical shift over the dipolar oscillation. This results in a two-dimensional map of the dipolar frequency versus chemical shift, as has been shown elegantly by the work of Opella and Waugh [20] on polyethylene. The correlation of the chemical shift and the dipole–dipole couping tensors has the potential not only of giving dipole–dipole interaction magnitudes but also of correlating the geometry of the principal axes of the chemical-shift tensor with those of the dipolar tensor that are usually more precisely known.

Another very important use of selective coupling of the rare and the abundant spin is the use of sequences like that of Fig. 6.23 [21]. It allows the acquisition of the resonance of I spins strongly coupled to rare spins out of a spectrum that consists primarily of I spins not coupled to rare spins. This sequence has potential to help evaluate spectra in complex systems such as catalysts, where one may be interested in studying the spectrum of only one proton while the preparation of the catalyst requires that many different protons be present on the catalyst. The sequence consists of two experiments alternately applied to the system. In the first experiment, the eight-pulse cycle discussed in Chapter 5 is applied to the system without any excitation of the S-spin system. The resulting time evolution of the abundant-spin I is

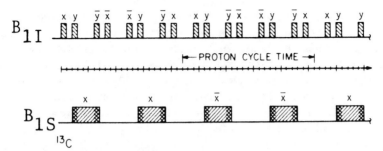

FIG. 6.23 A sequence to select the resonance of abundant spins that are coupled to a rare spin. (After Reimer and Vaughan [21]).

recorded. We note that this time decay is the sum of two parts, the first is due to spins not coupled to a rare spin and the second is from I spins that are coupled to rare spins. In the second experiment, the I spins are excited with the eight-pulse sequence, however, the S spins are simultaneously excited by a series of pulses strategically placed in the cycle so as to invert the *heteronuclear* dipole–dipole interaction. The I spin response to this experiment consists of two parts, one from I spins near rare spins and the other from I spins that do not come into contact with rare spins. However, the heteronuclear interaction having been refocused by the action of the sequence of π pulses, the first contribution to the spectrum is not broad but sharp. The second contribution is the same as it was for the first experiment. Hence, a subtraction of the results of the two experiments produces the decoupled decay of those spins affected by the heteronuclear dipole–dipole coupling, which can be analyzed by Fourier transformation to obtain the spectrum of only those protons.

By delaying the application of the sequence of π pulses a specified period of time in the second experiment, one may reintroduce a certain amount of dipole–dipole evolution into the observed spectrum of the coupled I spin. By studying this effect as a function of the amount of time the sequence is omitted, one can, in principle, observe the heteronuclear dipole–dipole oscillations in the spectrum of the abundant spin's spectrum. This experiment observes the transient oscillation of the rare-spin, but from the other side of the fence.

VII. Effect of Coupling of Quadrupolar Nuclei to Spin-$\frac{1}{2}$ Nuclei

In previous sections of this chapter, the emphasis has been on the existence, elimination, and use of heteronuclear dipolar interactions between spin-$\frac{1}{2}$ species. In this section we recognize that an important portion of matter of concern to the community utilizing NMR as a tool contains both spin-$\frac{1}{2}$ nuclei and quadrupolar nuclei. For example, the peptide linkage in biological systems contains ^1H, ^{13}C, and ^{14}N, as interacting species. In addition to these, some synthetic polymers contain ^1H, ^{13}C, and ^{35}Cl, and silica–alumina and zeolitic catalysts contain ^1H, ^{29}Si, and ^{27}Al. We present no detailed treatment of the problem of decoupling spin-$\frac{1}{2}$ nuclei from quadrupolar nuclei in solids, e.g., ^{13}C from ^{14}N in biopolymers; we simply quote the results and refer reader to the literature for details [22,23]. The basic facts to be appreciated are that, in the presence of strong quadrupolar splitting of a nucleus such as ^{14}N, the form of the $^{13}C–^{14}N$ dipolar interaction must take into account the fact that the states of ^{14}N are quantized along the electric-field gradient. Based on previous discussions, one might guess that there exist at least two methods of decoupling ^{13}C from ^{14}N. These are radio-frequency decoupling and magic-angle spinning. In a solid containing ^1H, ^{13}C, and ^{14}N, radio-frequency decoupling requires a triply tuned, single-coil system with good power efficiency at all frequencies concerned. These probes have been constructed (but are not standard) for use in double-cross-polarization experiments, in which polarization is first transferred from ^1H to ^{13}C and then from ^{13}C to ^{14}N [24]. The radio-frequency decoupling of one nucleus from another, however, implies an *intensity* of decoupling field that spans the spectral width of the nucleus being irradiated. For example, to decouple ^1H from ^{13}C, a case in which the homogeneous broadening of ^1H can be 20 kHz, a proton decoupling field of $\omega_{1H} \gg 20$ kHz is needed. We have seen that 10-G fields, implying $\omega_{1H} \simeq 42.65$ kHz, are easily obtainable with modest power (~ 160 W) in a 10-mm-i.d. rf coil. If ω_Q of ^{14}N is of the order of several megahertz, efficient decoupling of ^{14}N from ^{13}C would require rf powers that would melt NMR probes.

The use of magic-angle spinning (MAS) might appear to be viable because the decreased magnetic moment of ^{14}N relative to ^1H, resulting in a nearest-neighbor carbon–nitrogen dipolar interaction of roughly 4 kHz would appear to be attackable by MAS. In fact, this *is* the case, in a special sense. The dipolar Hamiltonian between ^{13}C and ^{14}N, in the weak-field case for the quadrupolar levels of ^{14}N, is not of a form that disappears under MAS. However, MAS does severely attenuate this effect, leaving broadened lines from which chemical information may be inferred, providing the carbons that are broadened are sufficiently separated from other carbon resonances. In the case of ^{14}N broadening ^{13}C under strong proton decoupling and MAS, the ^{13}C spectrum contains structure from which the geometry of the C–N bond may be inferred. For example, Fig. 6.24 shows the theoretical and experimental spectra of carbon-4 in *p*-nitrotoluene MAS and strong proton decoupling [22]. The ^{14}N splits the ^{13}C resonance into two bands, the shapes and splittings of which depend upon the orientations of the dipolar

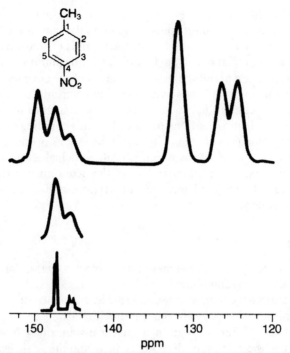

FIG. 6.24 The spectrum under magic-angle spinning of a spin-$\frac{1}{2}$ that is coupled by the dipole–dipole interaction to a quadrupolar nucleus that is not in a pure Zeeman state. (After Hexem *et al.* [22]. Reprinted with permission from *J. Am. Chem. Soc.*, Vol. 103, p. 224. Copyright 1981 American Chemical Society.)

and field-gradient tensors. Comparison of theoretical and experimental spectra may be used to infer bond distances, and orientations of dipolar and relevant interaction tensors.

VIII. Concluding Remarks

The study of rare spins and abundant spins coupled to rare spins introduces an additional term into the Hamiltonian—the heteronuclear dipolar Hamiltonian. The presence of this portion of the Hamiltonian has some remarkable effects on the time evolution of the system. Decoupling of its effects can be achieved by a variety of irradiation schemes on the abundant spins, allowing observation of the rare-spin chemical-shift spectrum. More importantly, the presence of the heteronuclear Hamiltonian provides a mechanism to enchance the signal of the rare spins by the cross-polarization process. This enhancement feature is an essential factor in achieving rare-spin spectra in a reasonable time. The process of magic-angle spinning may be applied to rare spins, as it is to abundant spins, to yield analytically valuable isotropic chemical-shift spectra in chemically complex solids. There exist techniques used in conjunction with magic-angle spinning (and spinning at other angles) to regain information about the principal values of the chemical-shift tensor, so that even this information is available. The heteronuclear dipole–dipole interactions also contain information about the structure of the molecule. Study of the transient oscillations produced by this coupling can yield information on the geometry of the environment, as well as produce information on the relative geometry of the chemical-shift principal axes to the dipole–dipole principal axis. Thus, the presence of the heteronuclear dipole–dipole interaction quite literally enhances our ability to examine chemical systems in a number of ways.

PROBLEMS

6.1. Give a rough order-of-magnitude value (in hertz) for the heteronuclear dipolar broadening for
   (a) a proton-^{13}C pair with an internuclear distance of 0.95 Å,
   (b) a proton-^{29}Si pair with an internuclear distance of 1.35 Å,
   (c) a ^{31}P–^{13}C pair with an internuclear distance of 1.70 Å.

6.2. What is the minimum decoupling field amplitude (in gauss) needed to suppress the effects of broadening in the pairs of Problem 6.1? What assumption is necessary, or what additional facts are needed?

6.3. (a) What is the form of Eq. (6.2) in the laboratory frame (assume rf irradiation at only ω_{0S} is present)?

(b) What is the transformation that results in Eq. (6.2) from the equation developed in part (a)?

6.4. Calculate $\bar{\mathscr{H}}_{D_{IS}}^{(0)}$ for the sequence illustrated in Fig. 6.4 and comment on the efficiency of heteronuclear decoupling under such a sequence compared to the sequence of Fig. 6.3.

6.5. (a) Will the Carr–Purcell sequence remove the effects of heteronuclear scalar coupling? Why? (b) Homonuclear scalar coupling? Why?

6.6 Common values of Q and V for resonant decoupling circuits supplied on commercial liquid-state NMR spectrometers are $Q = 100$, $V = 5$ cm^3. Transmitters with powers of 5 W are available for decoupling. Can such equipment be used for heteronuclear dipolar decoupling for the species given in Problem 6.1 (assume a dc field of 2.2 T)?

6.7. Derive Eq. (6.8) by using the appropriate interaction transformations indicated in Fig. 6.6.

6.8. Using the discussion following Eq. (6.8) as a guide, comment on decoupling ^{1}H and ^{13}C in a non-MAS experiment in which the resonance frequency for ^{13}C is 60 MHz. (Note: ^{13}C shielding anisotropies may be as large as 400 ppm.)

6.9. Starting with the form $\sigma = \sigma_{xx} \sin^2 \theta \cos^2 \phi + \sigma_{yy} \sin^2 \theta \sin^2 \phi + \sigma_{zz} \cos^2 \theta$, show that it can be written in form of Eq. (6.49).

6.10. Show that in the high-temperature approximation for a spin with quantum number I, in a field of magnitude B, $\mathscr{S} = CB^2/T$ using the relations

$$\mathscr{S} = -RT\rho \ln \rho$$

$$\text{Tr } \rho \mathscr{H} = E$$

$$E = -\mathbf{M} \cdot \mathbf{B} = -XB^2 = CB^2/T$$

where C is the Curie constant and $C = \gamma^2 n^2 I(I + 1)/3k$. (Hints: $\ln(1 + x) = x$; $\text{Tr } \rho = 2I + 1$.)

REFERENCES

1. L. R. Sarles and R. M. Cotts, *Phys. Rev.* **111**, 853 (1958).
2. A. Pines, M. G. Gibby, and J. S. Waugh, *J. Chem. Phys.* **59**, 569 (1973), and references contained therein.
3. P. Mansfield, P. K. Grannell, and M. A. B. Whittaker, *Phys. Rev.* **8B**, 4149 (1975).
4. J. Schaefer, E. O. Stejskal and R. Buchdahl, *Macromolecules* **10**, 384 (1977).
5. M. Mehring, "High-Resolution NMR Spectroscopy in Solids." Springer-Verlag, Heidelberg, 1976.
6. D. Demco, J. Tegenfeldt, and J. S. Waugh, *Phys. Rev.* **11B**, 4133 (1975).
7. D. A. MacArthur, E. L. Hahn, and R. F. Walstedt, *Phys. Rev.* **188**, 609 (1969).
8. P. D. Murphy, W. Stephens, T. T. P. Cheung, S. Lacelle, B. C. Gerstein, and D. M. Kertz, *J. Am. Chem. Soc.* **103**, 4400 (1981).
9. E. R. Andrew, Nuclear magnetic resonance in rapidly rotating solids. In *"Magnetic Resonance"* (C. K. Coogan, N. S. Ham, S. N. Stuart, J. R. Pilbrow, and G. H. H. Wilson, eds.), p. 163. Plenum, New York, 1970.
10. B. C. Gerstein, R. G. Pembleton, R. C. Wilson, and L. M. Ryan, *J. Chem. Phys.* **66**, 362 (1977).
11. J. Herzfeld and A. E. Berger, *J. Chem. Phys.* **73**, 6021 (1980).
12. M. M. Maricq and J. S. Waugh, *J. Chem. Phys.* **70**, 3300 (1979).
13. L. W. Jelinski, *Macromolecules* **14**, 1341 (1981).
14. W. B. Aue, D. J. Ruben, and R. G. Griffin, *J. Magn. Reson.* **43**, 472 (1981).
15. Ad Bax, N. M. Szeverenyi, and Gary Maciel, *J. Magn. Reson.* **55**, 494 (1983).
16. V. Bartuska, Chemagnetics, Inc., 208 Commerce Dr. #3C, Ft. Collins, CO 80524.
17. L. Müller, A. Kumar, T. Baumann, and R. R. Ernst, *Phys. Rev. Lett.* **32**, 1402 (1974).
18. R. K. Hester, J. L. Ackerman, V. R. Cross, and J. S. Waugh, *Phys. Rev. Lett.* **34**, 993 (1978).
19. M. E. Stoll, A. J. Vega, and R. W. Vaughan, *J. Chem. Phys.* **65**, 4093 (1976).
20. S. J. Opella and J. S. Waugh, *J. Chem. Phys.* **66**, 4919 (1977).
21. J. A. Reimer and R. W. Vaughan, *Chem. Phys. Lett.* **63**, 163 (1979).
22. J. G. Hexem, M. H. Frey, and S. J. Opella, *J. Am. Chem. Soc.* **103**, 224 (1981).
23. A. Naito, S. Ganapathy, and C. A. McDowell, *J. Chem. Phys.* **74**, 5393 (1981).
24. J. Schaefer, R. A. McKay, and E. O. Stejskal, *J. Magn. Reson.* **34**, 443 (1979).

THE FIELD OF A CURRENT LOOP:
A CLASSICAL MODEL

The magnetic field **B** at any point in space can be represented in terms of the vector potential **A**. The vector potential is determined by the current distribution **J(r)**. Thus the magnetic field **B** is also determined by **J(r)**. The Maxwell equations are the connecting link in relating the field to the current density, and they are the starting point for the calculation of the dipolar field.

One Maxwell equation requires the divergence of **B** to be zero. An immediate implication by Gauss's theorem is that

$$\int_{\mathscr{A}} \mathbf{B}(\mathbf{r}) \cdot d\mathscr{A} = 0 \tag{A1.1}$$

If **B(r)** satisfies the equation

$$\mathbf{B}(\mathbf{r}) = \nabla \times \mathbf{A}(\mathbf{r}) \tag{A1.2}$$

(A1.1) will always be satisfied. A second Maxwell equation connects the magnetic field to the current density:

$$\nabla \times \mathbf{B}(\mathbf{r}) = (4\pi/c)\mathbf{J}(\mathbf{r}) \tag{A1.3}$$

where c is the speed of light. An equivalent equation is found from substitution of (A1.2) in (A1.3):

$$\nabla \times (\nabla \times \mathbf{A}(\mathbf{r})) = (4\pi/c)\mathbf{J}(\mathbf{r}) \tag{A1.4}$$

The solution of this equation determines **A**. The double cross product in (A1.4) can be simplified by use of the vector identity

$$\mathbf{X} \times (\mathbf{Y} \times \mathbf{Z}) = \mathbf{Y}(\mathbf{X} \cdot \mathbf{Z}) - (\mathbf{X} \cdot \mathbf{Y})\mathbf{Z} \tag{A1.5}$$

Of the properties of the vector potential, we are only interested in the curl of A. Therefore, whatever the vector potential's curl may be, we may specify that the function A has zero divergence, without affecting the property of interest. Hence, the left-hand side of (A1.4) may be put in the form

$$\nabla(\nabla \cdot \mathbf{A}) - (\nabla \cdot \nabla)\mathbf{A} \tag{A1.6}$$

of which the first term is zero by the condition we impose on the vector potential. Equation (A1.4) then assumes the form

$$-\mathbf{V}^2\mathbf{A} = (4\pi/c)\mathbf{J} \tag{A1.7}$$

Equation (A1.7) is known as Poisson's equation and has the solution

$$\mathbf{A}(\mathbf{r}_2) = \frac{1}{c} \int_v \frac{\mathbf{J}(\mathbf{r}_1)}{|\mathbf{r}_2 - \mathbf{r}_1|} d^3\mathbf{r}_1 \tag{A1.8}$$

The relation of $\mathbf{r}_1$, $\mathbf{r}_2$ and the motion of the charge Q is indicated in Fig. A1.1.

The current loop is a useful model, for applications important to NMR spectroscopists, of (a) the nuclear dipolar field, (b) the chemical-shift field, or (c) the field in a solenoid. The spinning nucleus may be considered as a small rotating charge distribution. To simplify more, assume it to be a circular current loop of radius r_1, caused by the circulation of a charge Q. This is a crude model, but it is exemplary of the use of Maxwell's relations and will allow us to work towards an expression for the dipolar field of a nucleus. We may simplify (A1.8) by rewriting $|\mathbf{r}_2 - \mathbf{r}_1|$ as

$$|\mathbf{r}_2 - \mathbf{r}_1| = r_2\{1 + r_1^2/r_2^2 - (2\mathbf{r}_1 \cdot \mathbf{r}_2)/r_2^2\}^{1/2} \tag{A1.9}$$

Now consider $r_1 \ll r_2$; that is, assume that the extent of the region over which the charge is present is small compared to the distance to any other dipole. Since a nucleus has a radius on the order of 10^{-13} cm and the distance between nuclei is no smaller than 10^{-8} cm, this model is a reasonable approximation for nuclear moments producing fields that are experienced by other nuclear moments.

Thus, we can neglect, to 1 part in 10^5, all terms of higher order than r_1/r_2:

$$|\mathbf{r}_2 - \mathbf{r}_1| \approx r_2\{1 - 2[(\mathbf{r}_1 \cdot \mathbf{r}_2)/r_2^2]\}^{1/2} \tag{A1.10}$$

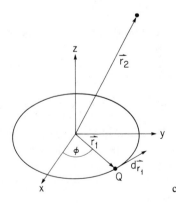

FIG. A1.1 Current loop of a circulating charge.

and

$$A(\mathbf{r}_2) \approx \frac{1}{cr_2} \int_v \frac{J(\mathbf{r}_1) \, d^3\mathbf{r}_1}{\{1 - [(2\mathbf{r}_1 \cdot \mathbf{r}_2)/r_2^2]\}^{1/2}} \tag{A1.11}$$

Since r_1/r_2 is small, the approximation

$$1/(1 - x)^{1/2} = 1 + \tfrac{1}{2}x + \cdots \tag{A1.12}$$

may be used:

$$A(\mathbf{r}_2) = \frac{1}{cr_2} \int_v d^3\mathbf{r}_1 \, J(\mathbf{r}_1) + \frac{1}{cr_2^3} \int_v (\mathbf{r}_1 \cdot \mathbf{r}_2) J(\mathbf{r}_1) \, d^3\mathbf{r}_1 \tag{A1.13}$$

For a circular loop

$$J(\mathbf{r}_1) \, d^3\mathbf{r}_1 = I(-\sin\phi \mathbf{i} + \cos\phi \mathbf{j}) \, d\phi \tag{A1.14}$$

and

$$A(\mathbf{r}_2) = \frac{I}{cr_2} \int_0^{2\pi} (-\sin\phi \mathbf{i} + \cos\phi \mathbf{j}) \, d\phi + \frac{1}{cr_2^3} \int_v (\mathbf{r}_1 \cdot \mathbf{r}_2) J(\mathbf{r}_1) \, d^3\mathbf{r}_1 \tag{A1.15}$$

where ϕ is the angle specifying the position of the charge on the loop (Fig. A1.1). The first term on the right-hand side of Eq. (A1.13) equals zero, then, from the integration over ϕ:

$$\int_0^{2\pi} \sin\phi \, d\phi = \int_0^{2\pi} \cos\phi \, d\phi = 0 \tag{A1.16}$$

This leaves

$$A(\mathbf{r}_2) = \frac{I}{cr_2^3} \int d\phi (\mathbf{r}_1 \cdot \mathbf{r}_2)(\mathbf{j}\cos\phi - \mathbf{i}\sin\phi) \tag{A1.17}$$

Changing the integration from $(\mathbf{j}\cos\phi - \mathbf{i}\sin\phi) \, d\phi$ to $d\mathbf{r}_1$ gives

$$A(\mathbf{r}_2) = \frac{I}{cr_2^3} \oint d\mathbf{r}_1 (\mathbf{r}_1 \cdot \mathbf{r}_2) \tag{A1.18}$$

where the symbol $\oint$ means to carry out a line integral along the loop. One may simplify the integrand of (A1.18) as

$$(\mathbf{r}_1 \times d\mathbf{r}_1) \times \mathbf{r}_2 + (\mathbf{r}_2 \cdot d\mathbf{r}_1)\mathbf{r}_1 = (\mathbf{r}_2 \cdot \mathbf{r}_1) \, d\mathbf{r}_1 \tag{A1.19}$$

The integral of the vector $\mathbf{r}_1(\mathbf{r}_2 \cdot d\mathbf{r}_1)$ about the circle is easily seen to be zero, since the positive and negative components of the scalar product cancel. The vector potential becomes upon rearrangement of (A1.19)

$$A(\mathbf{r}_2) = \left(\frac{I}{2c} \int \mathbf{r}_1 \times d\mathbf{r}_1 \right) \times \left(\frac{\mathbf{r}_2}{r_2^3} \right) \tag{A1.20}$$

Consider what has been accomplished: the vector potential has been separated into a product, the first term of which depends *only* on the structure of the loop and the last of which depends on the distance from the loop. Note that the vector $\mathbf{r}_1$ always points from the origin toward the loop and $d\mathbf{r}_1$ always points on a tangent to the curve. Thus, the cross product $\mathbf{r}_1 \times d\mathbf{r}_1$ is always perpendicular to the plane of the loop.

The vector

$$\frac{I}{2c} \int \mathbf{r}_1 \times d\mathbf{r}_1 = \mathbf{m} \tag{A1.21}$$

is the magnetic dipole moment of the loop; therefore

$$\mathbf{A}(\mathbf{r}_2) = \mathbf{m} \times (\mathbf{r}_2/r_2^3) \tag{A1.22}$$

The structure of the current loop has been replaced with the single quantity $\mathbf{m}$, which is taken as a fundamental property of the current loop. The field at $\mathbf{r}_2$ caused by this dipole at the origin can be calculated from Eq. (A1.2):

$$\mathbf{B}(\mathbf{r}_2) = \nabla \times \mathbf{A} = \nabla \times [\mathbf{m} \times (\mathbf{r}_2/r_2^3)] \tag{A1.23}$$

By expansion of the vector equation, the field is

$$\mathbf{B}(\mathbf{r}_2) = -\mathbf{m}/r_2^3 + 3[(\mathbf{m} \cdot \mathbf{r}_2)/r_2^5]/\mathbf{r}_2 \tag{A1.24}$$

This is the classical equation for the field $\mathbf{B}$ of a dipole at a point $\mathbf{r}_2$ away from a dipole. This expression will prove to be of enormous utility in discussions of the magnetic resonance of solids.

It should be noted that

(1) the form does not depend on the structure of the dipole, only on the fact that it is a dipole;

(2) more generally, for a whole assembly of dipoles disposed at various places $\mathbf{r}_i$, the field at a point far removed from any single dipole is

$$\mathbf{B}(\mathbf{r}_2) = -\sum_i \frac{\mathbf{m}_i}{|\mathbf{r}_2 - \mathbf{r}_i|^3} + 3\sum_i \frac{\mathbf{m}_i \cdot (\mathbf{r}_2 - \mathbf{r}_i)}{|\mathbf{r}_2 - \mathbf{r}_i|^5} (\mathbf{r}_2 - \mathbf{r}_i) \tag{A1.25}$$

(3) the formula for evaluating the dipole of a loop is

$$\mathbf{m} = \frac{I}{2c} \int \mathbf{r}_1 \times d\mathbf{r}_1 \tag{A1.26}$$

where the vector $\mathbf{r}_1$ is taken *from the center of the current loop.*

If a dipole is present at the point $\mathbf{r}_2$, it will have an interaction with the field of all the other dipoles, just as if the field were produced by a coil of wire in a magnet. The energy of this interaction, summed over all dipole–

dipole pairs, is the dipolar energy

$$E_D = -\sum_i \mathbf{m}_i \cdot \mathbf{B}_D(\mathbf{r}_i) \tag{A1.27}$$

or

$$E_D = -3 \sum_i \sum_{j \neq i} \frac{[\mathbf{m}_j \cdot (\mathbf{r}_i - \mathbf{r}_j)][(\mathbf{r}_i - \mathbf{r}_j) \cdot \mathbf{m}_i]}{|\mathbf{r}_j - \mathbf{r}_i|^5} + \sum_i \sum_{i \neq j} \frac{\mathbf{m}_i \cdot \mathbf{m}_j}{|\mathbf{r}_j - \mathbf{r}_i|^3} \tag{A1.28}$$

Although this result was discussed in the context that $\mathbf{m}$ is a current loop, the result (A1.28) is general. In particular, it is valid for the dipolar interactions of an assembly of nuclear spins. Quantum mechanically, the energy is the expectation value of some Hamiltonian $\mathscr{H}_D$. By the correspondence between classical and quantum mechanics, $\mathbf{m}_i = \gamma \hbar \mathbf{I}_i$. The form of the dipolar Hamiltonian is, in rad/sec,

$$\mathscr{H}_D = -3 \sum_{i,j}' \gamma_i \gamma_j \hbar \frac{[\mathbf{I}_j \cdot (\mathbf{r}_j - \mathbf{r}_i)][(\mathbf{r}_j - \mathbf{r}_i) \cdot \mathbf{I}_i]}{|\mathbf{r}_j - \mathbf{r}_i|^5} + \sum_{i,j}' \frac{\gamma_i \gamma_j \hbar}{|\mathbf{r}_j - \mathbf{r}_i|^3} \mathbf{I}_j \cdot \mathbf{I}_i \tag{A1.29}$$

where the prime indicates that the double sum is over all pairs of dipoles. This is the dipole–dipole interaction that causes such broad lines in certain NMR spectra of solids. In Chapters 5 and 6, the techniques used by NMR spectroscopists to eliminate selectively the effects of dipolar interactions on NMR spectra or to use them for specific purposes are discussed. We can, for example, infer geometrical information about randomly oriented molecules in solid or enhance the sensitivity of the NMR signal from a rare (e.g., ^{13}C, ^{15}N, or ^{29}Si) spin in a bath of ^1H spins by properly manipulating these dipolar effects.

UNITS AND PHYSICAL CONSTANTS

The functional form of all physical laws must be the same, no matter what system of units are used to describe them. However, the exact numerical values given to any physical quantity depend on the definitions of the various physical quantities involved. For example, the forms of electrostatic and electromagnetic forces in free space are

$$\mathbf{F}_e = k_e(q_1 q_2/r_{12}^3)\mathbf{r}_{12} \qquad \text{(A2.1)}$$

$$\mathbf{F}_m = k_m q\mathbf{V} \times \mathbf{B} \qquad \text{(A2.2)}$$

There are two commonly used systems of units: cgs (or Gaussian) and SI (System Internationale). The principal difference is that in cgs units, the units of k_e are chosen such that the first relation has the simple form

$$\mathbf{F}_e = (q_1 q_2/r_{12}^3)\mathbf{r}_{12} \qquad \text{(cgs)} \qquad \text{(A2.3)}$$

In addition, k_m is chosen such that, in free space, the magnetic induction equals the magnetic field (i.e., $\mathbf{B} = \mathbf{H}$). This requires $k_m = 1/c$,

$$\mathbf{F}_m = q(\mathbf{v}/c) \times \mathbf{B} \qquad \text{(cgs)}$$
$$= q(\mathbf{v}/c) \times \mathbf{H} \qquad \text{(A2.3a)}$$

In SI units, q is determined by a previous definition, such that in order to adjust the right-hand side of (A2.1) to correspond to mechanical forces,

$$F_e = (1/4\pi\epsilon_0)(q_1 q_2/r_{12}^3)\,\mathbf{r}_{12} \qquad \text{(SI)} \qquad \text{(A2.4)}$$

where the permittivity of free space $\epsilon_0 = 8.8542 \times 10^{-12}$ F m^{-1}. The magnetic field can then be defined such that

$$\mathbf{F}_m = q\mathbf{v} \times \mathbf{B} \qquad \text{(SI)}$$
$$= \mu_0 q\mathbf{v} \times \mathbf{H} \qquad \text{(A2.5)}$$

where the permeability of free space $\mu_0 = 1/\epsilon_0 c^2 = 4\pi \times 10^{-7}$ H m^{-1}.

The fundamental unit of force in SI is the newton ($\equiv 1$ kg m sec^{-2}), and the fundamental unit of charge is the coulomb (which is in turn related to the ampere, the unit of current). Hence, k_e and k_m are used to provide the bridge between mechanically derived units such as the newton and units such as the coulomb and the ampere, which are commonly used electrical units.

The fundamental unit of force in cgs is the dyne ($\equiv 1$ gm cm sec^{-2}), and the fundamental unit of charge, chosen to be congruent with this force unit, is the electrostatic unit (or esu). Hence, k_e is arbitrarily 1 in cgs units. This means that since most devices are calibrated in coulombs, amperes, and volts, any calculation in cgs must be converted to these units for comparison with most experimental measurements. Additionally, cgs units are chosen such a that, in free space **B** and **H** are the same; i.e., the field one applies, calculated from the current in some coil, *is* the field felt by magnetic dipoles elsewhere *in vacuo*.

In SI units, because of the maintenance of the definition of charge and because of the "simplified" force law (A2.5), it is required that **H** be multiplied by the permeability of free space μ_0 to determine the magnetic induction at the site of a dipole. Since electric and magnetic calculations in cgs do not involve a large number of conversions, in particular the one from applied field to magnetic induction, we have decided to use the cgs system, with conversion to volts, amperes, and coulombs after the fact, if necessary. The fundamental units and their relation to length, time, and mass are shown in Table A2.1. Table A2.2 gives values for important physical constants recommended in the *National Bureau of Standards Technical News Bulletin* **47**, 175 (1963).

The commonly used unit in cgs for energy is the erg ($\equiv 1$ gm cm^2 sec^{-2}). However, in magnetic resonance, most, if not all, of the processes involve

TABLE A2.1
SI and cgs Units

Quantity	SI	cgs
Length	Meter	Centimeter
Time	Second	Second
Mass	Kilogram	Gram
Force	Newton ($\equiv$ kg m sec^{-2})	Dyne ($\equiv$ gm cm sec^{-2})
Charge	Coulomb ($\equiv \sqrt{4\pi\epsilon_0}$ kg$^{1/2}$ m$^{3/2}$ sec^{-1})	Electrostatic unit ($\equiv$ gm$^{1/2}$ cm$^{3/2}$ sec^{-1})
Energy	Joule ($\equiv$ kg m^2 sec^{-2})	Erg ($\equiv$ gm cm^2 sec^{-2})
Electric field	Volt/meter	Dyne/electrostatic unit ($\equiv$ gm$^{1/2}$ cm$^{1/2}$ sec^{-1})
Electric potential	Volt $\left(\equiv \dfrac{1}{\sqrt{4\pi\epsilon_0}} \text{kg}^{1/2} \text{ m}^{1/2} \text{ sec}^{-1} \right)$	Statvolt (erg esu^{-1})
Magnetic field	Tesla ($\equiv$ V sec m^{-2})	Oersted ($\equiv$ dyne esu^{-1}) Gauss
Magnetic flux	Weber ($\equiv$ V sec)	Gauss square centimeter

TABLE A2.2
Values of Selected Physical Constants

Quantity	Symbol	Numerical value	Uncertainty (ppm)	Units
Speed of light in vacuum	c	299792458(1.2)	0.004	m sec^{-1}
Elementary charge	e	1.6021892(46)	2.9	10^{-19} C
Planck constant	h	6.626176(36)	5.4	10^{-34} J sec
Avogadro constant	N_A	6.022045(31)	5.1	10^{23} mole^{-1}
Electron rest mass	m_e	9.109534(47)	5.1	10^{-31} kg
Proton rest mass	m_p	1.6726485(86)	5.1	10^{-27} kg
Ratio of proton mass to electron mass	m_p/m_e	1836.15152(70)	0.38	
Faraday constant $N_A e$	F	9.648456(27)	2.8	10^4 C mole^{-1}
Rydberg constant $(\mu_0 c^2/4\pi)^2 (m_e e^4/\pi \hbar^3 c)$	R	1.097373177(83)	0.075	10^7 m^{-1}
Bohr radius $\alpha/4\pi R_e$	a_0	5.2917706(44)	0.82	10^{-11} m
Free electron g factor or electron magnetic moment in Bohr magnetons	$g_e/2 = \mu_e/\mu_\beta$	1.00115965567(35)	0.0035	
Bohr magneton $[e] (e\hbar/2m_e c)$	μ_β	9.274078(36)	3.9	10^{-24} J T^{-1}
Proton gyromagnetic ratio	γ_p	2.6751987(75)	2.8	10^8 sec^{-1} T^{-1}
Proton magnetic moment in Bohr magnetons	μ_p/μ_β	1.52103209(16)	0.011	10^{-3}
Ratio of electron and proton magnetic moments	μ_e/μ_p	658.2106880(66)	0.010	
Proton magnetic moment in nuclear magnetons	μ_p/μ_N	2.7928456(11)	0.38	
Nuclear magneton $|e| (e\hbar/2m_p c)$	μ_N	5.050824(20)	3.9	10^{-27} J T^{-1}
Compton wavelength of the electron, $h/m_e c$	λ_c	2.4263089(40)	1.6	10^{-12} m
Gravitational constant	G	6.6720(4)	615	10^{-11} m^3 sec^{-2} kg^{-1}
Boltzmann constant R/N_A	k	1.380662(44)	32	10^{-23} J K^{-1}

rotational energy. The Planck relation gives the corresponding energies for angular frequencies in a quantum system:

$$E \; (\text{ergs}) = hE \quad (\text{hertz}), \qquad E \; (\text{ergs}) = \hbar E \quad (\text{rad sec}^{-1})$$

Every formula in this book that has energy as a result has been adjusted to yield energies in radians per second. All that is necessary to convert to ergs is multiplication by $\hbar$. For example, an appropriate dipolar energy might be 20,000 Hz ($\cong 1.25 \times 10^5$ rad sec^{-1}). In cgs units, this value is

$$E \, (\text{ergs}) = 1.05 \times 10^{-27} \text{ erg sec} \cdot (1.25 \times 10^5 \text{ rad sec}^{-1})$$
$$= 1.313 \times 10^{-22} \text{ ergs}$$

VECTORS, TENSORS, AND TRANSFORMATIONS

It should be obvious from the text that the subject of magnetic resonance is based on an understanding of vectors such as the magnetic field, the magnetic moment, and the vector **r**, which represents the position of one particle with respect to the position of another. One of the dominant themes of magnetic resonance is how the mathematical description of a physical system composed of vectors is altered by the coordinate system to which these vectors are referenced. We have seen that this change of system can introduce such things as "fictitious fields," which alter the sense of motion in these systems. They may also (it is hoped) simplify the often complex description of the system.

In the three-dimensional space where we exist, any vector **a** can be represented by the components along the three normalized orthogonal vectors $\mathbf{e}_1$, $\mathbf{e}_2$, and $\mathbf{e}_3$ that span the space. (We call them $\mathbf{e}_i$ here, but **x**, **y**, and **z** can do just as well and are the most common symbols used.) Here

$$\mathbf{a} = a_1\mathbf{e}_1 + a_2\mathbf{e}_2 + a_3\mathbf{e}_3 \tag{A3.1}$$

and a_1, a_2, and a_3 together give all the information necessary to describe the vector **a** in the coordinate system. This mathematical description of **a** is, however, not unique because the choice of the vectors $\mathbf{e}_1$, $\mathbf{e}_2$, and $\mathbf{e}_3$ is not unique. Thus one could have equally well chosen a set of vectors $\mathbf{e}'_1$, $\mathbf{e}'_2$, and $\mathbf{e}'_3$ that also span the space. The components of the vector **a** in this description will be different from those of Eq. (A3.1) as indicated by

$$\mathbf{a} = a'_1\mathbf{e}'_1 + a'_2\mathbf{e}'_2 + a'_3\mathbf{e}'_3 \tag{A3.2}$$

However, since both descriptions represent the *same* vector, there must be some relationship between the two descriptions. We can use the properties of vector multiplication to get this relation. For example, for Eq. (A3.1) we can find the components of the vector **a**:

$$a_1 = \mathbf{a} \cdot \mathbf{e}_1 \tag{A3.3a}$$

$$a_2 = \mathbf{a} \cdot \mathbf{e}_2 \tag{A3.3b}$$

$$a_3 = \mathbf{a} \cdot \mathbf{e}_3 \tag{A3.3c}$$

In an analogous manner, one could find the components a_1', a_2', and a_3' by vector multiplication.

To established the functional relationship between the two descriptions of the vector $\mathbf{a}$, one must find how each component in one description depends on the components in the other description. One can use Eqs. (A3.2) and (A3.3) together to provide this connection:

$$a_1 = a_1'(\mathbf{e}_1' \cdot \mathbf{e}_1) + a_2'(\mathbf{e}_2' \cdot \mathbf{e}_1) + a_3'(\mathbf{e}_3' \cdot \mathbf{e}_1) \qquad \text{(A3.4a)}$$

$$a_2 = a_1'(\mathbf{e}_1' \cdot \mathbf{e}_2) + a_2'(\mathbf{e}_2' \cdot \mathbf{e}_2) + a_3'(\mathbf{e}_3' \cdot \mathbf{e}_2) \qquad \text{(A3.4b)}$$

$$a_3 = a_1'(\mathbf{e}_1' \cdot \mathbf{e}_3) + a_2'(\mathbf{e}_2' \cdot \mathbf{e}_3) + a_3'(\mathbf{e}_3' \cdot \mathbf{e}_3) \qquad \text{(A3.4c)}$$

The dot products in parentheses are scalar numbers that describe how the two coordinate systems used to describe the vectors are related to each other. Similarly, one could find each of the numbers a_1', a_2', and a_3' in terms of a_1, a_2, and a_3 and the same set of dot products. We shall give the dot products symbols in the following manner:

$$\mathbf{e}_i' \cdot \mathbf{e}_k = t_{ki} \qquad \text{(A3.5)}$$

The the set of Eq. (A3.4) could be rewritten as

$$a_1 = t_{11}a_1' + t_{12}a_2' + t_{13}a_3' \qquad \text{(A3.6a)}$$

$$a_2 = t_{21}a_1' + t_{22}a_2' + t_{23}a_3' \qquad \text{(A3.6b)}$$

$$a_3 = t_{31}a_1' + t_{32}a_2' + t_{33}a_3' \qquad \text{(A3.6c)}$$

We recognize this set of equations as the expansion of the multiplication of a matrix $\mathbf{T}$ by the column vector $\mathbf{a}'$. Thus, one could write a contracted notation for the three Eq. (A3.6) as

$$\mathbf{a} = \mathbf{T} \cdot \mathbf{a}' \qquad \text{(A3.7)}$$

where the vectors are column vectors. Here $\mathbf{T}$ is the transformation matrix that relates the two coordinate systems described by the sets of vectors $(\mathbf{e}_1, \mathbf{e}_2, \mathbf{e}_3)$ and $(\mathbf{e}_1', \mathbf{e}_2', \mathbf{e}_3')$.

One could similarly find $\mathbf{a}'$ from $\mathbf{a}$ by an analogous procedure. Let us, however, find this relationship from matrix algebra. One knows that for a unitary matrix of three-dimensional space like $\mathbf{T}$, there exists an inverse matrix, $\mathbf{T}^{-1}$, such that

$$\mathbf{T} \cdot \mathbf{T}^{-1} = \mathbf{T}^{-1} \cdot \mathbf{T} = \mathbf{1} \qquad \text{(A3.8)}$$

where $\mathbf{1}$ is a matrix with ones on the diagonal and zeroes elsewhere. Multiplying Eq. (A3.7) from the left by $\mathbf{T}^{-1}$ and using the fact that $\mathbf{1} \cdot \mathbf{a} = \mathbf{a}$, one gets

$$\mathbf{a}' = \mathbf{T}^{-1} \cdot \mathbf{a} \qquad \text{(A3.9)}$$

Thus, the inverse of the transformation matrix is a transformation in the reverse sense, allowing one to calculate $\mathbf{a}'$ when $\mathbf{a}$ is known.

Since in magnetic resonance one deals with vectors, it is not unsual that equations develop that relate one vector to another. An example is the relationship between the chemical-shift field $\mathbf{B}_{cs}$ and the applied field $\mathbf{B}_0$. Each component of $\mathbf{B}_{cs}$ in some coordinate system depends in some way on each of the components of $\mathbf{B}_0$ in that same coordinate system. One could write the equations, one for each component of $\mathbf{B}_{cs}$ as a set of equations, but one could also write the relationship in the contracted notation as well, as shown in

$$\mathbf{B}_{cs} = \sigma \cdot \mathbf{B}_0 \qquad (A3.10)$$

where σ is the *tensor* that contains information on *how* $\mathbf{B}_{cs}$ depends on all the components of $\mathbf{B}_0$—the chemical-shift tensor. Equation (A3.10) is a relationship between vectors expressed relative to the same coordinate system, whereas equations such as (A3.9) give a relationship between the descriptions of a vector relative to two different coordinate systems.

One question to be asked at this point about equations such as (A3.10) is "What would the form of the equation be if it were referred to some other coordinate system?" Our desire would be to have the equation invariant to the representation in which it presented; i.e., in the primed representation it would have the form

$$\mathbf{B}'_{cs} = \sigma' \cdot \mathbf{B}'_0 \qquad (A3.11)$$

In addition, it is also desirable to know how all of the various quantities in equations like (A3.10) and (A3.11) transform when they are referred to a new coordinate system. We already know how the vectors transform. We can use this example to find out how the tensors of rank two transform. First, we multiply (A3.11) from the left by $\mathbf{T}$; this gives an equation for $\mathbf{B}_{cs}$;

$$\mathbf{T} \cdot \mathbf{B}'_{cs} = \mathbf{B}_{cs} = \mathbf{T} \cdot \sigma \cdot \mathbf{B}'_0 \qquad (A3.12)$$

but, according to (A3.10), this is just equal to the product, as shown in

$$\sigma \cdot \mathbf{B}_0 = \mathbf{T} \cdot \sigma' \cdot \mathbf{B}'_0 \qquad (A3.13)$$

By Eq. (A3.9), we may substitute for $\mathbf{B}'_0$:

$$\sigma \cdot \mathbf{B}_0 = \mathbf{T} \cdot \sigma \cdot \mathbf{T}^{-1} \cdot \mathbf{B}_0 \qquad (A3.14)$$

The only way that this statement can be true is if the following equation is also true:

$$\sigma = \mathbf{T} \cdot \sigma' \cdot \mathbf{T}^{-1} \qquad (A3.15)$$

Hence, we have arrived at a transformation for tensors expressed with reference to different coordinate systems, whose vectors are related by the transformation matrix $\mathbf{T}$.

BIBLIOGRAPHY

A wide variety of books have been written on the subjects and sub-subjects of nuclear magnetic resonance. We have found a number of the following books to be extremely useful. We make no claim of completeness of this list, but it is hoped that the reader may find in these books and the references contained in them further information on NMR spectroscopy.

E. R. Andrew, "Nuclear Magnetic Resonance," Cambridge Univ. Press, Cambridge, 1955.

J. A. Pople, W. G. Schneider and H. J. Bernstein, "High-Resolution NMR Spectroscopy," McGraw-Hill, New York, 1959.

A. Abragam, "Principles of Nuclear Magnetism," Oxford Univ. Press, Oxford, 1961.

N. Bloembergen, "Nuclear Magnetic Relaxation," Benjamin, New York, 1961.

A. Carrington and A. D. McLachlan, "Introduction to Magnetic Resonance," Harper and Row, New York, 1967.

R. T. Schumacher, "Introduction to Magnetic Resonance," Benjamin, New York, 1970.

M. Goldman, "Spin Temperature and Magnetic Resonance in Solids," Oxford Univ. Press, Oxford, 1970.

T. C. Farrar and E. D. Becker, "Pulse and Fourier Transform NMR," Academic Press, New York, 1971.

J. B. Stothers, "Carbon-13 NMR Spectroscopy," Academic Press, New York, 1972.

D. Shaw, "Fourier Transform NMR Spectroscopy," Elsevier, Amsterdam, 1976.

U. Haeberlen, "High Resolution NMR in Solids: Selective Averaging," Academic Press, New York, 1976.

C. P. Slichter, "Principles of Magnetic Resonance," 2nd ed., Springer-Verlag, Heidelberg, 1978.

R. K. Harris and B. E. Mann, "NMR and the Periodic Table," Academic Press, New York, 1978.

D. Wolf, "Spin-Temperature and Nuclear-Spin Relaxation in Matter: Basic Principles and Applications," Oxford Univ. Press, Oxford, 1979.

G. C. Levy, R. L. Lichter, and G. L. Nelson, "Carbon-13 NMR Spectroscopy," 2nd ed., Wiley, New York, 1980.

E. D. Becker, "High-Resolution NMR—Theory and Chemical Applications," 2nd ed., Academic Press, 1980.

M. L. Martin, J. J. Delpeuch, and G. S. Martin, "Practical NMR Spectroscopy," Heyden, London, 1980.

M. Mehring, "High Resolution NMR in Solids," Springer-Verlag, Heidelberg, 1981.

E. Fukushima and S. B. W. Roeder, "Experimental Pulse NMR: A Nuts and Bolts Approach," Addison-Wesley, Reading, Massachusetts, 1981.

A. Abragam and M. Goldman, "Nuclear Magnetism: Order and Disorder," Oxford Univ. Press, Oxford, 1982.

R. K. Harris, "Nuclear Magnetic Resonance Spectroscopy," Pitman, London, 1983.

C. A. Fyfe, "Solid-State NMR for Chemists," CFC Press, Guelph, Ontario, 1983.

J. W. Akitt, "NMR and Chemistry," 2nd ed., Chapman and Hall, London, 1983.

A. Bax, "Two-Dimensional Nuclear Magnetic Resonance in Liquids," D. Riedel, Dordrecht, 1982.

INDEX

A

Absorption, 1, 20, 21, 28
  resonant, 1
  magnetic, 20
Adiabatic demagnetization in the rotating
  frame, 242, 248–249, 251
ADRF, *see* Adiabatic demagnetization in
  the rotating frame
Alternating-current circuit theory, 35–36
Angular momentum, 2, 3, 4, 6, 17, 46–47,
  49, 91, 104, 111, 155, 211, 258
  nuclear-spin, 46, 123
  operator, 155, 156
  orbital, 108, 119, 120, 123
  spin, 103, 213
  states, 45
  vector, 50
Aue sequence, 263–265

B

Baker–Campbell–Hausdorff formula, 137,
  138, 139, 141, 151
BCH formula, *see* Baker–Campbell–
  Hausdorff formula
Bloch equations, 2, 12–16, 22, 81, 87, 88
Bohr magneton, 94
Broadening
  dipolar, 199, 255, 256
  homogeneous, 227, 238, 256
  inhomogeneous, 92, 238
  quadrupolar, 126
BR-24 cycle, *see* Burum–Rhim cycle
Burum–Rhim cycle, 203, 207, 214

C

Carr–Purcell sequence, 121, 137, 172–174,
  229, 230, 231, 274

Carr–Purcell–Meiboom–Gill sequence,
  121, 174, 193, 194, 195, 196, 224, 225
Chemical shift, 92, 103–119, 134, 135, 199,
  201, 206, 229, 230, 231, 233, 236, 238,
  245, 255, 256, 263, 266, 267
  anisotropy, 134, 199
  diamagnetic, 108
  dispersion, 257
  isotropic, 265
  paramagnetic, 108
  tensor, *see* tensor, chemical shift
Constants, physical (t), 280–285
Continuous-wave NMR, 18, 20, 29, 30,
  162, 234–239
Coulomb's law, 125
Counterclockwise, 3
CP, *see* Cross-polarization
Cross-polarization, 227, 228, 239, 240–254,
  255, 265, 267, 273
Curie constant, 58, 244
Curie-Weiss law, 58, 59, 245
Current loop, 277–281
Cycle
  Burum–Rhim, 203, 207, 214
  BR-24, *see* Burum–Rhim
  Eight-pulse, *see* MREV-8
  MREV-8, 204, 207
  WAHUHA, 168, 169, 175, 201, 204, 225

D

Decoupling
  continuous-wave, 162, 234–239
  dipolar, 255, *see also* heteronuclear and
    homonuclear
  field, 234, 239
  heteronuclear, 174, 229, 239, 240, 255
  homonuclear, 199, 206, 207, 213, 214,
    225, 255
  magic-sandwich, 249

pulse, 229–234, 238
    radio-frequency, 213, 256, 272
    spin-lock, 239
    spin–spin, 238
Demodulation, 21
Density matrix, *see* Density operator
Density operator, 54, 56, 57, 59–64, 65, 66,
    68, 69, 75, 80, 88, 89, 137, 144, 157,
    164, 169, 196, 197
Diamagnetism, 108
Dipole
    electric, 3
    magnetic, 3
    moment, 3
Dipolar coupling tensor, 85, 125
Dipolar-echo sequence, 168
Dipole–dipole coupling, 255
    constant, 135
    interaction, 91, 94, 95, 96, 97, 103, 133,
    169, 227, 238, 240
    tensor, 95, 125
Dirac notation, 43, 44
Dispersion, 20, 21, 28
Double-quantum coherence, 60
Dyson
    expression, 68, 137, 138, 147, 150–151,
    152, 154, 163, 164, 169
    time-ordering operator, 68

E

Echo
    dipolar, 199, 200
    rotational, 265
    solid, 199, 200
    spin, 193, 196, 198
Electric dipole, 3
Electric-field gradient, 135
Electronic moments, 94
Electron spin resonance, 21
Equation
    Bloch's, 2, 12–16, 22, 81, 87, 88
    Heisenberg's, 71
    Laplace's, 122, 124
    Liouville–von Neumann, 56, 57, 61–64,
    87, 167, 169, 208
    Planck's, 285
    Poisson's, 278
    Schroedinger's, 43, 51, 54, 66, 71, 72, 74,
    91, 258
    van Vleck's, 99

Errors, phase, *see* Phase errors
Expectation value, 44, 45, 59, 63, 155, 160
Exponential
    approximation, 137
    operator, 65

F

Fermi contact interaction, 120
FID, *see* Free-induction decay
Field
    effective, 157
    electrostatic, 131
    local, 93
    magnetic, 4, 31, 104
    nuclear, 104
    of a current loop, 277
    radio-frequency, 35
    static, 2, 20, 103, 104, 195
    time-varying, 20
Force, 5
Fourier transformation, 21, 27, 28, 29, 101
Frame
    interaction, 17, 71, 72, 77, 161
    rotating, *see* Rotating frame
Free-induction decay, 25, 26–29, 31, 39,
    100, 155, 177, 178, 187, 195, 245, 246,
    247, 248, 249
    cross-polarized, 246
Frequency
    Larmor, 7, 8–11, 18, 27, 71, 79, 93, 133,
    134, 193, 222
    NMR (*t*), 8–11
    precessional, 2
    rotational, 199

G

Gauge transformation, 104
Gyromagnetic ratio, 2, 4, 18, 89, 95, 133,
    134, 199, 245

H

Hamiltonian, 43, 44, 46, 49, 53, 54, 57, 59,
    61, 63, 64, 68, 71, 74, 75, 83, 87, 88,
    89, 91, 92, 93, 94, 100, 104, 105, 109,
    111, 121, 127, 132, 133, 135, 138, 150,
    151, 156, 157, 160, 162, 164, 166, 171,
    172, 181, 182, 183, 186, 191, 203, 204,

207, 208, 209, 210, 211, 213, 215, 220,
228, 229, 235, 236, 245, 253, 263, 273
average, 164, 166–170, 185–187, 190,
   197, 209, 210, 211, 224, 225, 226,
   230–236
chemical-shift, 74, 167, 203, 225, 231,
   257, 265
dipole–dipole, 95, 138, 167, 199, 201,
   202, 203, 207, 208, 225, 229, 231, 273
heteronuclear, 231
internal, 91, 92, 93, 164, 167, 174, 189,
   196, 201, 213
offset, 146, 167, 182, 186, 188
phase-error, 181, 189
quadrupole, 123, 126, 136
randomly modulated, 132
radio-frequency, 74, 93, 153
spin, 199
time-dependent, 75, 257
time-independent, 57, 198
Zeeman, 49, 51, 79, 92, 93, 95, 111, 126,
   136, 161, 173, 182, 235
Hartmann–Hahn
   condition, 244, 245, 249
   matching, 245, 248
Heisenberg representation, 71
Helmholtz free energy, 88
Hermitian operator, 44, 49, 55, 66, 67, 142
Herzfeld-Berger technique, 260, 261, 265,
   267
Hilbert space, 66
Homonuclear pulse experiments, *see* Pulse,
   homonuclear

I

Interaction frame, 17, 42, 69, 71, 72, 73, 74,
   77, 151–160, 161, 211
Isotope, NMR-active (*t*), 8–11

J

J coupling, 91, 120, 121, 238

K

Kramers
   rule, 126
   degenerate levels, 132

L

Laplace's equation, 122, 124
Larmor frequency, *see* Frequency, Larmor
Lattice, 61, 83, 242
Levi-Civita symbol, 46
Linear momentum, 3, 46, 104
Liouville operator, 63, 64, 65, 258
Liouville–von Neumann equation, 56, 57,
   61–64, 87, 167, 169, 208
Longitudinal relaxation, *see* Spin-lattice
   relaxation
Lorentzian line, 20, 21, 28, 39, 102, 118
Lowe-Tarr scheme, 222
Lurie-Slichter experiment, 245

M

Maclaurin series, 122
Macroscopic moment, 11
Magic-angle, 135, 202, 213
   flipping, 265, 266
   spinning, 199, 213, 215, 222, 228,
      255–267, 272, 273, 274
Magnetic moment, 1, 2, 3, 4, 5, 6, 7, 8, 9,
   10, 11, 18, 20, 24, 34, 39, 45, 71, 87
Magnetization
   absorption, 20
   dispersion, 20
   equilibrium, 58
   macroscopic, 59
   total, 11
Magnetogyric ratio, *see* Gyromagnetic ratio
Magnus expansion, 137, 138, 139,
   143–156, 164, 167, 168, 169, 170, 182,
   196, 207, 209
MAS, *see* Magic-angle spinning
Meiboom–Gill–Carr–Purcell sequence, *see*
   Carr–Purcell–Meiboom–Gill sequence
Mixer, 21, 22
Moment
   dipole, 3
   electronic, 94
   magnetic, 1, 2, 3, 4, 5, 6, 7, 8–16, 18, 20,
      24, 34, 39, 45, 71
   macroscopic, 11
   pseudo-first and pseudo-second, 102
   quadrupole (*t*), 8–11
   second, 103
   transverse, 30
   zeroth, 99

Momentum, *see* Angular momentum, Linear momentum
MREV-8 cycle, 204, 207
Multiple-pulse NMR, 164, 166–170, 171, 188, 199, 201, 202, 207, 213, 215, 222, 223, *see also* Sequence *and* Cycle
Multiple-quantum coherence, 60, 61

N

Natural abundance (*t*), 8–11
Newton's second law, 3
Nuclear quadrupole resonance, 132
NQR, *see* Nuclear quadrupole resonance
Nuclear spin angular momentum, 46, *see also* Angular momentum
Nuclei (*t*), 8–11

O

Off-magic-axis spinning, 215, 260, 267
Operator
    angular-momentum, 46, 155, 156, 166
    average-Hamiltonian, 185
    density, *see* Density operator
    Dyson time-ordering, 68, 151
    evolution, 137, 211
    exponential, 65, 70, 81, 139, 150, 156, 211
    Hermitian, 44, 49, 55, 66, 67, 142
    Linear-momentum, 46
    Liouville, 63, 64, 65, 79, 258
    offset, 211
    probability, 53, 55, 56, 79
    projection, 44
    rotation, 69, 138
    space, 83
    spin, 50, 51, 68, 69, 70, 93, 138, 253
    time-dependent, 151
    time-evolution, 143
    time-ordering, 68, 151
    transition, 208
    unitary, 66
    Zeeman, 79, 161
OMAS, *see* Off-magic-axis spinning

P

Pake doublet, 60, 97
Pauli
    spin matrices, 50, 51, 85, 88, 89, 90, 155

spin vector, 50
Parameters for NMR-active nuclei (*t*), 8–11
Phase errors, 178, 181, 188, 191, 209, 213
Physical constants, 282–285
Planck's relation, 285
Poisson's equation, 278
Polarization transfer, 240, 241, 243, *see also* Cross-polarization
Powder spectrum, 60, 97, 113, 115, 118, 128, 131
Principal-axis system, 109, 110, 111
Principal values, 109
Probability operator, 53, 55, 56, 79
Probe ringdown, 221, 222
Pulse, 21, 153, 156, 159, 164, 198–207
    decoupling, 229–234
    delta-function, 182, 187, 193, 201
    heteronuclear, 227–275
    homonuclear, 164–171
    multiple, 166, 171, *see also* Cycle *and* Sequence

Q

Quadrupolar
    interaction, 91, 92, 121–132, 134–135
    nuclei, 8–11, 60, 111, 271, 272, 273
Quadrupole moment, 8–11, 126, 131

R

Receiver recovery, 221
Relaxation time
    cross, 253
    longitudinal, *see* Spin-lattice relaxation
    spin-spin, 2, 13
    transverse, 2, 13, 193
Resonance, quadrupole, *see* Nuclear quadrupole resonance
Rhim sequence, 202, *see also* Burum–Rhim cycle
Ringdown
    probe, 221, 222
    transmitter, 221
Rotary saturation, 240
Rotating frame, 16–20, 23, 24, 26, 30, 31, 34, 39, 40, 57, 71, 74, 75, 76, 89, 100, 155, 156, 157, 158, 160, 162, 163, 167, 178, 181, 186, 208, 229, 237, 244, 250
Rotation, magic-angle, *see* Magic-angle spinning

S

Scalar coupling, *see* J coupling
Schröedinger's equation, 43, 51, 54, 66, 71, 72, 74, 91, 258
Sensitivity (t), 8–11
Sequence
  Aue, 263, 264, 265
  Carr-Purcell, *see* Carr–Purcell sequence
  Carr-Purcell-Meiboom-Gill, *see* Carr–Purcell–Meiboom–Gill sequence)
  dipolar-echo, 168
  Reimer-Vaughan, 271
  Rhim, 202
  solid-echo, 200
  Stoll–Vega–Vaughan, 269
Shielding anisotropy, 206, 215
  tensor, 107, 108
Single-quantum coherence, 60, 61
Solenoid, 31–35
Spectrometer (schematic), 22, 165
Spin, 8–11(t), 49, 50, 69, 121, 127, 271, 272, 273
Spin-lattice relaxation, 2, 13, 133, 227, 242, 246, 251, *see also* T_1, $T_{1\rho}$
Spin lock, 30, 31, 239, 240, 242, 243, 248, 253, 269
Spin quantum number, 133, 135
Spin–spin coupling, 135
Spin–spin relaxation, 2, 13, 31, 156, *see also* T_2, T_2^*
Spin temperature, 241, 242, 248
Spur, 54
Stationary states, 43
Superoperator, 68, 141
Suppression, 92

T

T_1, 2, 13, 14, 39, 40, 60, 61, 79–84, 92, 132, 133, 134, 135, 137, 156, 247, 253
$T_{1\rho}$, 30, 31, 40, 253
T_{cp}, 251, 252, 255
T_2, 2, 13, 14, 15, 39, 61, 155, 174, 238, 247
T_2^*, 92, 164, 177, 178, 213
Tensor, 120, 286–288
  chemical-shift, 111, 112, 113, 115, 263, 267, 288
  dipole–dipole coupling, 95, 125
  shielding, 107, 108, 113, 125, 227

Time, relaxation, *see* Relaxation time, T_1, $T_{1\rho}$, T_2, T_2^*
Time constant of probe circuit, 221
Torque, 5, 6
Trace, 59
Transformation, 71, 72, 73, 75, 286–288
  Fourier, 21, 27, 28, 29
  to interaction frame, 17
  to rotating frame, 16–20
Transverse relaxation, *see* T_2, T_2^*
Tuning, 215–220
Two-dimensional NMR, 263, 265, 266, 267

U

Units, 282

V

van Vleck's equation, 97
Vector, 286–288
  polarization, 42
  potential, 277–288

W

WAHUHA cycle, 168, 169, 175, 201, 204, 225
Wigner rotation matrices, 258

Z

Zeeman
  energy levels, 132, 134
  energy reservoir, 82, 95, 134, 242, 248
  field, 71, 93, 119, 126, 132, 160, 225
  Hamiltonian, 49, 51, 79, 92, 93, 95, 111, 136, 161, 173, 182, 235
  interaction, 13, 16–20, 31, 49, 51, 52, 54, 57, 63, 71, 74, 75, 79, 91, 92, 93, 94, 96, 98, 104, 126, 135, 153, 160, 162, 163, 229, 241
  operator, 79, 161
  splitting, 126, 199
  states, 91, 111, 248
Zero-quantum coherence, 60
Zero-time-resolution NMR, 103